NON-NEUTRAL PLASMA PHYSICS II

AIP
CONFERENCE
PROCEEDINGS 331

NON-NEUTRAL
PLASMA PHYSICS II

BERKELEY, CA JULY 1994

EDITORS: **JOEL FAJANS**
UNIVERSITY OF CALIFORNIA,
BERKELEY

DANIEL H. E. DUBIN
UNIVERSITY OF CALIFORNIA,
SAN DIEGO

American Institute of Physics   **New York**

In memory of our colleague and friend

John Malmberg

1927–1992

CONTENTS

Introduction

Non-Neutral Plasmas
The Fourth State of Matter Re-Visited

C. W. Roberson

Physical Science & Technology Division
Office of Naval Research
Arlington, VA 22217

The concept that a plasma is the fourth state of matter comes from considering what happens when we add energy to matter. If, for example, we add energy to ice in the form of heat it transforms to water, a liquid. Adding more energy transforms it to steam, a gas. If we continue to add energy we ionize the gas. This energetic fourth state of matter we call a plasma. If we remove energy from a plasma, the electrons and ions recombine to form a gas, the gas condenses into a liquid and the liquid turns into a solid. That is the normal order. Naturally occurring plasmas on earth are uncommon because of the high temperatures required. The most common man made plasma is the "neon" sign. In this case the energy to ionize the gas is provided by energetic electrons driven by an electrical circuit. Spectroscopic data from such plasmas provided much of the early information about atomic structure and these plasmas are the gain media of many lasers. Today these low temperature, collision dominated plasmas have found wide spread industrial applications in the plasma processing of materials.

Plasmas are common in space, in fact they are by far the dominant form of matter in space. The space plasma we are most familiar with is the sun. This plasma, which is driven by nuclear fusion, is the primary energy source for the earth, making life possible. Reproducing of the sun's fusion energy process in a controlled way occupies a major fraction of the international plasma research and technology effort. Since the advent of spacecraft, we have discovered the space between the Sun and Earth is filled with plasma. Plasma is the natural environment of spacecraft. Space plasma provides the basis of over the horizon radar and low frequency communication systems. The ionosphere and magnetosphere plasmas affect our high frequency communications and navigation systems. The activity in plasma research has grown to the point that the Plasma Physics Division of the American Physical Society is the second largest in the Society.

The Berkeley Workshop on Non-Neutral Plasmas in Traps is devoted to a different approach to plasmas. Suppose we separate the electrons and ions after we make the plasma but before we cool it down. The states of matter we get then will be quite different because it can not recombine. The plasma will remain a single component non-neutral plasma through the liquid and solid phases.

From the papers in the proceedings of the Berkeley Workshop on Non-Neutral Plasmas in Traps it is clear that this area of research has evolved to the status of a new paradigm[1]. These single component non-neutral plasmas have most of the essential properties of neutral plasmas and significant differences. This paradigm has been approached from two directions. On the one hand it came about by plasma physicists trying to create simpler plasmas in order to improve our understanding and predictability of this complex media. From the other direction it has been approach by atomic physicists interested in designing the transition from the simplicity of single particle dynamics to collective behavior in a predictable way.

This Berkeley workshop is the third in a series and is a timely snap shot of the status of the field. The first was an Office of Naval Research Symposium and was held in Washington DC[2]. The second meeting was run as a workshop, and was held at the National Academy of Science's Beckman Center in Irvine, California. That workshop was part of the Plasma Science Committee's Opportunities in Plasma Physics study. The results will be integrated into their report. This workshop's proceedings is particularly important since it is the first time this group has published such a conference proceedings since the original ONR Symposium.

During this period the American Physical Society has recognized investigators working in this field with two Maxwell prizes, an Excellence in Plasma Physics award, and a Davidson-Germer prize. There have been several plenary session review talks and numerous invited talks at American Physical Society meetings. One individual has been elected to the National Academy of Sciences and one is a member of both the National Academy of Sciences and Engineering.

A new text book has been written on the subject. In addition to the work on non-neutral microplasmas in traps which this workshop has focused on, it covers non-neutral beams in the limit where self field effects are important[3].

On the one hand we have simplified the plasma to its most elementary state. This elementary plasma is to plasma physics what the hydrogen atom is to atomic physics. Because most of the free energy has been removed from these plasmas and they are near thermal equilibrium, the experiments are precise and reproducible. The result has been a series of experiments that has set a new standard of excellence in plasma physics. These pure electron microplasmas, which consist of a few hundred million particles(i.e. the population of the US), exhibit the continuum properties of a fluid. The phenomena which are studied are connected with ideal, incompressible flow and appear in aeronautics, in meteorology and oceanography. For example, research in fluid vortex dynamics with pure electron plasmas has blossomed since the ONR Symposium. The most carefully diagnosed fluid vortex experiments are now carried out in pure electron plasmas. Experiments that follow the detailed evolution of the pure electron plasma from its turbulent beginning to equilibrium are under way. There are surprises along the way. Long lasting "crystal" lattices of 3–10 vortices arising spontaneously from turbulence are observed. These metaequilibria are a exciting new area for both experiment and theory. They challenge each other. We are forced to follow the internal logic of the science. This leads to a new depth of understanding of this unique media which lays the ground work for new technology.

With the technique of laser cooling, radiation pressure from a laser can be used to control the temperature of trapped ions. These ions can be used in atomic clocks. To improve the signal to noise ratio a large numbers of ions are required, hence the need for pure ion plasma traps. However, a large number of ions increases the time dilation shift due to plasma rotation, reducing the potential accuracy of the clocks. It is important to be able to minimize and predict these shifts. Inaccuracies of atomic clocks based on stored ions may eventually be less than 1 part in 10^{18}. On the road to making better atomic clocks, fascinating new states of matter have been discovered. Pure ion liquid and solid plasmas have been created in the process of laser cooling ions in a trap. It is now possible to store and image one ion in a trap. The number of ions can be increased in a controlled way and the structure that evolves can be predicted based on an energy minimization principle. A new ion trap has become operational which is capable of trapping and cooling more than 10^5 ions. Theory predicts this plasma should make the transition to a bcc lattice, a solid ion plasma. Ions have been laser cooled in a miniature storage ring. They form a helical structure, a natural rotating electrostatic quadrupole.

A pure positron plasma has been created for the first time. The positrons come from a sodium radioactive emitter with a broad energy spectrum which passes through a moderator to slow down the positrons and narrow the energy spectrum, which can then be trapped. Although we do not expect the plasma dynamics of these pure positron plasmas to differ in any significant way from pure electron plasmas, they make possible some fundamental experiments on the stability of equal mass electron-positron plasmas. Such positron plasmas can be combined with antiprotons in a trap to form antihydrogen, the fundamental antimatter atom. As with electrons, non destructive diagnostics of positrons are very difficult. However experiments with ion plasmas where laser diagnostics are possible have confirmed our theoretical understanding of plasma modes in these traps. The mode structure of these plasmas can in turn be used as a diagnostic in positron and electron plasmas. The temperature and density of these plasmas can be inferred from measurements of the plasma modes. Positrons make good diagnostic particles since they produce a distinct 0.5 MeV radiation when annihilated by electrons. The strongest radiation line from astrophysical plasmas comes from electron- positron annihilation. Annihilation rates on large molecules have been carried out in positron traps as a data base for diagnostics in neutral plasmas and modeling of the strong astrophysical radiation from positron annihilation. And of course, antimatter plasmas stimulate the Sci-Fi imagination.

A new trap that uses positrons from a linear accelerator, thus avoiding the inefficiencies of the moderator has just become operational. About 10^4 positrons per pulse can be trapped. Since it is a high duty cycle Linac it should be possible to test the density limits of positron plasmas by pulse stacking.

Electrons or positrons are difficult to cool. Laser cross sections are too small to make any kind of laser cooling practical. One must rely on radiation cooling. One way to overcome these problems is to use a cylindrical microwave Penning trap which is constructed to be an approximation to an ideal microwave cavity. The cavity walls are split so that sections form the electrodes of a Penning trap in which from one electron to more than 10^5 electrons can be suspended and studied. Electrical techniques have been refined to the point where it is now possible to observe and study a single particle with good signal-to-noise without ejecting it from the trap. Coupling of the electron cyclotron motion to a dissipative load via the electromagnetic cavity modes provides a means of cooling the electrons or positrons. These traps provide a means of precision measurements of fundamental constants of elementary particles such as the magnetic-moment. In the opposite limit these traps also make it possible to experimentally study in detail the transition from single particle behavior of one electron to the collective plasma dynamics of many electrons, i. e. the "plasma physics hydrogen atom."

A new positron trap designed to be used in conjunction with an antiproton trap to make antihydrogen was reported. This trap uses a injection technique that preserves the hard vacuum and can be cooled to 4 degrees Kelvin. The positron-antiproton recombination rate is predicted to change by more than six orders of magnitude as one lowers the plasma temperature from room temperature to 4 degrees Kelvin. In addition positron-antiproton recombination rates are significantly different in the strong magnetic field limit.

At present antiprotons are decelerated from a GeV to 6 MeV and injected into a Penning trap partially filled with 4 degree Kelvin electrons. The antiprotons are cooled via electron cooling by a factor of 10^{10}. These traps offer the possibility of an elementary particle bucket. One can go to the accelerator facility, fill your trap with an elementary particle plasma and take it to your lab to do the experiments.

Particle transport across magnetic field lines is considered by many to be the most important problem in plasma physics. It is key to the design of a confinement system for thermonuclear fusion plasmas. At the temperatures and densities of laboratory thermonuclear plasmas the transport is primarily due to wave-particle collisions. The rapid grow of electron cyclotron reactors for plasma processing of microelectronics has created a need to understand the transport from the collision dominated low temperature gas discharge region through the sheath where wave-particle collisions dominate to the surface.

Transport experiments are much more difficult than measurements of plasma wave properties. In fact some people think they are so difficult, that to do a definitive experiment one must resort to single component non-neutral plasmas. One great advantage is the existence of a rigid rotor thermal equilibrium. This is a Maxwellian in the rotating frame. There is no thermal equilibrium for neutral plasmas. They are, at best, steady state plasmas with a local thermal equilibrium approximation.

Another great advantage of like particle transport is the constraint imposed by the conservation of canonical angular momentum. If one particle moves out 10 radii, then 10 particles must move in one radii. As a result, wave-particle transport due to instabilities in the plasma is driven towards thermal equilibrium, not to loss of confinement as in neutral plasmas. In one set of experiments carried out at the kinetic level, the approach to thermal equilibrium was studied by looking at the equilibration of the parallel and perpendicular components of temperature. The data covered eight orders of magnitude in the ratio of electron cyclotron radius to impact parameter. There was close agreement between traditional theory and experiment in the warm plasma low magnetic field regime. In the cold high magnetic field limit were the plasma approached a pure electron liquid a new theory based one $\mathbf{E} \times \mathbf{B}$ collisions had to be developed. A new many-particle adiabatic invariant was discovered during the course of the work. This work represents a new standard of excellence in plasma physics.

A new magnesium ion plasma with in situ Laser Induced Fluorescence measurements of temperature and density profiles has become operational. A novel feature of this experiment is the addition of a rotating electrostatic quadrupole to compensate for losses due to azimuthal asymmetries. Confinement times of 10 days are routinely observed.

Electron streams have been a source and means of amplifying coherent electromagnetic radiation since the invention of electron tubes. Current research in this area focuses on high power (Megawatt) short radiation (microwave to X ray) sources such as free electron lasers and gyrotrons. The intensity of the beams in these devices is sufficiently high that the self fields affect the particle dynamics to the extent they should be considered as streaming electron plasmas. The beam quality of high energy accelerators is determined in large part by the low energy injection and transport section where the beam is a streaming plasma. In these traps we can do experiments on "beams at rest." That is we can carry our detailed experiments that are related to the dynamics of beams in the rest frame in a regime where self field effects are important.

New insights into the high frequency cyclotron modes of an electron plasma have been obtained. Landau damping, nonlinear trapping similar that observed in space charge waves has been observed for the cyclotron modes. Electron Bernstein modes have also been observed, and used to infer the temperature. This work is related to crossfield microwave devices.

New and better diagnostics of non-neutral have been developed since the ONR

Symposium which has created new challenges to our understanding. This includes the new experiment designed to use Laser Induced Fluorescence and new cameras to image the profile in a single shot have made for some interesting new experimental discoveries has led to challenges of assumptions in the field. It has been discovered there is a limit in which asymmetric electron plasmas of almost any shape are confined very well. These asymmetric plasmas are locally stable if the energy is a maximum when compared to the energy of all the nearby accessible states.

The assumption of the scaling of confinement time with length and magnetic field due to collective effects has been challenged. That is, confinement time is independent of density and decreases with length. Particle losses appear to be more characteristic of accelerators than plasma confinement experiments.

Our ideas about Debye shielding are being called in to question. A recent experiment has created a spherical ion plasma at the Brillouin limit by passing an electron beam through a low pressure gas. Dusty plasmas have played a role in the workshop. These are inexpensive ways to study strongly coupled plasmas. There is an apparatus under construction to take advantage of the ideal spherical symmetry that can be achieved in these traps to carry out an inertial electrostatic fusion experiment.

Computer simulations are playing an increasing role in the field. Examples include predictions of the transition to an ion crystal lattice, confirmation of the asymmetric maximum energy states, analysis of the reflection at the ends and boundary effects on the diocotron instability.

Because we can make a pure electron plasma in a known equilibrium state with a loss rate proportional to the pressure, then with gas of known cross section we have the foundations of a new absolute pressure standard for the measurement of vacuum in a region where none now exists.

The similarity of these traps to cyclotron mass spectrometers was pointed out at the ONR Symposium. Speakers from that community have been invited to both the Irvine and Berkeley workshops. This year one of the recent graduates of the program who has been working in the cyclotron mass spectrometer community gave a talk on the relationships of the two cultures. Given the quality of research in this field and needs of the other it will be a surprise if the non-neutral plasma community does not have a significant impact on cyclotron mass spectrometer research.

The field has benefited and grown as a result of cross fertilization. It is starting to impact a variety of other areas of science as well as grow as a unique field. The study of non-neutral plasmas in traps is the keystone that connects parts of atomic and molecular physics, fluid dynamics, condensed matter physics, astrophysics and antimatter physics with a variety of technologies such as atomic clocks, cyclotron mass spectrometers and a new standard for a pressure gauge, coherent radiation sources, accelerators, and fusion.

References

[1] Thomas S. Kuhn, *The Structure of Scientific Revolutions*, Chicago Press, Chicago, (1970).

[2] C.W. Roberson and C.F. Driscoll, *Non-Neutral Plasma Physics*, American Institute of Physics, New York, (1988).

[3] R.C. Davidson, *Physics of Non-Neutral Plasmas*, Frontiers in Physics Series, Addison Wesley, Reading, Massachusetts, (1990).

Long Ion Plasma Confinement Times with a "Rotating Wall"

F. Anderegg, X.-P. Huang, C.F. Driscoll, G.D. Severn,*
and E. Sarid,†

Institute for Pure and Applied Physical Sciences and Physics Department
University of California at San Diego, La Jolla, CA 92093

*University of San Diego, San Diego, CA 92110

†NRCN, P.O. 9001, Beer Sheva 84190, Israel

Abstract. Static field errors in a Penning-Malmberg trap exert a drag on confined non-neutral plasmas, causing radial expansion and loss. We suppress this transport by applying an electrostatic wall asymmetry rotating faster than the plasma. This results in inward radial transport and plasma compression. The experiments are performed on a magnesium ion plasma in a magnetic field of 4 Tesla, with *in situ* Laser Induced Fluorescence(LIF) measurement of density and temperature profiles. Confinement of ions for up to 10 days is routinely observed.

INTRODUCTION

It is well known that static azimuthal asymmetries and collisions with neutral particles exert a drag on rotating non-neutral plasmas, reducing the plasma angular momentum and consequently causing radial expansion and losses. The transport due to collisions with neutrals is well understood, and is negligible for electrons at pressures below 10^{-7} Torr [1]. Anomalous transport presumably caused by static azimuthal asymmetries in the electric or magnetic fields has been experimentally characterized and extensively documented [2], but the underlying physics is still unclear. One noticeable exception is the case where the asymmetry is near the column ends [3,4].

The present work is based on the idea that angular momentum can be "injected" into the plasma by a rotating electrostatic asymmetry on the wall; we will refer to this as a "rotating wall." Here, the applied perturbation has azimuthal mode number $m = 2$ and is phased to rotate somewhat faster than the maximal plasma rotation frequency f_R. A similar concept has been applied to small ion clouds: when a laser beam torques on the plasma, compression through the Brillouin limit has been observed [5]. The simplicity of the rotating wall concept may be attractive for applications requiring control of the size of an ion cloud, such as in high precision measurements using ion plasmas

[6]. Earlier work on rotating asymmetries reported only small increases in the central density [7] or small increases in the angular momentum limited by heating of the electron plasma [8]. These early works used $m = 1$ rotating asymmetries at high frequencies compared to the plasma rotation frequency, typically near a plasma mode. More recently, the $m = 2$ low frequency rotating wall technique has been successfully applied to electron plasmas where the heating was balanced by cyclotron radiation [9].

APPARATUS

A new ion machine, denoted IV, shown schematically in Fig. 1, has been recently built and is now operational. The goal of this apparatus is to measure test particle transport in a variety of transport regimes. IV is a standard Penning-Malmberg trap with a wall radius $R_w = 2.86$ cm and for the present work a uniform magnetic field $B_z = 4$T and a plasma length $L_p = 11.4$ cm. The choice of the magnesium ion was dictated by optical considerations for the LIF diagnostic and by scaling laws from electron to ion plasmas [10].

The Mg ions are produced with a Metal Vapor Vacuum Arc source [11], creating a neutral plasma from the magnesium of the cathode. The positive confining potentials applied to the end gates then trap the ions, while the electrons escape freely. In order to avoid creating other "gratuitous" electron traps, the potentials are made monotonic at each end of the confining region. During plasma injection, the background pressure (97% H_2) is increased to $\sim 10^{-8}$ Torr for 2 minutes to cool the ions from ~ 20 eV down to ~ 0.1 eV. This allows better LIF measurements.

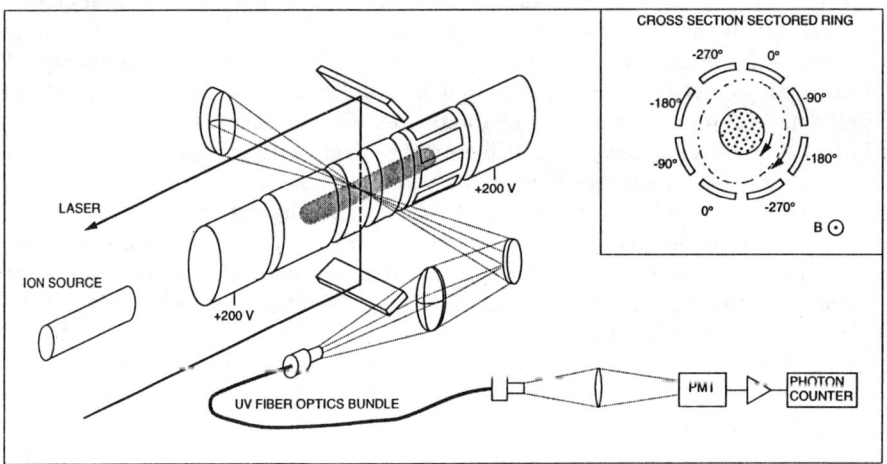

Figure 1: Experimental setup with LIF diagnostic and "rotating wall." A uniform magnetic field B = 4T is aligned with the trap axis. The plasma length is $L_p \cong 11.4$ cm.

One of the confining ring electrodes is sectored, consisting of eight 27° sectors positioned every 45°. Electrostatic asymmetries varying as $\cos(2\theta + 2\pi f t)$ are created with small forward and reverse-propagating harmonics also created from the discrete sectors. Following the argument that asymmetries applied at the end of the column can affect all the ions due to the axial bounce motion, the sectored ring was placed at one end of the column [3]. The frequency of the rotating wall asymmetry is fixed at 20 kHz (0.5V pp), chosen to be more than m times the plasma rotation frequency at all radii.

The main diagnostic of the apparatus is LIF [12]. We chose a CW laser based on signal to noise considerations and for optimum velocity resolution. Transitions from the ground state ($3s^2 S_{1/2}$) to the lowest excited states ($3p^2 P_{1/2}$, or $3p^2 P_{3/2}$) of a Mg^+ ion are in the 280 nm range, obtained by frequency doubling of the CW ring dye laser. This setup uses a Bergquist scheme [13]: A nonlinear crystal (BBO) in an external stabilized cavity converts the tunable fundamental beam (560 nm, ~ 0.7W) into an ultra-violet beam (280 nm, ~ 15 mW) with a bandwidth of 1 MHz. The diagnosed volume of plasma is determined by the intersection of the laser beam and the detection volume, typically a cylinder of ~ 1 mm diameter by ~ 3 mm long. The beam is scanned across the entire ring diameter. At each radial location, the laser frequency is scanned to obtain the ion velocity distribution, and the density, temperature and plasma rotation frequency are derived from this. A complete scan with 31 radial positions takes ~ 90 sec. The laser-measured charge densities are as yet only roughly calibrated, from dumped charge measurements.

RESULTS

The evolution of the central density with (triangle) and without (circle) the rotating wall asymmetry is shown in Fig. 2. For the first 400 sec, the central density decays essentially the same with or without the rotating wall. With the rotating wall, the decay of the central density is drastically reduced after 400 sec, and experiences little change after 5000 sec. In contrast, without the rotating wall, all the ions are lost by 2000 sec.

The rotating wall perturbation causes substantial heating, which appears to be balanced by cooling from collisions with the neutral background. Note that at $P \sim 7 \times 10^{-11}$ Torr and Ti ~ 4 eV, the ion cooling time is about 400 sec.

We have established that the rotating asymmetry can either increase or decrease the angular momentum (*i.e.* mean-square radius $\langle r^2 \rangle = \langle R^2 \rangle / R_w^2$) of the plasma, depending on the direction of rotation. We performed the following experiment: after the plasma was confined for ~ 5 days with a wall signal of 20 kHz (0.4V pp) we reversed the rotation direction of the wall perturbation, leading to a rapid loss of angular momentum at a rate $d\langle r^2 \rangle / dt$ of $\gamma_{rev} = 5.3 \times 10^{-4}$ sec^{-1} as shown in Fig. 3. Density and temperature profiles (A) and (B) reflect these changes; note that a reversed rotating wall signal does not create any substantial heating. After 2400 sec, ions start to escape (B), and the positive rotation direction is re-established resulting in a slow recompression of the plasma ($\gamma_{fwd} = -1.9 \times 10^{-4}$ sec^{-1}). Interestingly, the

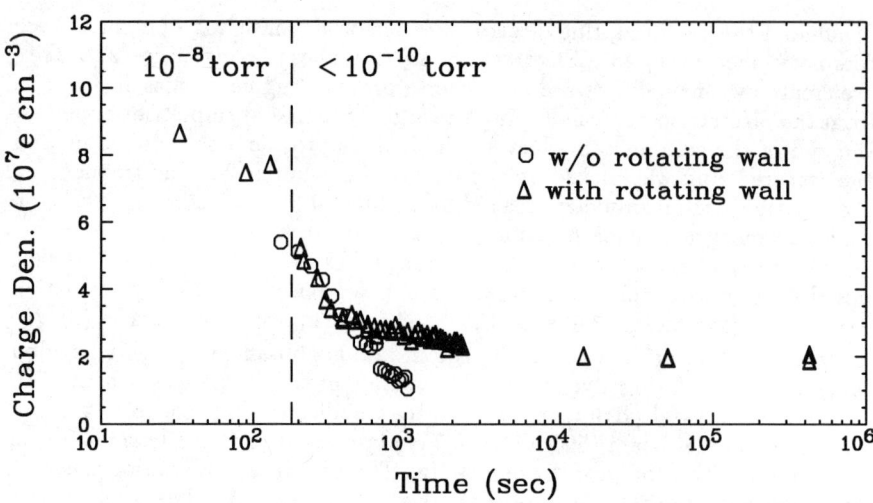

Figure 2: Central charge density versus time with and without the rotating wall asymmetry.

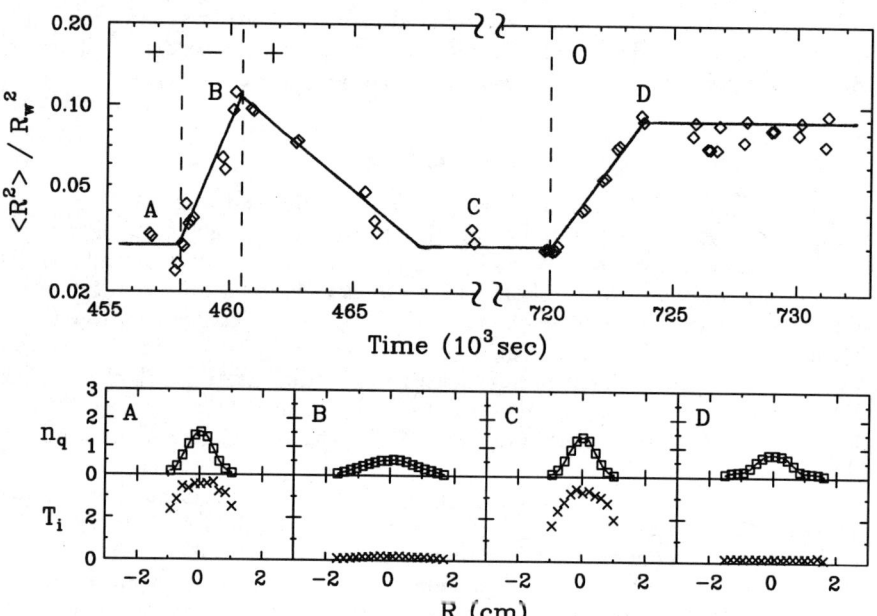

Figure 3: Plasma mean square radius $\langle r^2 \rangle = \langle R^2 \rangle / R_w^2$ versus time during forward rotating (+), reversed (−) and no (0) wall perturbations of 20 kHz and 0.4V pp. Also shown are ion density n_i [10^7 cm^{-3}] and temperature T_i [eV] versus radius at 4 times A-D.

plasma profile after recompression (C) is essentially the same as before expansion (A). Eventually, after 8 days, we let the plasma evolve without signal applied to the wall, and angular momentum is lost ($\gamma_{\text{free}} = 3.0 \times 10^{-4}$ sec^{-1}) resulting in ions being lost to the wall (D). From these rates, one would conclude that the rotating wall torque in the forward direction ($\gamma_{\text{fwd}} - \gamma_{\text{free}} = -4.9 \times 10^{-4}$ sec^{-1}) is about twice the torque in the reverse direction ($\gamma_{\text{rev}} - \gamma_{\text{free}} = 2.3 \times 10^{-4}$ sec^{-1}).

Theoretically [3], one expects that compression will result only if the asymmetry rotation rate is greater than the plasma rotation rate, *i.e.* $f/m > f_R$. The steady-state profile A in Fig. 3 has a laser-measured rotation frequency of $f_R = 7$ kHz, from an applied frequency $f = 20$ kHz. Reducing the applied frequency results in lower charge density, but to date the use of higher frequency has not lead to higher density. Calculation by Crooks and O'Neil [3] predicts a radial flux which can be directed inward or outward depending on the difference between f/m and f_R. Future experiments may provide quantitative comparison to theory, including measurement of the local fluxes and dependencies on drive and plasma rotation frequencies.

CONCLUSIONS

We have experimentally demonstrated that an electrostatic "rotating wall" can be used to confine ion plasmas for an arbitrary long time. This perturbation changes the angular momentum of the plasma and modifies the equilibrium density profile. This technique might have applications to control the size of an ion cloud in precision measurements, if the heating effect can be balanced by cooling.

ACKNOWLEDGMENTS

We gratefully acknowledge the support of the Office of Naval Research (ONR N000144-89-J1714). One of us (G.D.S.) was supported by a National Science Foundation Research Opportunity Award. We also thank T.M. O'Neil for suggesting the theoretical concept of a "rotating wall." One of us (F.A.) would like to thank J.C. Bergquist for advice on laser frequency doubling.

REFERENCES

1. Malmberg, J.H., and Driscoll, C.F., *Phys. Rev. Lett.* **44**, 654 (1980).
2. Driscoll, C.F., Fine, K.S., and Malmberg, J.H., *Phys. Fluids* **29**, 2015 (1986).
3. Crooks, S.M., and O'Neil, T.M., "Transport due to Rotational Pumping," in this conference proceedings.
4. Cluggish, B., and Driscoll, C.F., "Transport and Damping from Rotational Pumping," in this conference proceedings.
5. Heinzen, D.J., Bollinger, J.J., Moore, F.L., Itano, W.M., and Wineland, D.J., *Phys. Rev. Lett.* **66**, 2080 (1991).

6. Bollinger, J.J., private communication.

7. Eggleston, D.L., O'Neil, T.M., and Malmberg, J.H., *Phys. Rev. Lett.* **53**, 982 (1984).

8. Mitchell, T.B., Ph.D. dissertation, UCSD (1993), p. 58.

9. Pollock, R.E., and Anderegg, F., "Spin up of an Electron Plasma-First Results," in this conference proceedings.

10. Driscoll, C.F., "Containment of Single-Species Plasmas at Low Energies," in *Low Energy Antimatter*, Singapore: World Scientific, 1986, pp. 184-195.

11. Brown, I.G., Galvin, J.E., MacGill, R.A., and Wright, R.T., *Appl. Phys. Lett.* **49**, 1019 (1986).

12. Hill, D.N., Fornaca, S., and Wickham, M.G., *Rev. Sci. Inst.* **54**, 309 (1983).

13. Bergquist, J.C., Hemmati, H., and Itano, W.M., *Optics Comm.* **43**, 437 (1982).

Electrostatic Normal Modes in Nonneutral Plasmas

DAVID L. BOOK

Enigmatix, Inc., P.O. Box 11834, Washington, DC 20008

ABSTRACT

A fluid description is employed to derive the dispersion relation for modes near the cyclotron frequency Ω in a nonuniform cylindrical nonneutral plasma of radius R with finite temperature confined by a uniform magnetic field $\mathbf{B} = B_0 \mathbf{e}_z$. In contrast to the theory of Gould and LaPointe, the model includes the diamagnetic drift but omits finite-Larmor-radius effects. The eigenfrequencies for high-frequency electrostatic modes with wavevectors satisfying $\mathbf{k} \cdot \mathbf{B} = 0$ (Bernstein modes) are found in the form $\omega = -\Omega + \Delta\omega$. Solutions are obtained and compared with experiment and the theory of Gould and LaPointe. The present theory predicts that at a given temperature modes with $m > 1$ propagate only when the density is less than a critical value that increases with m, and that $\Delta\omega$ normalized by the diocotron frequency depends only on the ratio of the Debye length to the plasma radius and hence is independent of B and the particle mass.

I. BACKGROUND

Gould and LaPointe[1] observed electrostatic azimuthally propagating [i.e., varying as $\exp(im\theta - i\omega t)$] eigenmodes for ω near the cyclotron frequency $\Omega = qB/Mc$ in a cylindrical electron plasma. They found that the $m = 1$ mode is downshifted from Ω by the diocotron frequency, while for $m > 1$ families of modes appear, upshifted by varying amounts. In connection with their measurements they employed a theoretical description[1,2] which included the effect of finite temperature by retaining the lowest-order finite-Larmor-radius (FLR) correction in the perpendicular dielectric function. Although the theory of Gould and LaPointe agrees well with their experiment, particularly for $m = 1$, their neglect of pressure forces in the equilibrium condition and the treatment of the perturbations may be unjustified. Moreover, their derivation relies to some extent on analysis by Pearson[3] which assumed that the unperturbed plasma is in an exponential rigid-rotor equilibrium, whereas the plasma in the experiments of Gould and LaPointe rotated nonrigidly and probably never reached a state of global thermodynamic equilibrium. Numerous distinct kinetic (Vlasov) equilibria can be constructed that are consistent with the observed density profile to within the accuracy of the measurements.

The present theory is intended to complement that of Gould and LaPointe. Having in mind eventual applications to the ion experiments of Sarid et al.[4] and ion cyclotron spectrometry, we regard the charge as positive. We make no use of the microscopic description in terms of single-particle constants of motion,[5] but instead start from the ideal fluid equations with a finite transverse pressure, which presuppose only local thermodynamic equilibrium. This allows the radial dependence of the density and temperature to be specified arbitrarily in the unperturbed state. The r-dependent rotation frequency ω_r and the electric field follow from radial force balance and the Poisson equation, respectively. Like Gould and LaPointe,[1] we make the experimentally validated approximations that the plasma frequency is small compared with the gyrofrequency and the particle Larmor radius is small compared with the plasma radius.

We linearize the fluid equations, assuming that all quantities are subjected to a small flutelike ($k_z = 0$) perturbation. The assumption that the equilibrium is isentropic simplifies treatment considerably. Solution of the linearized equations then yields a fourth-order differential equation in r for the amplitude of the perturbed electrostatic potential (in contrast to the cold-plasma case,[5] which leads to a second-order equation). For modes with $\omega \sim -\Omega$, this equation reduces approximately to second order. If the density and temperature profiles are parabolic, the eigenvalue problem closely resembles that treated by Gould and LaPointe.[1] The main difference is that the temperature has to exceed a certain threshold for Bernstein modes with $m > 1$ to be trapped, and the frequency shift normalized by the diocotron frequency is a function only of the Debye length divided by the plasma radius.

II. THE MODEL

The number density n, radial and azimuthal velocities u and v, temperature T (expressed in units of energy), and electrostatic potential ϕ, satisfy the continuity equation, the inviscid momentum equation with a Lorentz force term, the adiabatic law, and the Poisson equation. In equilibrium the radial component of the momentum equation yields

$$-\frac{v^2}{r} + \frac{1}{nM}\frac{d}{dr}(nT) = \frac{q}{M}\left(E + \frac{1}{c}vB\right),\tag{1}$$

where we assume that $n(r)$ and $T(r)$ are specified, and E is found by solving the Poisson equation

$$\frac{1}{r}\frac{d}{dr}(rE) = 4\pi nq.\tag{2}$$

This is identical with the single-species form of the radial force balance condition for a general rigid-rotor equilibrium in the *kinetic* (Vlasov) description [Eq. (4.41) of Ref. 5b]. For slow rotation $|\omega_r| \ll \Omega$, where $\omega_r(r) = v/r$, Eq. (1) simplifies to

$$\omega_r \approx \frac{1}{r\Omega}\left[\frac{1}{nM}\frac{d}{dr}(nT) - \frac{qE}{M}\right],\tag{3}$$

From Eqs. (2) and (3) we see that a negative pressure gradient gives rise to a diamagnetic drift in the same direction as the $\mathbf{E} \times \mathbf{B}$ drift, whereas FLR corrections tend to reduce the effective electric field strength.[1]

If first-order quantities (labeled with subscript 1) are taken to vary as $\exp(im\theta - i\omega t)$, the linearized equations are

$$-i\omega' n_1 + \frac{1}{r}\frac{d}{dr}(rnu_1) + \frac{im}{r}nv_1 = 0; \tag{4}$$

$$-i\omega' u_1 - \Gamma v_1 + \frac{1}{nM}\left[\frac{d}{dr}(n_1 T + nT_1) - \frac{n_1}{n}\frac{d}{dr}(nT)\right] = -\frac{q}{M}\frac{d\phi_1}{dr}; \tag{5}$$

$$-i\omega' v_1 + \Gamma' u_1 + \frac{im}{nMr}(n_1 T + nT_1) = -\frac{q}{M}\frac{im}{r}\phi_1; \tag{6}$$

$$-i\omega'\left[T_1 n^{1-\gamma} - (\gamma - 1)Tn^{-\gamma}n_1\right] + u_1\frac{d}{dr}(Tn^{1-\gamma}) = 0; \tag{7}$$

$$\frac{1}{r}\frac{d}{dr}\left(r\frac{d\phi_1}{dr}\right) - \frac{m^2}{r^2}\phi_1 = -4\pi q n_1, \tag{8}$$

where we have introduced the first-order electrostatic potential ϕ_1, $\Gamma = \Omega + 2\omega_r$, $\Gamma' = \Gamma + rd\omega_r/dr$, and γ is the adiabatic index.

III. SIMPLIFICATION OF THE LINEARIZED EQUATIONS

The equations simplify considerably in the case of isentropic equilibrium ($Tn^{1-\gamma}$ independent of r). Solving for u_1 and v_1 and substituting in Eq. (4), we find the first-order density in the form

$$-\omega' n_1 = \frac{\omega'}{\Omega^2}\left[\frac{1}{r}\frac{d}{dr}\left(\frac{rn}{D}\frac{d\Upsilon_1}{dr}\right) - \frac{m^2 n}{r^2 D}\Upsilon_1 - \frac{m\Upsilon_1}{r\omega'}\frac{d}{dr}\left(\frac{n\Gamma}{D}\right)\right], \tag{9}$$

where $\omega' = \omega - m\omega_r$, $\Upsilon_1 = q\phi_1/M + v_0{}^2 f^\nu n_1/n_0$, $D = (\omega'^2 - \Gamma\Gamma')/\Omega^2$, $v_0 = (\gamma T_0/M)^{1/2}$ is the thermal velocity, n_0 and T_0 are the values of the unperturbed density and temperature at $r = 0$, $f(r) = n(r)/n_0$, and $\nu = \gamma - 2$.

It is convenient to use $x = r^2/R^2$ in place of r as the independent variable. Considerable simplification results if we take $\delta \ll 1$, $\varepsilon \ll 1$, $|\eta| \ll 1$, where $\delta = 4\pi n_0 q^2/M\Omega^2$, $\varepsilon = 4v_0{}^2/\Omega^2 R^2$, and $\eta = 2\Delta\omega/\Omega = 2(\omega + \Omega)/\Omega$. Now the slow rotation frequency (2) becomes $\omega_r \approx -[\delta g(x) + \varepsilon h(x)]\Omega/2$, where $g(x) = x^{-1}\int_0^x dx' f(x')$ and $h(x) = -f^\nu(df/dx)$. (The functions f and g are the same as those used by Gould and LaPointe.[1]) The determinant in Eq. (9) can then be approximated as

$$D = -\eta + \delta[f(x) + (1 - m)g(x)] + \varepsilon[(2 - m)h(x) + x(dh/dx)]. \tag{10}$$

This reduces to the denominator of Eq. (1) in Ref. 1 in the limit $\varepsilon \to 0$ if due allowance is made for signs and normalization. Everywhere else we take $\Gamma \approx \Gamma' \approx -\omega' \approx \Omega = \text{const.}$

We introduce a new dependent variable

$$\chi_1 = \frac{qx}{M}\left(\frac{d\phi_1}{dx} + \frac{m\phi_1}{2x}\right) = x^{1-m/2}\frac{q}{M}\frac{d}{dx}\left(\phi_1 x^{m/2}\right). \tag{11}$$

In terms of x and χ_1, Eq. (9) becomes

$$-\frac{n_1}{n_0} = \frac{4}{\Omega^2 R^2}\left(\frac{d}{dx} - \frac{m}{2x}\right)\left[\frac{f\chi_1}{D} + v_0{}^2\frac{xf}{D}\left(\frac{d}{dx} + \frac{m}{2x}\right)\left(f^\nu\frac{n_1}{n_0}\right)\right], \qquad (12)$$

and Eq. (8) can be written

$$\frac{4}{\Omega^2 R^2}\left(\frac{d}{dx} - \frac{m}{2x}\right)\chi_1 = -\delta\frac{n_1}{n_0}. \qquad (13)$$

Combining (13) and (14) and integrating we get

$$\varepsilon x f\left(\frac{d}{dx} + \frac{m}{2x}\right)\left[f^\nu\left(\frac{d}{dx} - \frac{m}{2x}\right)\chi_1\right] + (D - \delta f)\chi_1 = C^{(i)}Dx^{m/2}, \qquad (14)$$

where $C^{(i)}$ is constant. This inhomogeneous second-order equation for χ_1 is similar to Eq. (4) of Ref. 1a. The most important difference is that the second derivative term, which arises from the diamagnetic drift, has the opposite sign here from that in Ref. 1, where it is a consequence of FLR effects.

IV. THE EIGENVALUE PROBLEM FOR m = 1

Following Gould and LaPointe,[1] we specialize to the case of a parabolic (in r) density profile, $f(x) = 1 - x$. We introduce a new dependent variable $\psi_1 = x^{-m/2}f^\nu\chi_1$. For $\nu = 0$ the isentropicity condition implies that the temperature profile is also parabolic, and in terms of ψ_1 the potential equation becomes

$$\varepsilon(1 - x)\left(x\frac{d^2\psi_1}{dx^2} + m\frac{d\psi_1}{dx}\right) + [D - \delta(1 - x)]\psi_1 = C^{(i)}D, \qquad (15)$$

where

$$D = -\eta + (\delta + \varepsilon)(2 - m) + \delta(m - 3)x/2. \qquad (16)$$

The corresponding equation in the case $\nu = -1$ is

$$\varepsilon\left[x\frac{d^2\psi_1}{dx^2} + \left(m - \frac{x}{1 - x}\right)\frac{d\psi_1}{dx}\right] + \left[D - \delta(1 - x) - \frac{m}{1 - x} - \frac{x}{(1 - x)^2}\right]\psi_1 = \frac{C^{(i)}D}{1 - x}, \qquad (17)$$

where

$$D = -\eta + \delta[2 - m + (m - 3)x/2] + \varepsilon\left[\frac{2 - m}{1 - x} + \frac{x}{(1 - x)^2}\right]. \qquad (18)$$

We seek solutions of Eq. (15) or (17) for which ϕ_1 is finite at $x = 0$ and vanishes on a conducting wall at $x = x_0 = (R_0/R)^2 \geq 1$. In the case $x_0 > 1$ the potential in the region $1 < x \leq x_0$ is the solution of Laplace's equation that vanishes at $x = x_0$:

$$(q/M)\phi_1^{\text{ext}} = C^{(e)}\left[(x/x_0)^{m/2} - (x_0/x)^{m/2}\right], \qquad (19)$$

where $C^{(e)}$ is constant. The corresponding form of ψ_1 is

$$\psi_1^{\text{ext}} = mC^{(e)}x_0^{-m/2}. \tag{20}$$

The matching condition is the requirement that ϕ_1 and its first derivative (or equivalently, ψ_1) be continuous at $x = 1$. Another condition, found by integrating Eq. (4), is that nu_1 go to zero at $x = 1$.

For $m = 1$ the requirement of finiteness or the continuity of nu_1 implies that ψ_1 vanishes at $x = 1$, which in turn implies $C^{(ii)} = 0$. Hence the conditions for continuity of the first-order potential and electric field at $x = 1$ reduce to

$$C^{(i)}(\delta/2 - \eta) = C^{(e)}\left(x_0^{-1/2} - x_0^{1/2}\right) \tag{21}$$

and

$$(1/2)C^{(i)}(\delta - \eta - 3\delta/2) = (1/2)C^{(e)}\left(x_0^{-1/2} + x_0^{1/2}\right). \tag{22}$$

Taking the ratio of Eqs. (22) and (23) we obtain

$$\frac{\delta + 2\eta}{\delta - 2\eta} = \frac{x_0 + 1}{x_0 - 1}, \tag{23}$$

or

$$\frac{\Delta\omega}{\Omega} = \frac{\pi n_0 q^2}{M\Omega^2}\left(\frac{R}{R_0}\right)^2 = \frac{\langle\omega_p^2\rangle}{2\Omega^2}, \tag{24}$$

where the angle brackets denote the average over the cross section of the chamber, i.e., $0 \leq r \leq R_0$. Hence there is a shift in frequency toward smaller magnitudes by an amount equal to the diocotron frequency. This agrees with the result obtained by Gould and LaPointe.[1]

V. THE EIGENVALUE PROBLEM FOR m > 1

For $m > 1$ we assume $\nu = -1$. We follow Gould and Lapointe[1] in looking for modes which are radially trapped close to the axis, i.e., at $x \ll 1$. Writing $(1 - x)^{-1} \approx 1 + x$ we have

$$\varepsilon\left[x\frac{d^2\psi_1}{dx^2} + (m - x)\frac{d\psi_1}{dx}\right] - \left[\eta + (\delta + 2\varepsilon)(m - 1)\right.$$
$$\left. -(x/2)(\delta - 4\varepsilon)(m - 1)\right]\psi_1 = C^{(i)}D(1 + x). \tag{25}$$

If we write the homogeneous part of ψ_1 as $\tilde{\psi}_1\exp(\kappa x)$ and take

$$\kappa = \frac{1}{2}(1 \pm \Delta), \tag{26}$$

where

$$\Delta = [8m - 7 - 2(m - 1)\delta/\varepsilon]^{1/2}, \tag{27}$$

then $\tilde{\psi}_1$ satisfies

$$x\frac{d^2\tilde{\psi}_1}{dx^2} + [m + (2\kappa - 1)x]\frac{d\tilde{\psi}_1}{dx} - \left[\frac{\eta}{\varepsilon} + \left(\frac{\delta}{\varepsilon} + 2\right)(m-1) - \kappa m\right]\tilde{\psi}_1 = 0. \qquad (28)$$

Taking the lower sign in (26) to ensure that $\tilde{\psi}_1$ decreases for large x and writing $\alpha = [\eta + (m-1)(\delta + 2\varepsilon) - m\kappa\varepsilon]/(\varepsilon\Delta)$, $\beta = m$, and $z = \Delta x$, we see that $\tilde{\psi}_1(z)$ satisfies the confluent hypergeometric equation

$$z\frac{d^2y}{dz^2} + (\beta - z)\frac{dy}{dz} - \alpha y = 0. \qquad (29)$$

The solution of Eq. (29) which is well behaved at the origin is the Kummer function[6] $\Phi(\alpha, \beta; z)$. Its power-series expansion terminates if α equals a nonpositive integer $-l$, so that ψ_1 is well behaved for large argument also. Hence the dispersion relation is

$$\eta = -(m-1)(\delta + 2\varepsilon) + m\varepsilon/2 - \varepsilon\Delta(l + m/2), \qquad (30)$$

$l = 0, 1, 2, \ldots$. We see that the condition for $\Delta > 1$ ($\kappa < 0$) is $\varepsilon > \delta/4$.

VI. DISCUSSION

For $m = 1$ the present model unsurprisingly yields the same result (a down-shift equal to the diocotron frequency) as that of Gould and LaPointe.[1] One would expect this behavior, which represents a center-of-mass displacement of the plasma as a whole, to be insensitive to the shape of either temperature or density profile. For $m > 1$, however, we have seen that trapped modes to exist (in the case of a uniform temperature profile) only if the nondimensional temperature ε exceeds a certain minimum value proportional to the density. Note that Eq. (31) says that the scaled frequency shift $\lambda = \eta/\delta$ is a function only of $(\varepsilon/\delta)^{1/2}$, i.e., the Debye length divided by the plasma radius, and so does not depend on B or M. Since Eq. (31) contains terms that are not proportional to δ, the quantity λ *increases* as the density falls. This is inconsistent with the observations unless the temperature is decreasing faster than the density. The theory of Gould and LaPointe [Eq. (4) of Ref.1a; Eq. (6) of Ref. 1b] predicts that λ should be independent of density. Both the the present theory and that of Gould and LaPointe[1] predict that the spacing between adjacent modes, which is constant within a given family, should increase with m. The observations of Ref. 1 do not appear to support this. Equation (31) predicts that the frequency shift increases with l, the number of nodes in the radial profile of the eigenmode, while the theory of Ref. 1 makes the opposite prediction. Observations may conceivably decide which is correct.

One attraction of a simple soluble model like the present one is that it can provide a better understanding of the physics than a more elaborate one that has to be solved numerically. The use of the adiabatic law to describe the pressure, however, may be an oversimplification. Usually the adiabatic law is invoked for processes that are fast compared to outside influences and losses (which is true here) but slow compared with microscopic processes (which is not).

The justification for omitting finite-Larmor-radius effects is more convincing. To lowest order these reduce the effective electric field by a relative amount $k^2 r_L^2/2$, where k is the local wavenumber. This is equivalent to replacing ϕ_1 by $\phi_1 - (M/q)\delta v_0^2 f^\nu (n_1/n_0)$ in the linearized momentum equations, which leads to an equation differing from (17) by terms of order $\delta\varepsilon$. Consequently, the FLR corrections in λ found in Ref. 1 are smaller by $\varepsilon^{1/2}$ than the Doppler shift. In contrast, the diamagnetic drift introduces terms in the expression for the eigenfrequency that are the same size as the Doppler shift if we take δ and ε to be formally of the same order.

A possible extension of this work is to the perturbative treatment of multiple species with similar masses, at least when one of them dominates the others. This prospect, which was one motivation for studying the present model, has potentially important applications to the experiments of Sarid et al.[4] and ion mass spectrometry, but it can only be realized when the eigenvalues and eigenstates of the single-species case are throughly understood.

ACKNOWLEDGEMENTS

I am grateful to Profs. R. W. Gould and E. Sarid for many helpful discussions and for allowing me to read their manuscripts prior to publication. This work was supported by ONR under contract N00014-94-C-0012.

REFERENCES

1. R. W. Gould and M. A. LaPointe, (a) *Phys. Rev. Lett.* **67**, 3685 (1991); (b) *Phys. Fluids* B4, 2038 (1992).

2. R. W. Gould (to be published).

3. G. A. Pearson, *Phys. Fluids* **9**, 2454 (1966).

4. E. Sarid, F. Anderegg, and C. F. Driscoll, *Bull. APS* **38**, 1971 (1993); F. Anderegg, X.–P. Huang, G. D. Severn, and C. F. Driscoll, *Bull. APS* **39**, 1971 (1994).

5. R. C. Davidson, (a) *Theory of Nonneutral Plasmas*, Benjamin, Reading, MA (1974); (b) *Physics of Nonneutral Plasmas*, Addison–Wesley, Redwood City, CA (1990).

6. M. Abramowitz and I. A. Stegun (Eds.), *Handbook of Mathematical Functions*, National Bureau of Standards (U. S. Govt. Printing Office, Washington, 1972).

Measurement of Transport and Damping from Rotational Pumping

B.P. Cluggish and C.F. Driscoll

Physics Department, University of California at San Diego, La Jolla, CA 92093

Abstract. Radial transport and the damping of the $m = 1$ diocotron mode from "rotational pumping" has been measured on a magnetized electron column. Rotational pumping is the collisional dissipation of axial compressions caused by the $\mathbf{E} \times \mathbf{B}$ rotation of the column through asymmetric confining potentials. The observed transport rates are in close agreement with theory.

INTRODUCTION

We present measurements of radial particle transport and mode damping from "rotational pumping" of a magnetized electron column displaced from the axis of a cylindrical trap. Rotational pumping is the collisional dissipation of axial compressions caused by $\mathbf{E} \times \mathbf{B}$ rotation of the column through asymmetric confinement potentials. Here, the confinement potentials appear asymmetric only because of the displacement of the column away from the symmetry axis of the trap.

We find that this transport conserves particle number; conserves angular momentum by moving the column back to the trap axis as the column expands; and conserves total energy by dissipating electrostatic energy into thermal energy. The transport rate is proportional to the electron-electron collision rate, which drops precipitously in the cryogenic, strongly magnetized regime; surprisingly, the transport is otherwise independent of the magnetic field strength. The observed transport rates are in close agreement with a new theory by Crooks and O'Neil (1), described elsewhere in these proceedings.

EXPERIMENT

We confine the plasmas in a Penning-Malmberg trap (2,3), shown schematically in Fig. 1. Electrons from a tungsten filament are confined in a series of conducting cylinders of radius $R_w = 1.27$ cm, enclosed in a vacuum can at $4.2°$K. The electrons are confined axially by negative voltages $V_c = -200\ V$ on cylinders 1 and 4; radial confinement is provided by a uniform axial magnetic field, with $10 < B < 00$kG. The trapped plasma typically has initial density $10^9 \leq n \leq 10^{10}$ cm^{-3}, RMS radius $R_p \sim 0.04$ cm, length $L_p \sim 3$ cm, with a characteristic expansion time $10^2 \leq \tau_m < 10^3$ sec for a centered plasma.

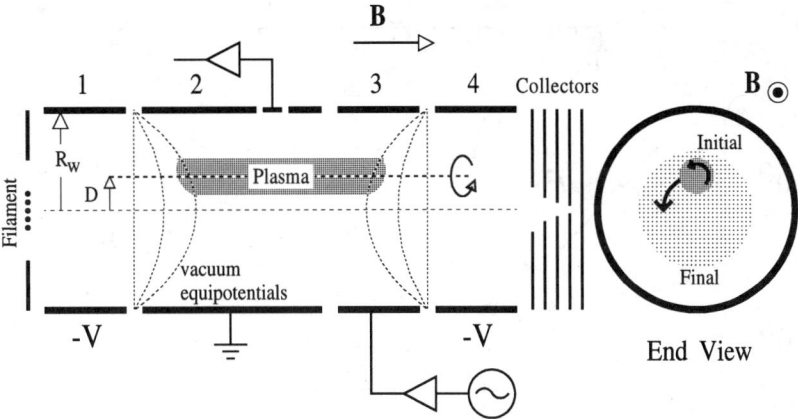

Figure 1. Schematic of the cylindrical apparatus and electron plasma.

The self electric field of the electrons causes an $\mathbf{E} \times \mathbf{B}$ drift rotation around an axis through the center of charge, at a rate $500 \leq f_E \leq 3000$ kHz. When the column is displaced from the center of the trap, image charges in the conducting walls cause the column to orbit around the trap axis in the $m = 1$ "diocotron" mode, at frequency $5 \leq f_d \leq 20$ kHz. The image charge signal received on a wall sector is proportional to the displacement, D, of the column from the trap axis.

The z-integrated density of the plasma, $q(\rho)$, is obtained by measuring the charge dumped onto the 5 end collectors when cylinder 4 is grounded. Here, ρ refers to the radius from the plasma axis, *i.e.* $\mathbf{r} = \mathbf{D} + \boldsymbol{\rho}$. The parallel plasma temperature, $T_{\parallel}$, is measured by slowly ramping the voltage on cylinder 4 to ground, and measuring the number of electrons which escape as a function of the confining voltage. We find $0.003 \leq T_{\parallel} \leq 20\text{eV}$, giving axial electron bounce frequencies $4 \times 10^5 \leq f_b \leq 3 \times 10^7$ Hz, and $T_{\parallel}$ to $T_{\perp}$ collisional equilibration rates $10^3 \leq \nu_{\perp\parallel} \leq 10^5$ sec^{-1}.

We calculate the 3D plasma density $n(x, y, z)$ and potential $\phi(x, y, z)$ from the measured $q(\rho)$ and D, by numerically solving Poisson's equation on a $125 \times 125 \times 200$ grid. Here, we assume

$$n(x, y, z) = n_0(x, y) \exp\{e\phi(x, y, z)/kT_{\parallel}\} , \qquad (1)$$

where $n_0(x, y)$ follows from $\int dz \, n(x, y, z) = q(\rho)$, where $\rho^2 = (x - D)^2 + y^2$.

Figure 2 shows a typical evolution of the density profile $n(\rho, z = 0)$ of an off-axis column. The RMS radius increases from $R_p = 0.044$ cm at $t = 0$ to $R_p = 0.22$ cm at $t = 10$sec, while the central density decreases a factor of 8. However, the total number of particles, N, is conserved to within 1%.

The displacement D decreases as the plasma expands, damping the $m = 1$ diocotron mode. In these strongly magnetized plasmas, the total angular

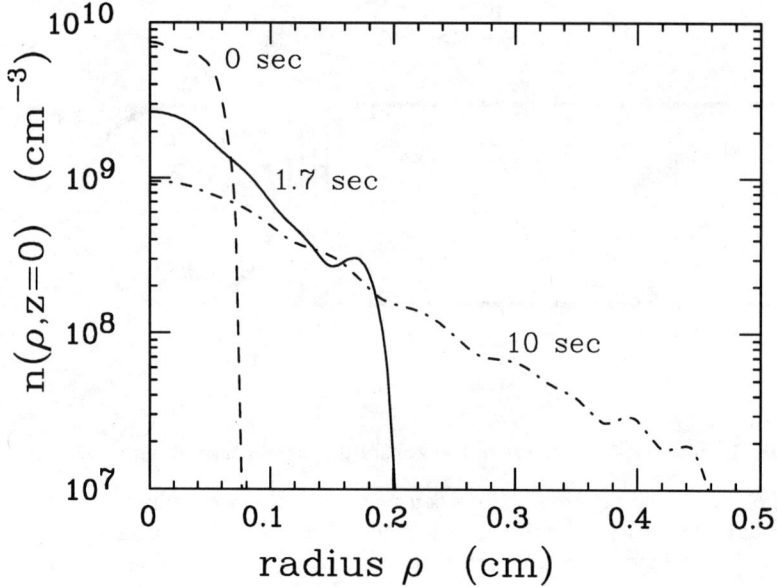

Figure 2. Measured radial density profiles at 3 different times.

momentum is dominated by the electromagnetic component, i.e.

$$P_\theta \equiv \int d^3\mathbf{r}\, n\left(mv_\theta r - \frac{eB}{2c}r^2\right) \approx \frac{eB}{2c}N[D^2 + R_p^2]\,, \qquad (2)$$

where $R_p^2 \equiv (1/N)\int 2\pi\rho d\rho\ \rho^2 q(\rho)$ is the mean square radius of the plasma. Figure 3(a) shows $(D/R_w)^2$, $(R_p/R_w)^2$, and their sum for the evolution of Fig. 2. Initially, 97% of P_θ is in $(D/R_w)^2$; after 10 seconds, expansion and mode damping leave only 1% of P_θ in $(D/R_w)^2$. The sum of the two P_θ components remains constant, implying that the forces causing the transport are azimuthally symmetric around the trap axis.

This expansion and symmetrization process converts electrostatic energy, H_ϕ, into kinetic energy, H_k. We also calculate the total radiated energy, H_{rad}, and the work done on the power supplies which provide the confining potentials, W_{ps}, as

$$H_\phi = -(\tfrac{1}{2})\int d^3x\, n(x,y,z)\, e\phi(x,y,z)/N\,, \quad H_k = \tfrac{1}{2}\,k(T_\| + 2T_\perp)\,,$$

$$H_{rad} = \tfrac{3}{2}\int_0^t dt'\, kT_\perp/\tau_{rad}\,, \qquad W_{ps} = -(\tfrac{1}{2})\,\Delta Q\, V_c/N\,.$$

Here, $T_\perp$ is the perpendicular temperature, $\Delta Q(t)$ is the change in the charge on the end cylinders, and τ_{rad} is the energy loss time of the electrons due to cyclotron radiation. At $B = 40$ kG, we measure $\tau_{rad} = 0.29$ sec (2). Because $\tau_{rad} \gg \nu_{\perp\|}^{-1}$, we assume that $T_\perp \approx T_\| \equiv T$ in calculating H_k and H_{rad}.

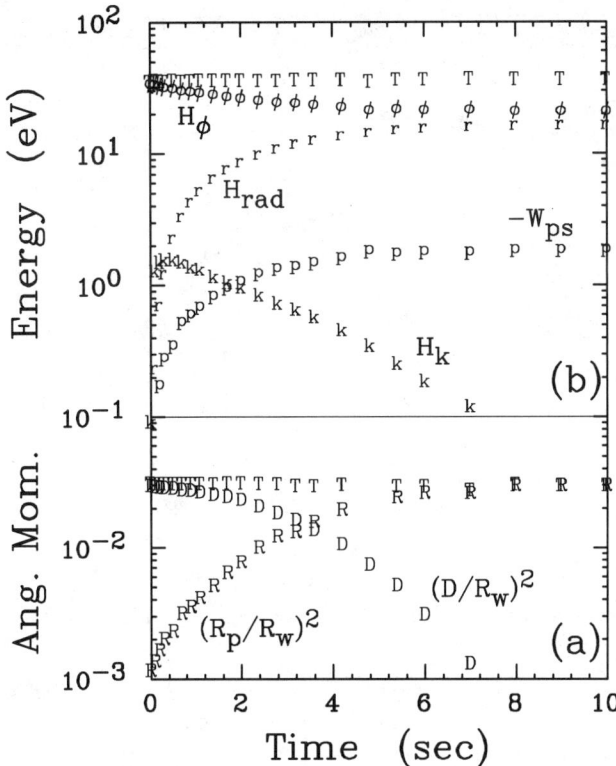

Figure 3. Evolution of (a) plasma diameter and displacement components of P_θ; (b) electrostatic, radiated, kinetic, and power supply components of total energy.

Figure 3(b) shows the evolution of H_ϕ, H_t, W_{ps}, H_{rad}, and their sum. Over 10 seconds, H_ϕ decreases by 40% from its initial value. H_k increases over the first 0.4 seconds, until radiative cooling dominates. The dissipation of electrostatic into kinetic energy indicates that $\mathbf{E} \times \mathbf{B}$ drift dynamics alone cannot be responsible for the observed transport.

We characterize this transport by the damping rate of the $m = 1$ diocotron mode, γ, i.e. by the rate of decrease of the column displacement D, which we measure non-destructively. For $D_0 \leq R_p$, the displacement decreases as $D(t) = D_0 \exp(-\gamma t)$.

We find that the transport rate exhibits several striking parameter dependences. One is that γ is essentially independent of B. For moderate temperatures, as B is increased from 10 to 60 kG, γ decreases only 30%. This is in contrast with "conventional" transport scaling as B^{-2} (4,5) or B^{-1} (6), generally due to dependence on the cyclotron radius, r_c, or on $\mathbf{E} \times \mathbf{B}$ drift velocities.

Even more striking is the dependence of γ on plasma temperature for small

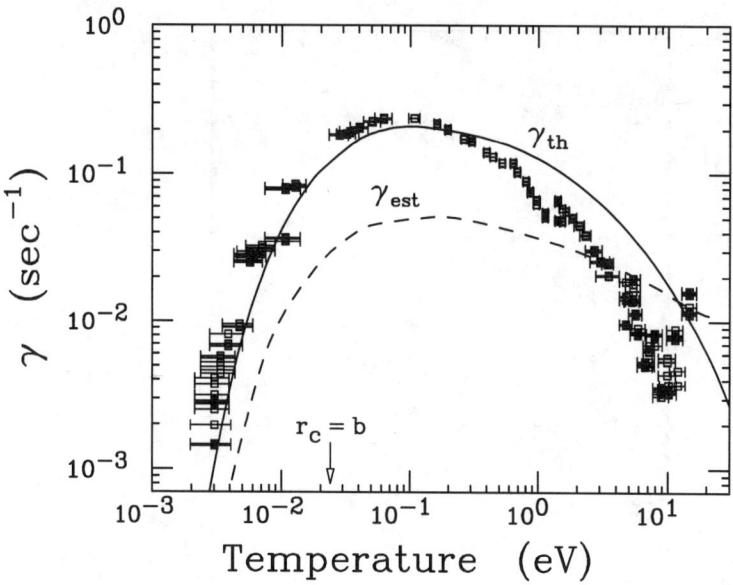

Figure 4. Measured damping rate γ versus plasma temperature, with estimated and calculated theory predictions.

T, as shown in Fig. 4. As T is decreased from 0.01 to 0.003 eV, we find that γ decreases by 2 orders of magnitude. In this strongly magnetized regime, r_c is smaller than the distance of closest approach, $b \equiv e^2/kT$, and $\nu_{\perp\parallel}$ becomes exponentially small (3). This is a strong indication that $\gamma \propto \nu_{\perp\parallel}$. Experimentally, we maintain a constant plasma temperature by applying a 2 MHz oscillation to cylinder 3, thus balancing the radiative cooling; γ is independent of the frequency and amplitude of this heating oscillation, except through the plasma temperature.

COMPARISON WITH THEORY

The recent theory of rotational pumping by Crooks and O'Neil (1) analyzes the cyclic variation in the length of a tube of plasma as it $\mathbf{E} \times \mathbf{B}$ drifts around the plasma axis. The length at radius ρ varies as $L(\rho, t) = L_0(\rho) + \delta L(\rho) \sin \omega_R t$, where $\delta L \ll L_0$ and $\omega_R(\rho)/2\pi = f_E(\rho) - f_d$ is the rotation frequency of the plasma in the dioctron mode frame. If we assume the plasma has uniform n and T, and approximate δL by $\delta L = \kappa(D/R_w)\rho$ (7), where $\kappa \sim 2$, the theory gives an estimated damping rate

$$\gamma_{est} \equiv -\frac{1}{D}\frac{dD}{dt} = 4\kappa^2 \nu_{\perp\parallel} \frac{\lambda_D^2 R_p^2}{L_p^2 R_w^2} \frac{1}{1 - 2R_p^2/R_w^2}, \qquad (3)$$

where $\lambda_D = \sqrt{kT/4\pi ne^2}$ is the Debye length.

The estimated rate γ_{est} is independent of D and depends on B only through the Coulomb logarithm in $\nu_{\perp\parallel}$. We have plotted γ_{est} in Fig. 4 as a dashed curve. It agrees with our data within a factor of 6. A more accurate theoretical prediction γ_{th} is obtained using δL from the measured $q(\rho)$ and Eq.(1). This gives the solid curve in Fig. 4, which shows closer agreement with the data. The difference in slope between γ_{est} and γ_{th} for $T > 1$ eV is due to a decrease in the actual δL with temperature: higher energy electrons penetrate farther towards the end cylinders, where the vacuum equipotential surfaces have less curvature, as shown in Fig. 1.

In addition to D, B, and T, we have measured the dependence of γ on V_c, n, R_p, and L_p. In all cases, agreement between γ_{th} and experiment is comparable to the agreement in Fig. 4.

This rotational pumping mechanism should damp the $m \geq 2$ diocotron modes. It may also cause the dissipation of several otherwise stable nonneutral plasma configurations, including 2 electron vortex equilibria (8), vortex crystals (9), asymmetric equilibria (10), and toroidal electron plasmas (11). Rotational pumping should damp azimuthal rotation in non-axisymmetric systems such as tandem mirrors (12), just as the analogous process of "magnetic pumping" (13) strongly damps poloidal rotation in tokamaks.

ACKNOWLEDGMENTS

The authors would like to acknowledge experimental suggestions by the late J.H. Malmberg, and enlightening discussions with T.B. Mitchell and K.S. Fine. This work was supported by NSF PHY91-20240 and ONR N00014-89-J-1714.

REFERENCES

1. Crooks, S.M., and O'Neil, T.M., *Phys. Plasmas*, accepted (1994).
2. Beck, B.R., Ph.D. Dissertation, UCSD (1990).
3. Beck, B.R., Fajans, J., and Malmberg, J.H., *Phys. Rev. Lett.* **68**, 317 (1992).
4. Malmberg, J.H., and Driscoll, C.F., *Phys. Rev. Lett.* **44**, 654 (1980).
5. Driscoll, C.F., Fine, K.S., and Malmberg, J.H., *Phys. Fluids* **29**, 2015 (1986).
6. Driscoll, C.F., Malmberg, J.H., and Fine, K.S., *Phys. Rev. Lett.* **60**, 1290 (1988).
7. Peurrung, A.J., and Fajans, J., *Phys. Fluids B* **5**, 4250 (1993).
8. Mitchell, T.B., Driscoll, C.F., and Fine, K.S., *Phys. Rev. Lett.* **71**, 1371 (1993).
9. Fine, K.S., Driscoll, C.F., and Cass, A., *Bull. Am. Phys. Soc.* **37**, 1736 (1994).
10. Notte, J., and Fajans, J., *Phys. Plasmas* **1**, 1123 (1994).
11. Khirwadkar, S.S., et al., *Phys. Rev. Lett.* **71**, 3443 (1993).
12. Price, H.D., et al., *Nucl. Fusion* **23**, 1043 (1983).
13. Stacey, W.M., and Jackson, D.R., *Phys. Fluids B* **5**, 1828 (1993).

Linear Response of the 2-d Pure Electron Plasma; Quasi-Modes for Model Profiles

Noel R. Corngold

Watson Laboratories of Applied Physics,
California Institute of Technology,
Pasadena, California 91125

The following is a summary of a poster presentation at the Workshop. A full discussion of the matters touched on here will appear elsewhere.

Introduction

We are concerned with diocotron waves in a non-neutral plasma. In a recent experiment[1], Pillai and Gould excited the m=2, diocotron- mode of a column of pure-electron plasma, and observed its subsequent, damped, "ringing". When the amplitude of the short, exciting burst is relatively small, one expects a linear analysis of the response to be fruitful. (deGrassie and Malmberg, in an earlier study of diocotron waves[2], emphasized the non-linear regime.) Linear "ringing" is usually associated with a normal mode of oscillation of a system, or with a discrete eigen-value, or a point in the point-spectrum of an operator describing the evolution of the system. But it has been known for some time[3] that such modes do not exist in the collision-less dynamical model most commonly used to describe the pure-electron plasma, under the conditions believed to prevail in these experiments. There is, however, a continuous spectrum associated with the system, and it brings what appears to be a paradox: the evolution of a system governed by a continuous spectrum is generally non-exponential, but the experimenter sees exponential behavior. Two responses are available. One can remark that the integrand in the integration over the continuous range of frequencies is peaked sharply at the observed frequency. Or, one can ask why? - is the peak the "shadow" of an interesting singularity sitting nearby? - a "quasi-mode"?

And might the quasi-mode be viewed as a signature, or used as a diagnostic?

The situation is not new. It resembles that encountered in the study of Landau-damping. To understand it fully one must -in the language of the theory - examine a multi-valued Green's function which has been continued analytically onto a sheet adjacent to the "physical sheet". This problem was discussed generally and thoroughly, two decades ago, by Briggs, Daugherty, and Levy[3] (henceforth BDL). This paper supplements BDL in showing that such an analysis may be carried out analytically for all angular modes - not merely m=2, or the peculiar m=1 - in plasmas characterized by convex, power-law profiles . Then, the quantities of interest may be expressed in terms of hypergeometric functions and the continuation is straightforward. One finds the "quasi-mode" in an appropriate place. One obtains formulae for the dependence of frequency and damping of the mode upon profile, mode-number, and gap-size. We shall illustrate by considering three examples (cases): in the first the plasma fills the interior of a perfectly conducting cylindrical shell; in the second the plasma fills a "good", but not perfect shell, and in the third the plasma occupies the central portion of the interior of a perfect shell; a vacuum gap separates the plasma from the wall. In addition, we discuss some ideas about special solutions for profiles more complicated than power-law, the latter being somewhat unnatural in their edge-behavior.

Analysis

The model we consider describes the plasma density $n(r,\theta,z)$ and the electrostatic potential $\phi(r,\theta,z)$ by the equations

$$\frac{\partial n}{\partial t} + \frac{c}{B}\left(\nabla\phi \times \nabla n\right)\cdot \hat{z} = 0 \qquad\qquad \nabla^2\phi = -4\pi q n \ .\qquad\qquad (1)$$

The plasma is contained in a very long, grounded circular cylinder, whose radius is denoted by "a". The system is immersed in an uniform magnetic field, $B\,\hat{z}$. We shall consider only distributions independent of z.

An arbitrary, smooth profile, $n_0(r)$, accompanied by a related $\phi_0(r)$, or by an angular velocity profile $\omega_0(r) = \frac{c}{B_0}\frac{1}{r}\frac{d\phi_0}{dr}$, will satisfy the equations of motion. Imagine a small perturbation about this steady state. If we write $n(r,\theta,t) = n_0(r) + n_1(r,\theta,t)$, and a similar expression for ϕ, we are led to

$$\frac{\partial n_1}{\partial t} + \omega(r)\,\frac{\partial n_1}{\partial \theta} - \gamma(r)\,\frac{\partial \phi_1}{\partial \theta} = 0 \quad \text{and} \quad \nabla^2 \phi_1 = -4\pi q n_1 , \qquad (2)$$

where $\gamma(r) = \dfrac{c}{B_o}\dfrac{1}{r}\dfrac{dn_o}{dr}$. The main idea of this paper is that special profiles

$$\hat{n}_o(r) = \left(1 - \left(\tfrac{r}{b}\right)^{2p}\right) \qquad 0 \le r \le b \le a \qquad \hat{\omega}_o(r) = \left(1 - \frac{1}{1+p}\left(\tfrac{r}{b}\right)^{2p}\right)$$

which are not unreasonable on physical grounds, lead to solutions of the linearized equations in terms of known functions. (The profiles are scaled by their values "on-axis".) Consider, for example, the initial value problem for functions proportional to $\exp(im\theta)$. Solve it by Laplace transformation, writing the transform variable as $s = -i\omega m$. The associated Green's function has two ingredients: solutions to a homogeneous equation and the Wronskian denominator, a function of ω whose zeros yield poles in the transform, and signal collective modes. All of these quantities may be expressed in terms of hypergeometric functions $F(\bar{a}, \bar{b}; \bar{c}; z)$, with $\bar{c} = \bar{a} + \bar{b} + 1 = \dfrac{p+m}{p}$, $-\bar{a}\,\bar{b} = \dfrac{p+1}{p}$, and $z = \dfrac{1}{(1-\omega)(1+p)}\left(\tfrac{r}{b}\right)^{2p}$. In a typical application, a boundary condition at $r = b$

requires that F vanish there, that

$$z = z_e = \frac{1}{(1-\omega)(1+p)}$$

be a zero of the hypergeometric function. (The corresponding frequency, ω, is the mode frequency.) But one can show that the fundamental branch of our multi-valued function F has no zeros. We are led to search - successfully- for zeros on a neighboring branch. Deformation of the contour for Laplace inversion then extracts the associated "quasi-mode". The mode frequency is complex; its real part is usually equal to $m\omega(r)$ for some r, $0 \le r \le b$; its imaginary part produces a small damping. There is one mode of overwhelming dominance. It can be located quite well by approximate analytic techniques, very accurately by easy computation.

A Sampling of Results

Some examples follow: (Note that our definition of "frequency", ω, causes it to be <u>wave speed</u>, rather than frequency. The true frequency is $m\omega$.)

1) In Case 1, the complex frequency of the quasi-mode lies on the curve

$$\omega = 1 - \frac{1}{1+p \, (1+ \frac{\sin\Theta}{\Theta} e^{i\Theta})} \; , \qquad \pi \leq \Theta < 2\pi \; ,$$

particular values of Θ depending on m and p. This result is based on an accurate, but approximate, estimate of the location of the zero . It is borne out by detailed computation.

2) The same approximation gives

$$\frac{\omega}{\omega_e} = 1 + \frac{1}{1+p} \, (\frac{m\text{-}1}{2p+1}) + i \, \pi \, \frac{1}{1+p} \, (\frac{m\text{-}1}{2p+1})^2 + \dots \qquad (m\text{-}1)/p \to 0 \; , \tag{3}$$

$$= 1 - \frac{1}{1+p} \, (\frac{1}{\ln \frac{m}{p}}) + i \, 2\pi \, \frac{1}{1+p} \, (\frac{1}{\ln \frac{m}{p}})^2 + \dots \qquad (m\text{-}1)/p \to \infty,$$

where $\omega_e = p/(1+p)$, the "edge-frequency", is $\hat{\omega}_0(b)$. Thus, for large m/p, the waves are dispersion-less; the damping increases with m, almost linearly. For very flat profiles and moderate m, so that $(m\text{-}1)/p \to 0$, the modes cluster at the edge-frequency. Indeed, ω_e is the limit point for the wave-speed in both limiting cases.

3) The quasi-mode frequencies which emerge are all "reasonable" when compared with experiment[1,2] or with other computations. A typical result, for m=2, a quadratic profile (p=1), and the "egg" of Fig.(2) is $\omega =$ (0.47, 0.023), which gives a true frequency of (0.94, 0.046) $\omega(0)$. Flatter profiles will produce less damping, larger "Q's". Then, when we study the m-dependence, we are struck by the lack of dispersion.Were it not for the mode-dependent damping, one would expect an initial disturbance containing many angular modes to evolve into a characteristic shape, which would endure for a perceptible interval of time. Since the damping is mode-dependent, though small, the shape would be altered slowly, its smaller features disappearing more rapidly. The estimates in Eq.(3) prove to be quite reasonable; more precise numerical calculations appear to show the logarithmic drifts which appear in Eq.(3).

4) Since flat, or "top-hat" profiles produce real modes, the study of the profile-dependence of quasi-modes contains little that is surprising. As p, and therefore "flatness" increases, so damping -which, ultimately, is phase-mixing- decreases. For a given profile, creating a gap or, equivalently,

compressing the plasma, lowers the frequency of the quasi-mode. Small gaps do not affect the damping, while large gaps cause it to increase. It appears that the larger the density - or frequency gradient in the profile, the larger is the phase-mixing (damping). (That the damping is proportional to the gradient of the density at the resonant radius appears in the perturbation calculations of BDL.)

Acknowledgments

This work has benefitted greatly from valuable discussions with my colleagues Roy Gould and Sateesh Pillai.

References

1. S. Pillai and R.W. Gould, "Collisionless Damping and Particle Trapping in a Pure Electron Plasma", Bull. Am. Phys. Soc., 38, No. 10, 1970 (1993).

2. J.S. deGrassie and J.H. Malmberg, "Waves and Transport in the Pure Electron Plasma", Phys. Fluids 23, 63 (1980).

3. R.J. Briggs, J.D. Daugherty, and R.H. Levy, "Role of Landau Damping in Cross-Field Electron Beams and Inviscid Shear Flow", Phys. Fluids 13, 421 (1970).

4. J.D. Jackson, "Classical Electrodynamics", Wiley, New York, 1962.

5. R.A. Smith and M.N. Rosenbluth, "Algebraic Instability of Hollow Electron Columns and Cylindrical Vortices", Phys. Rev. Lett. 64, 649 (1990).

6. C.F. Driscoll and K.S. Fine, "Experiments on Vortex Dynamics in Pure Electron Plasmas", Phys. Fluids B2, 1359 (1990).

LLNL Pure Positron Plasma Program

J.H. Hartley§, B.R. Beck†, T.E. Cowan†, J. Fajans‡, R. Gopalan‡,
R.H. Howell†, J.L. McDonald†, and R.R. Rohatgi†

†Physics Department and §U.C. Davis Department of Applied Science, Lawrence Livermore
National Laboratory, Livermore, CA 94551
‡Physics Department, University of California, Berkeley, CA, 94720

Abstract. Assembly and initial testing of the Positron Time-of-Flight Trap at the Lawrence Livermore National Laboratory (LLNL) Intense Pulsed Positron Facility has been completed. The goal of the project is to accumulate a high-density positron plasma in only a few seconds, in order to facilitate study that may require destructive diagnostics. To date, densities of at least 6×10^6 positrons per cm^3 have been achieved.

TRAP DESIGN

The LLNL pure plasma trap has an unusual configuration for a non-neutral plasma device, which was driven by its intended use as a high-density positron gas target for particle physics experiments (1). The positron trap consists of a 76 cm long cylindrical-electrode Penning-Malmberg trap in a high-uniformity solenoid capable of producing a 62 kG axial magnetic field. The apparatus is situated on the positron beam line of the LLNL 100 MeV electron Linac.

At the time of the design, the conventional wisdom in non-neutral traps was that azimuthal asymmetries of the field and electrodes were a major limiting factor to the plasma confinement time (2). For this reason, great care was taken to ensure that the solenoid field was very uniform, and that the electrodes were aligned with great precision. The measured field strength varies by less than 0.1% axially over the length of the trap, and ~20 ppm peak-to-peak azimuthally at the electrode radius of 1 cm (Fig. 1). The size and radial dependence of the asymmetries in the field can be accounted for by a horizontal 0.125 mm bow in the 1 m long coil, resulting in a toroidal field with a radius of curvature of 1 km.

The electrodes are gold-plated copper cylinders aligned in a ceramic V-block cradle. All components are machined to 2.5 μm tolerance over the length of the trap. The trap structure is situated in a copper bore whose temperature can be varied from 300 K to 4.2 K.

The primary plasma diagnostic consists of a micro-channel plate (MCP) in front of a phosphor screen, imaged by a CCD camera. Camera images are captured on a VAX Workstation.

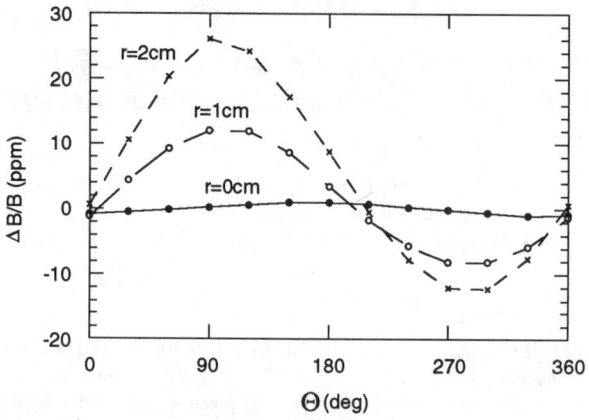

Figure 1. Magnetic field variation with azimuthal angle at three radii. 0° and 180° define the vertical plane.

OPERATION WITH ELECTRON PLASMAS

Preliminary testing of the trap utilized electrons from a spiral filament as the plasma particles. The trap region was 6 cm long, with B=30 kG and T=77 K. Plasma density measurements were made by dumping the plasma from the trap onto a phosphor screen. The radial density profile was determined from the image intensity and a capacitive pick-off of the total charge on the screen.

At the time of our initial measurements, summarized in Figures 2 and 3, the CCD camera imaging system was not yet operational, so human eyes and a ruler were used to measure the diameter of the plasma. Work done at UC San Diego (3) and UC Berkeley (4), and the qualitative appearance of the phosphor image, suggested that assuming flat profile would be a reasonable approximation and that the calculated average density was adequate to characterize the plasma expansion.

The plasma underwent an early stage of rapid transport, followed by a much slower expansion. This was consistent with the behavior of other non-neutral plasma devices, and suggested a quick evolution of the radial profile to an equilibrium flat distribution. The data was fit as two exponentials, and the t=0 equilibrium density was defined to be the intercept of the second, slower exponential with the y axis. Confinement time was defined as the time it took for the average density to drop to one-half of the equilibrium density.

With the bore at room temperature, confinement times were on the order of ten seconds. With the bore cooled to 77 K, lifetimes improved to as much as several thousand seconds. This factor of 100 difference is presumed to result from the improvement in vacuum pressure due to cryopumping. Pressure measured well outside the cryopumping region improved from 2×10^{-9} to 2×10^{-10} Torr with cooling, and the improvement was likely much greater in the trap region.

Earlier work (5) has suggested that the lifetime scaling should follow:

$$\tau_m \propto \frac{B^2}{L^2} \cdot \frac{1}{P} \cdot \frac{f(T)}{n^a}, \tag{1}$$

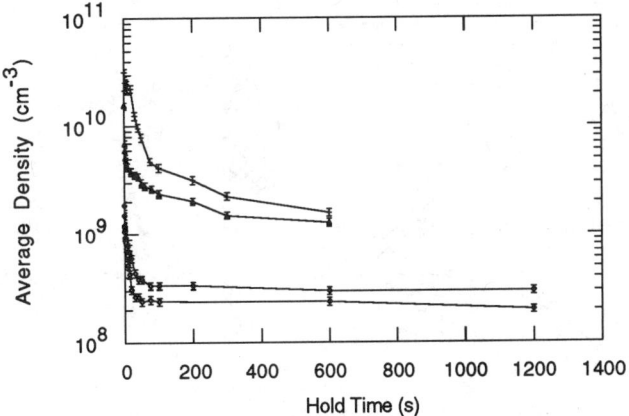

Figure 2. Electron average density as a function of hold time. (L=6 cm, B=30 kG, T=77 K)

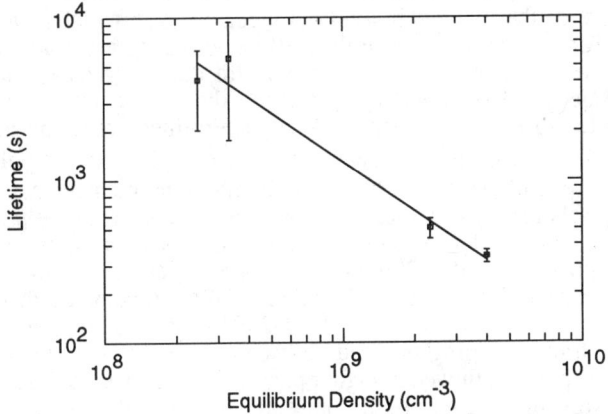

Figure 3. Electron plasma lifetime as a function of equilibrium density, determined from data in Fig. 2 for hold times greater than 100 s.

where B is the magnetic field, L is plasma length, P is neutral pressure, n is plasma density, and f is an undetermined function of plasma temperature. A fit to the LLNL data (Fig. 3) indicates a value of $a \approx 1$. Interpolating over several orders of magnitude in density and radius between UCSD data (5) and these preliminary LLNL data suggests $a \sim 1.5$. The CCD camera diagnostic is now fully on-line, and work is underway to further refine the trapping characteristics and scaling laws of this device.

POSITRON BEAM AND PULSED INJECTION

Positrons are injected into the trap from the LLNL Linac intense pulsed positron beam. (6) The positron beam is produced from the photon-electron-positron shower

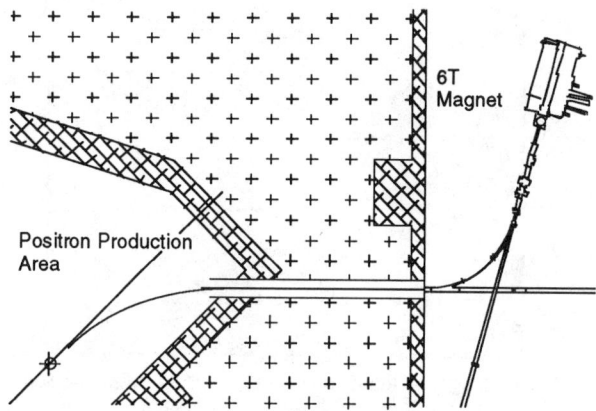

FIGURE 4. Diagram of the positron transport system.

generated by the 100 MeV electron beam in a tungsten converter target. High energy positrons are thermalized in a set of tungsten vanes, which have a negative work function for positrons, so those that diffuse to the vane surfaces are ejected with 1-2 eV. To preserve the 20 ns pulse width during transport through the solenoidal guide field to the trap (Fig. 4), the positrons are accelerated electrostaticly to ~6 kV. The positron flux, determined from the annihilation radiation at a beam stop just upstream of the trap, is 3×10^6 e+/pulse. At the maximum Linac repetition rate of 1440 pulses/s, the beam intensity is 4.4×10^9 e+/s.

The trapping scheme for accumulating positrons makes use of the pulsed nature of the positron beam. A 1000 Å tungsten remoderation foil at the front of the trap thermalizes the positrons, with ~20% efficiency, greatly enhancing the phase-space brightness of the beam by removing the large axial energy spread resulting from injection into the high magnetic field. The 1-2 eV positrons are then more easily captured in the trap. Trapping proceeds in two phases, the first by simple TOF (time-of-flight) capture, followed at higher positron densities by collisional cooling into the axial well. In the first phase, the remoderation foil, which is used as an electrostatic end cap, is dropped to ground potential for a time $\tau_g \geq \tau_p$, where τ_p is the width of the positron pulse. The positrons drift into the trap and spread evenly throughout the trapping region before the next pulse arrives. The fraction of trapped particles lost when a new pulse is injected is τ_g/τ_r, where τ_r is the round-trip travel time of positrons in the trap. The number of pulses trapped in this phase then asymptotically approaches τ_r/τ_g, at which point positrons are lost at the same rate at which they are injected.

The second phase of trapping occurs when the density of particles in the trap is high enough for collisions to equilibrate the transverse and longitudinal temperatures before the next pulse arrives (7). The trap is held at a negative bias, so as longitudinal energy is scattered into the transverse degrees of freedom the positrons cool into the axial potential well. Cyclotron radiation cools the transverse energy in less than a second. Once the positrons are in the potential well they can no longer escape when the front barrier is lowered. The transition to this second phase of trapping can be clearly seen in numerical simulations. For example, Figure 5 shows positron accumulation for the conditions in our first positron injection

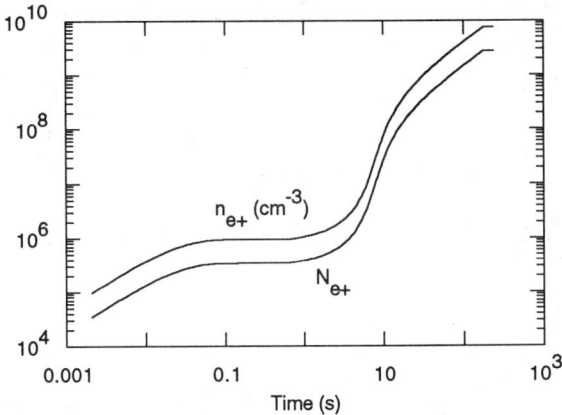

FIGURE 5. Numerical simulation of positron accumulation scheme. (L=46 cm, B=30 kG, well depth = 60 V, Linac rep. rate = 480 pps.)

experiments, described below. Under ideal conditions of injection and remoderation, ~9x10^7 positrons could be trapped per second.

FIRST RESULTS OF POSITRON INJECTION

The two major technical difficulties encountered in the first experiment were transport into the high-field region of the trap and remoderation of the positrons. As the positrons passed from the 150 G guide field into the 30 kG field used in the experiments presented here, magnetic mirroring gave an acceptance angle of less than five degrees. Irregularities in the guide field provided a transverse impulse to the positrons, resulting in a "spinning up" of the beam and consequent mirroring of the positrons. Refined techniques for operating the transport magnets improved transmission efficiency from an initial value of 1% to greater than 35%.

The unannealed tungsten foil used in the initial experiment gave only a 0.8% remoderation efficiency, instead of the expected 20%, when cooled to 77 K with the rest of the trap. Crystal defects in an unannealed foil inhibit positron diffusion to the surface. Cryopumping of impurities on the foil would inhibit re-emission. A recent experiment with an annealed foil at 300 K has improved performance to 5-6% remoderation. Cooling the bore with the annealed remoderation foil in place will show whether cryopumping is actually a problem requiring redesign of the remoderation stage.

These improvements resulted in 3.5x10^4 remoderated positrons injected into the trap per pulse. The round-trip flight time was 240 ns in this trap configuration. With an injection gate width of 25 ns, the TOF capture limit was about 3.5x10^5 positrons, achieved in a small fraction of a second. The actual accumulation of positrons as a function of hold time is shown in Figure 6. The fact that positrons continued to accumulate for up to 20 s indicates that particle cooling into the well was occurring. The time distribution of the charge signal as the trap was dumped indicates a significant accumulation of positrons near the bottom of the potential

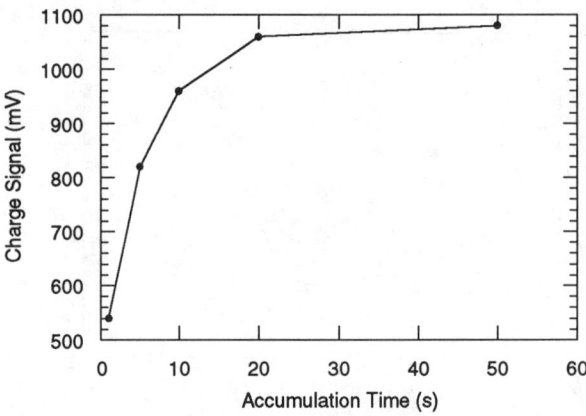

Figure 6. Charge dump signal on phosphor as a function of accumulation time.

well. Both the accumulation rate and the maximum density achieved scaled linearly with the repetition rate of the positron beam.

Calibration of the total number of positrons with the observed charge signal was complicated by saturation of the MCP. The relatively low repetition rate of one charge dump per 20 s made statistical counting from the annihilation gamma-ray signal also difficult. Three attempts at indirect calibration were made, using electron plasma diocotron measurements, the annihilation signal of single positron pulses, and the signal from a charge-sensitive pre-amp attached to the MCP with no voltage across the plate. Different calibrations give results ranging from 1.8×10^6 to 2×10^7 positrons trapped. Direct measurement of the number of trapped particles will be made in the next run by looking at the diocotron frequency of the positron column. The large number of positrons captured also allows use of the MCP at lower gain, which should reduce the problem of charge saturation.

An image of the dumped positron plasma is shown in Figure 7. The measured FWHM radius was 0.45 mm. The trapping region was 46 cm long, which gave a volume of approximately 0.29 cm^3, and a density of at least 6×10^6 cm^{-3} and possibly as high as 6×10^7 cm^{-3}.*

A comparison of the data of Figure 6 and calculations of Figure 5 reveals several important results regarding positron accumulation and confinement. Initially, the trap was filling too quickly to be accounted for by positron collisions alone. For example, the calculated density at 1s is 10^6 cm^{-3}, but the actual accumulation was at least three times that value. The accumulation should continue linearly throughout this range, but the results show saturation, while CCD images indicated no radial transport. This implies a large effect from collisions with neutral particles, enhancing early collisional cooling into the well and causing positron loss due to annihilation or positronium formation on a time scale of a few seconds. This data is consistent with the early results of Surko *et. al.* (8). Experiments to improve

---

* An abstract for a poster session at the annual meeting of the APS Division of Plasma Physics mistakenly suggested that design-goal densities of 10^{10} cm^{-3} had actually been achieved. We apologize for the error.

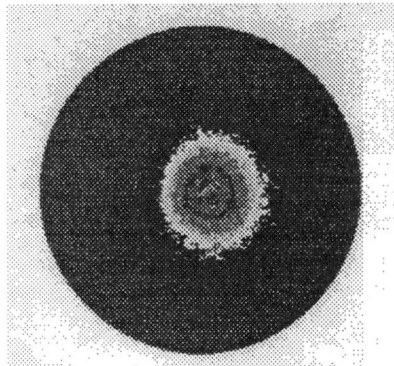

Figure 7. CCD Image of positron column. FWHM radius is 0.45 mm

confinement by cooling the trap are in preparation. If surface contamination of the remoderation foil proves not to be a limiting factor, we expect to achieve significantly higher densities in the near future.

ACKNOWLEDGMENTS

We wish to acknowledge the invaluable assistance of W. Patterson and H. Koberle in the development of the experiment and the operation of the Linac. This work was funded by the LLNL Laboratory Directed Research and Development program and the LLNL Plasma Physics Research Institute. Work was performed under the auspices of the U.S. Department of Energy by Lawrence Livermore National Laboratory under contract No. W-7405-ENG-48.

REFERENCES

(1) Cowan, T. E., Howell, R. H., Rohatgi, R. R., and Fajans, J., *Nuc. Inst. and Methods B*, **56-7**, 599–603 (1991).
(2) Driscoll, C. F., and Malmberg, J. H., *Phys. Rev. Let.*, **50**, 167–170 (1983).
(3) Malmberg, J. H., and Driscoll, C. F., *Phys. Rev. Let.*, **44**, 654–657 (1980).
(4) Peurrung, A. J., and Fajans, J., *Phys. Fluids B*, **1**, 2073–2084 (1989).
(5) Cass, A., contribution, this conference.
(6) Howell, R. H., Rosenberg, I. J., and Fluss, M.J., *Applied Physics A*, **43**, 247–255 (1987).
(7) Beck, B. R., Fajans, J., and Malmberg, J. H., *Phys. Rev. Let.*, **68**, 317–320 (1992).
(8) Surko, C.M., Leventhal, M., Passner, A., and Wysocki, F.J., "A Positron Plasma in the Laboratory – How and Why", *Proceedings of the ONR Conference on Non-Neutral Plasmas*, 1988, pp. 185–200.

Transport Due to Rotational Pumping

S. M. Crooks and T. M. O'Neil

Physics Department
University of California at San Diego, La Jolla, CA 92093

Abstract. An effect which we call rotational pumping (by analogy with magnetic pumping) causes cross-field transport in nonneutral plasmas when the end confinement potentials are non-axisymmetric. Because the Debye length is small the asymmetries are screened out within the plama, but cause the surface of the plasma to distort. As a flux tube of plasma undergoes $\mathbf{E} \times \mathbf{B}$ drift rotation about the center of the column, the length of the tube oscillates about some mean value and the $P_{\parallel} dV$ work produces a corresponding oscillation in $T_{\parallel}$. In turn the collisional relaxation of $T_{\parallel}$ toward $T_{\perp}$ produces a slow dissipation of electrostatic energy into heat and a consequent radial expansion (cross-field transport) of the plasma. Detailed comparisons between theory and experiment have been made for the case where the asymmetry is produced by displacing the column off-axis, that is, by creating an $m = 1$ diocotron mode (see paper by Cluggish and Driscoll in these proceedings). The theory is generalized to include time dependent asymmetries. For the case where the asymmetry is a traveling wave that rotates faster than the $\mathbf{E} \times \mathbf{B}$ drift rotation of the plasma the particle flux is directed radially inward.

Recent experiments have involved the confinement of pure electron plasmas in Penning traps [1-4]. A schematic diagram for such a trap is shown in Fig. 1. A conducting cylinder is divided axially into three sections, the two end sections being held at a negative potential relative to the central section. There is a uniform magnetic field, B, directed along the axis of the cylinder. The electron plasma resides in the central section, with axial confinement provided by the negatively biased end sections and radial confinement by the magnetic field. The Larmor radius is typically small, so the cross field motion may be described by $\mathbf{E} \times \mathbf{B}$ drift dynamics [3,5].

The most important principle needed for understanding transport in these plasmas is that external torques drive radial particle transport. In guiding center theory the canonical angular momentum of the plasma is approximately [6]

$$P_\theta \simeq \frac{eB}{2c} \sum_j R_j^2 = \frac{eB}{2c} \frac{1}{N} < R^2 > , \qquad (1)$$

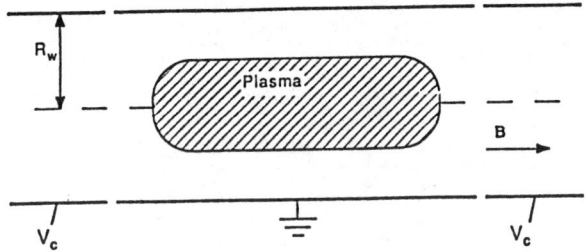

Figure 1: The confinement geometry.

where R_j is the radial position of the jth particle and $< R^2 >$ is the mean square radius of the plasma. In the absence of torques, P_θ and $< R^2 >$ are constant. In the experiments torques may arise from neutrals which are stationary in the lab frame [2,7,8] or from azimuthal asymmetries in the confining fields [9-11]. In this paper we consider transport driven by field asymmetries localized at the ends of the plasma.

When the Debye length is small the plasma has a well defined edge and the non-axisymmetric end potentials cause the end shape of the plasma to become non-axisymmetric. This asymmetry can be characterized by the length of the plasma parallel to the magnetic field,

$$L(r, \theta) = L_0(r) + \delta L(r, \theta) \tag{2}$$

where (r, θ) is a cylindrical coordinate system centered on an axis through the center of charge.

A simple transport equation can be derived by considering a single flux tube of plasma as shown in Fig. 2. The flux tube has length $L(r, \theta)$ as given by Eq. (2), cross section area δA, and contains δN particles. The dominant cross field motion of the flux tube is the $\mathbf{E} \times \mathbf{B}$ drift $\mathbf{v}_D = (c/B)\hat{z} \times \nabla \Phi$. Assuming that the plasma column has a circular cross section the electric potential is of the form $\Phi = \Phi(r)$, and the flux tube drifts in a circular orbit about the center of the column with the frequency

$$\omega_R = \frac{c}{Br} \frac{\partial \Phi}{\partial r} . \tag{3}$$

Setting $\theta = \omega_R t$ in Eq. (2) then implies that the length of the flux tube varies temporally as $L(r, t) = L_0(r) + \delta L(r, \omega_R t)$. The cyclic axial compression and expansion produces a cyclic variation in the parallel temperature, and this is coupled collisionally to the perpendicular temperature. The full temperature evolution is governed by the equations

$$\frac{dT_\parallel}{dt} = -T_\parallel \frac{2}{L} \frac{dL}{dt} + \nu_{\perp,\parallel}(T_\perp - T_\parallel) \tag{4}$$

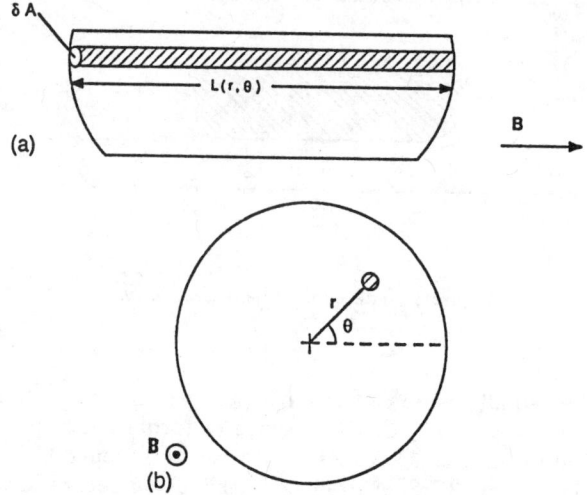

Figure 2: (a) Side view of flux tube. (b) End view of flux tube.

and

$$\frac{dT_\perp}{dt} = -\nu_{\perp,\|}(T_\perp - T_\|) , \tag{5}$$

where $\nu_{\perp,\|}$ is the collisional equipartition rate. We have used the fact that δN is constant in deriving these equations. The first term on the right hand side of Eq. (4) describes the compressional heating (or expansion cooling) of the parallel degrees of freedom, and the second term describes the collisional coupling to the perpendicular degrees of freedom. The perpendicular degrees of freedom are not directly affected by the change in length, so the R.H.S. of Eq. (5) contains only the collisional coupling term. The factor of two difference in the collisional coupling term for Eq. (5) relative to Eq. (4) simply reflects the fact that there are 2 perpendicular degrees of freedom and 1 parallel.

A two time scale analysis of Eqs. (4) and (5) based on the frequency ordering $\omega_R \gg \nu_{\perp,\|}$ yields the result

$$\frac{d<T_\|>}{dt} = 8\nu_{\perp,\|}\frac{<T_\|>}{L_0^2} <\delta L^2(r,t)> + 2\nu_{\perp,\|}(<T_\perp> - <T_\|>) \tag{6}$$

$$\frac{d<T_\perp>}{dt} = -\nu_{\perp,\|}(<T_\perp> - <T_\|>) , \tag{7}$$

where $< \cdot >$ indicates an average over the fast time scale, that is over one rotation period. In addition to the energy conserving terms, the first term on the right hand side of Eq. (6) represents a secular increase in $T_\|$. Physically, this term arises because collisions cause a small phase shift in the parallel temperature fluctuations so that the parallel temperature and pressure are slightly larger in the compression stage than in the expansion stage. More work is done on the plasma during compression than is done by the plasma during expansion. The result is that the plasma in the flux tube is heated.

This effect is similar to magnetic pumping [14], and by analogy, we refer to it as rotational pumping.

Since the confinement potentials are time independent, the total energy in the plasma is conserved, and the increase in thermal energy must be balanced by a corresponding decrease in the electrostatic energy. The particle flux is found by equating the increase in the thermal energy of the flux tube to local Joule heating. That is,

$$n\frac{d}{dt}\left(\tfrac{1}{2} < T_{\parallel} > + < T_{\perp} >\right) = -e\frac{\partial\Phi}{\partial r}\Gamma_r \, , \tag{8}$$

where Γ_r is the radial particle flux and n is the density. The R.H.S. of this equation is the Joule heating per unit volume, and again we have used the fact that $\delta N = \text{CONST}$. Equations (6)-(8) are solved for the flux and yield

$$\Gamma_r = 4\nu_{\perp,\parallel}n(r)\frac{T}{-e\partial\Phi/\partial r}\frac{< \delta L^2 >}{L_0^2} \, , \tag{9}$$

where

$$< \delta L^2 > = \frac{1}{2\pi}\int_0^{2\pi} d\theta \, \delta L^2(r,\theta) \, . \tag{10}$$

This result may also be obtained by a formal kinetic treatment [17] in the adiabatic limit where the bounce frequency for a thermal particle, $\bar{\omega}_B = \pi/L_0\sqrt{T/m}$, is much larger than the $\mathbf{E}\times\mathbf{B}$ drift rotation frequency, ω_R. When $\bar{\omega}_B < \omega_R$ resonant particle flux can be the dominant effect [17]. This theory is closely related to the work of Ryutov and Stupakov on transport in magnetic mirror traps [12].

It is striking that the flux depends on magnetic field strength only through $\nu_{\perp,\parallel}$. In the regime of weak magnetization (i.e. $r_c \gg b$, where $r_c = \bar{v}/\Omega_c$ and $b = e^2/m\bar{v}^2$) this dependence is very weak, $\nu_{\perp,\parallel} \propto \ln(r_c/b)$. In the regime of strong magnetization (i.e. $r_c \ll b$) $\nu_{\perp,\parallel}$ becomes exponentially small (15,16) and our theory predicts that Γ_r becomes exponentially small.

The theory has been used successfully to explain the observed damping of the $m = 1$ diocotron mode [17,18]. One can think of this mode as a rigid displacement of the plasma column away from the central axis of the trap. The displaced column sees non-axisymmetric end potentials and these cause the end shape of the plasma to become non-axisymmetric. Since the total canonical angular momentum is conserved the column must move back to the axis of the trap as it expands, i.e. rotational pumping transport causes the mode to damp. A simple calculation [17] in which we take $\delta L = \kappa \, D/R_w \, r\sin\theta$ [13], where κ is a numerical constant, D is the displacement of the column off axis, and R_w is the radius of the conducting wall yields a damping rate

$$\gamma = -2\kappa^2\nu_{\perp,\parallel}\frac{\lambda_D^2 r_p^2}{L_0^2 R_w^2}\frac{1}{(1 - r_p^2/R_w^2)} \tag{11}$$

where r_p is the radius of the plasma and λ_D is the Debye length. Good agreement between theory and experiment has been found by Cluggish and Driscoll

[18]. In particular, they have observed the unusual scalings with magnetic field strength (see paper by Cluggish and Driscoll in these proceedings).

Another case where we expect rotational pumping to be important is where non-axisymmetric time dependent potentials are directly applied at one end of the main confinement cylinder. For the special case of a plane wave potential, the time dependent length of the plasma will have the form

$$L(r, \theta, t) = L_0(r) + \delta L(r) \, \cos(m\theta - \omega t) \, . \tag{12}$$

If we work in a frame rotating at frequency ω/m, the length fluctuations are static and the simple transport calculation yields

$$\Gamma_r = 4\nu_{\perp,\parallel} n(r) \frac{T}{-e\frac{\partial \Phi_{rot}}{\partial r}} \frac{<\delta L^2>}{L_0^2} \tag{13}$$

where Φ_{rot} is the potential in a frame rotating at frequency ω/m. The transformation to the rotating frame gives a $1/c \, \mathbf{v} \times \mathbf{B}$ contribution to the electric field so that

$$-\frac{\partial \Phi_{rot}}{\partial r} = -\frac{\partial \Phi}{\partial r} + \frac{1}{c} Br \frac{\omega}{m} \, . \tag{14}$$

The flux can now be written as

$$\Gamma_r = 4\nu_{\perp,\parallel} n(r) \frac{T}{-\frac{eBr}{c}[\omega_R - \omega/m]} \tag{15}$$

where we have used $\omega_R = c/Br \, \partial \Phi/\partial r$. Note that this calculation neglects resonant particles and assumes $|m\omega_r - \omega| \gg \nu_{\perp,\parallel}$.

By changing ω one can change the magnitude of the flux. For $|\omega| > m|\omega_R|$ the flux is negative and the plasma is compressed. This corresponds to a field error that rotates faster than the plasma so that the torques act to decrease P_θ. This type of experiment has been performed by Anderegg et al. [19]. Although the transport mehanism is still in question, they have used the technique to demonstrate exceptionally long confinement times (plasma contained for eight days).

ACKNOWLEDGMENTS

We wish to thank Professor Chuan Liu for his suggestion that magnetic pumping may be important in non-neutral plasmas. This work was supported by ONR N00014-89-J-1714 and NSF PHY91-20240.

REFERENCES

1. Malmberg, J.H., and deGrassie, J.S., *Phys. Rev. Lett.* **35**, 577 (1975).

2. deGrassie, J.S., and Malmberg, J.H., *Phys. Fluids* **23**, 63 (1980).

3. Driscoll, C.F., and Fine, K.S., *Phys. Fluids B* **2**, 1359 (1990).

4. Tan, J., and Gabrielse, G., *Phys. Rev. Lett.* **67**, 3090 (1991).

5. Nicholson, D.R., *Introduction to Plasma Theory*, New York: John Wiley & Sons, 1983.

6. Malmberg, J.H., and O'Neil, T.M., *Phys. Rev. Lett.* **39**, 1333 (1977).

7. Douglas, M.H., and O'Neil, T.M., *Phys. Fluids* **21**, 920 (1978).

8. Malmberg, J.H., and Driscoll, C.F., *Phys. Rev. Lett.* **44**, 654 (1980).

9. Eggleston, D.L., O'Neil, T.M., and Malmberg, J.H., *Phys. Rev. Lett.* **53**, 982 (1984).

10. Eggleston, D.L., and Malmberg, J.H., *Phys. Rev. Lett.* **59**, 1675 (1987).

11. Notte, J., and Fajans, J., *Phys. Plasmas* **1**, 1123 (1994).

12. Ryutov, D.D., and Stupakov, G.V., *Fiz. Plazmy* **4**, 521 (1978).

13. Peurrung, A.J., and Fajans, J., *Phys. Fluids B* **5**, 11 (1993).

14. Berger, J.M., Newcomb, W.A., Dawson, J.M., Frieman, E.A., Kulsrud, R.M., and Lenard, A., *Phys. Fluids* **1**, 301 (1958).

15. Glinsky, M.E., O'Neil, T.M., Rosenbluth, M.N., Tsuruta, K., and Ichimaru, S., *Phys. Fluids B* **4**, 1156 (1992).

16. Beck, B.R., Fajans, J., and Malmberg, J.H., *Phys. Rev. Lett.* **68**, 317 (1992).

17. Crooks, S.M., and O'Neil, T.M., *Phys. Plasmas*, "Rotational Pumping and Damping of the $m = 1$ Diocotron Mode," *Phys. Plasmas*, to appear.

18. Cluggish, B., and Driscoll, C.F., "Transport and Damping from Rotational Pumping," in this conference proceedings.

19. Anderegg, F., Huang, X.-P., Driscoll, C.F., Severn, G.D., and Sarid, E., "Long Ion Plasma Confinement Times with 'Rotating Wall'," in this conference proceedings.

Turbulence and Relaxation in 2D Non-Neutral Plasmas

C.F. Driscoll, X.-P. Huang, T.B. Mitchell, and K.S. Fine

Department of Physics and Institute for Pure and Applied Physical Sciences
University of California at San Diego, La Jolla, CA 92093

Abstract. Magnetically confined electron columns evolve as near-ideal 2D fluids, allowing quantitative study of instabilities, turbulence, relaxation, and self-organization. We find that rapid global symmetrization of a distorted column can occur by a decay instability due to nonlinear beat wave damping. In the free relaxation of vortex turbulence, we find that the relaxation rate is limited by vorticity holes which persist for hundreds of rotations even in strong background shear. Finally, we observe that turbulence self-organizes to a meta-equilibrium state which is accurately predicted by minimization of enstrophy for a range of unstable initial conditions.

1. INTRODUCTION

Magnetically confined electron columns evolve as near-ideal 2D fluids, allowing quantitative study of turbulence and relaxation. Here, we describe several experiments which characterize this process from the seemingly disparate perspectives of near-linear waves and fully nonlinear vortices. Interestingly, a number of similarities are seen.

A "generic" experimental apparatus is shown schematically in Fig. 1; the experiments described here were performed on the "EV", "V'", and "CamV" apparatuses (1–3). The electron columns are confined radially by a uniform magnetic field, B_z, and contained axially by voltages applied to end sections of the cylindrical wall. The confined plasma is sensed and manipulated by antennas in the wall, and the z-averaged electron density $n(r, \theta, t)$ is accurately measured by dumping the column onto a phosphor screen imaged by a $512 \times 512 \times 16$ bit CCD camera. The axial bounce frequency of an electron is large compared to the $\mathbf{E} \times \mathbf{B}$ drift rotation frequency ($\omega_B \gg \omega_E$), allowing a 2D fluid description of the system.

The (r, θ) flow of the electrons is described by the 2D drift-Poisson (4) equations,

$$\frac{\partial n}{\partial t} + \mathbf{v} \cdot \nabla n = 0 \,, \quad \mathbf{v} = -\frac{c}{B} \nabla \phi \times \hat{z} \,,$$

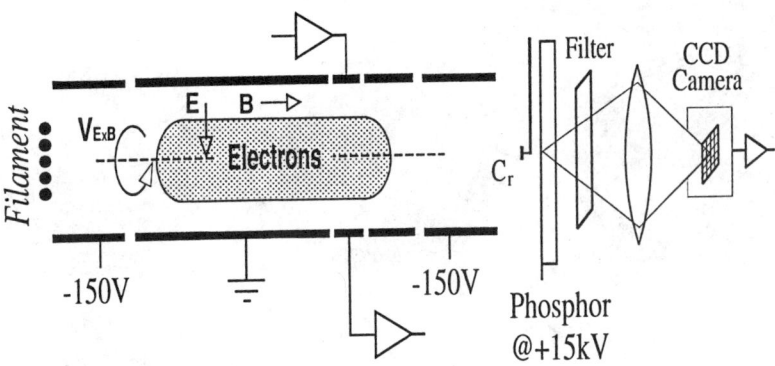

Figure 1. The cylindrical experimental apparatus with phosphor screen/CCD camera diagnostic.

$$\nabla^2 \phi = 4\pi en \,, \tag{1}$$

where $n(r, \theta)$ is the (z-averaged) electron density, $\mathbf{v}(r, \theta)$ is the $\mathbf{E} \times \mathbf{B}$ drift velocity, and $\phi(r, \theta)$ is the electrostatic potential. The vorticity of the flow, $\zeta \equiv \hat{z} \cdot \nabla \times \mathbf{v} = (4\pi ec/B)n$, is proportional to the electron density, which is directly measured. These equations are isomorphic to the Euler equation (5,6). A column of electrons in vacuum surrounded by a conductor thus evolves as would a 2D vortex in an incompressible inviscid fluid surrounded by a circular free-slip boundary. Of course, there are also various "viscous" plasma effects (7) on small spatial scales, which are not modelled by the Navier-Stokes equation. We believe the effects described here do not depend on the details of the fine-scale dissipation.

We study instabilities and relaxation by creating well-controlled but unstable initial conditions and then observing the free evolution. If we start with a smooth, symmetric density profile $n(r)$ and add a small perturbation in θ, we are able to study near-linear modes and instabilities. As discussed in the next section, the global modes (stable, damped, and growing) are reasonably well understood. The alternate perspective of nonlinear vortices and turbulence is less well understood. Here, shear-flow instabilities on an initially hollow profile lead to the formation of vortices, which then merge, shed filaments, and eventually relax to an axisymmetric, sheared meta-equilibrium state (MES). An example of such an evolution is shown in Fig. 2, labeled in bulk rotation times τ_R.

In this paper, we discuss three processes which affect this relaxation of turbulence. From the near-linear perspective, global symmetrization of the vortex column can occur due to direct or beat wave damping of surface modes; both effects depend on a spatial resonance which is ignored in step-profile theory and in contour dynamics simulations. From the vortex perspective,

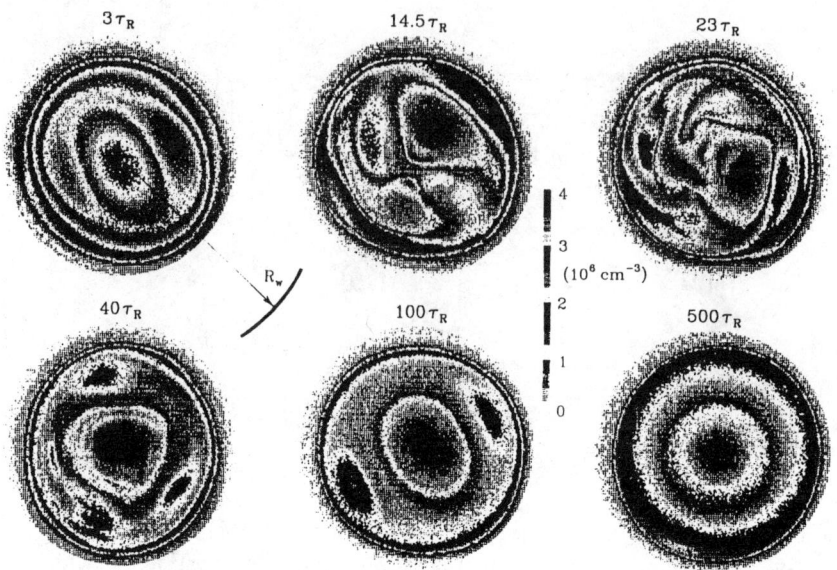

Figure 2. False-color camera images of electron density (*i.e.* vorticity) showing the evolution of an initially hollow electron column.

turbulence generally relaxes by vortex merger and fine-scale filamentation; here, energy flows to long spatial scales, and enstrophy and entropy cascade to fine scales where they are dissipated. We find that the rate of this relaxation can be limited by the longevity of negative vorticity holes which are stable in the background shear flow. Interestingly, these holes often form a tripole pattern which could be considered to be the extreme nonlinear limit of the saturated state of a partially damped $m = 2$ wave. Finally, we consider the axisymmetric meta-equilibrium state to which the turbulence eventually relaxes. Here, we find that a range of initially unstable configurations relax to the minimum enstrophy vortex. For initial conditions with larger excess energy, the observed meta-equilibrium states are more peaked, and theory predictions are presently being developed.

2. MODES AND INSTABILITIES

Our $n(r, \theta, t)$ data allows us to completely characterize the $k_z = 0$ flute modes, varying as $\exp(im\theta - i\omega_m t)$. It is well known that hollow density profiles can have diocotron (*i.e.* Kelvin-Helmholtz) instabilities, whereas monotonically decreasing profiles have only stable or damped modes. More precisely, Rayleigh's inflection-point theorem (8) as extended by Arnold (9) shows that the system is nonlinearly stable if $n(r)$ is monotonically decreasing. When $n(r)$

and $\omega_E(r)$ are sufficiently non-monotonic, $m \geq 2$ exponential instabilities are predicted by inviscid 2D fluid theory (5). These $m \geq 2$ shear-flow instabilities are readily observed on hollow beams and trapped electron columns (6,10,11), generating the turbulence of Fig. 2.

It is important to emphasize that the stable and unstable diocotron modes are separate and apparently mutually orthogonal, and may co-exist on hollow profiles (6,12). This distinction is missed by the usual "step-profile" vorticity patch analysis, which gives modes occurring in complex conjugate pairs (4). Experimentally, the stable and unstable modes for $m \geq 2$ have completely different real frequencies, and one mode can be unstable even when the other is not significantly damped. Both the frequencies and growth rates of the $m \geq 2$ modes are reasonably well characterized by computational solution of the eigenvalue equation using the measured density profiles. For short wavelengths (*e.g.* $m \geq 15$), however, experiments have observed stabilization by plasma effects outside 2D fluid theory (11).

For $m = 1$, a strong exponential instability is also observed experimentally (12). This exponentially growing mode does not exist within 2D fluid theory, although an initial-value analysis demonstrates that perturbations can grow algebraically with time (13). More recent analyses suggest that finite Larmor radius and finite length effects outside 2D fluid theory can result in exponential growth of $m = 1$ (14,15).

In neutral plasmas, where electric fields arise from regions of non-neutrality, similar flute modes have been analyzed in various approximations (16) and have recently been observed (17,18). One interesting question is the net charge of the plasma and the effects of image charges in the walls. For cylindrical geometry, the image charges from a net plasma charge give rise to the stable $m = 1$ diocotron mode, wherein the entire displaced plasma column orbits around the center of the confinement cylinders. For toroidal geometry, this image charge mode would be more complicated.

3. DIRECT AND BEAT-WAVE DAMPING

Small amplitude surface waves on the idealized circular, constant vorticity patch are all stable, with frequencies $\omega_m = \omega_E \left[m - 1 + (R_p/R_w)^{2m} \right]$. However, this vortex patch idealization precludes resonance between the wave and the fluid flow at a radius where the vorticity is not spatially constant (19). This resonance for mode m occurs at a radius r_s where

$$\omega_m = m\, \omega_E(r_s)\,, \tag{2}$$

where $\omega_E(r)$ is the $\mathbf{E} \times \mathbf{B}$ rotation frequency of the vortex. This "direct" resonance can give rise to inviscid spatial Landau damping of the wave (1), *i.e.* symmetrization of the vortex. For even moderate wave amplitudes, this

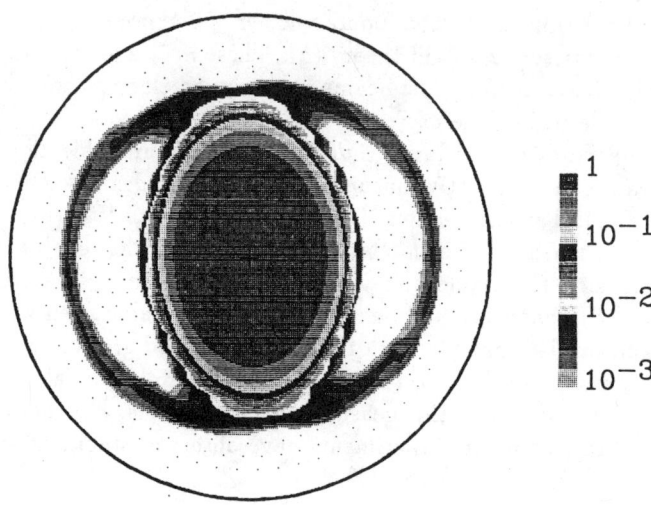

Figure 3. Measured density $n(r, \theta)$ phase coherent with a partially damped $m = 2$ diocotron mode.

damping is typically nonlinear, and the damping may decrease (20,21) or cease (6) when the resulting "cat's-eye" flows generate fine-scale filaments inside the vortex. For reasonably sharp-edged vorticity profiles, the resonant radii can be completely outside the vortex, in which case no direct resonance damping occurs.

Of perhaps more interest for real plasmas and flows are the large amplitude modes. The Kirchoff ellipse is an exact nonlinear solution (22) for $m = 2$, and nonlinear solutions for $m \geq 3$ have been found numerically and analytically (23). For large amplitude distortions, $\omega_E(r)$ no longer adequately describes the flow, and direct wave/flow resonances can occur even for vortex patches. In this case, the filaments generated at the resonance "X-points" may be partially outside the vortex (23,24). Again, the evolution of this process will be highly nonlinear, and the resonant damping may cease with the formation of steady-state filamentary structures. Figure 3 shows the nominally steady-state filamentary structure which results from the direct resonance acting on a large $m = 2$ mode. This structure is also described as a "tripole," since the "holes" surrounded by the filaments are negative vorticity relative to the center (25). Furthermore, this state seems to be the "linear wave" analogue of the long-lived "holes" discussed in the next section.

Even when no direct resonance occurs, we find (1) that a single excited wave varying as $\sin(m\theta)$ will decay into a daughter wave varying as $\sin[(m-1)\theta]$, through "beat wave" resonance damping of the nonlinear beat at frequency $\omega_m - \omega_{m-1}$ (26). The direct and beat wave damping processes are shown

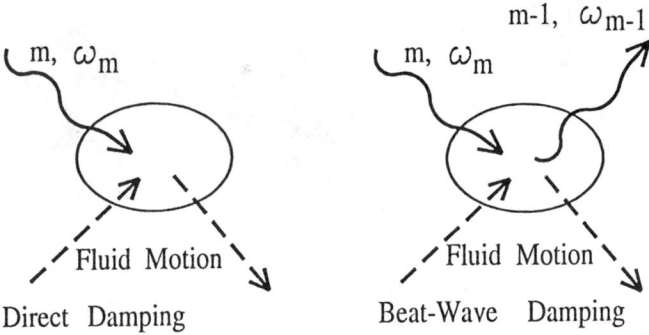

Figure 4. Diagrammatic representation of (a) direct spatial Landau damping and (b) beat-wave damping.

diagrammatically in Fig. 4. When this beat-wave resonant coupling exists, a single surface wave is an unstable equilibrium, and no equilibrium exists with two waves.

This beat-wave damping is seen to give global symmetrization of an asymmetric vortex, while presumably also generating fine-scale resonance filaments within the vortex. For many vortex profiles, this beat wave damping is observed to be the fastest mechanism for symmetrization. Figure 5 shows the measured density (*i.e.* vorticity) distribution before and after beat-wave damping of an $m = 3$ perturbation on an otherwise symmetric column. This $m = 3$ parent mode decays rapidly to an $m = 2$ daughter mode plus a diffuse halo, on a time scale of 10 rotations of the vortex. The resultant $m = 2$ mode is then unstable to decay into an $m = 1$ mode. We note that the initial condition exhibits slight filamentation due to direct resonance damping of the large $m = 3$ mode.

Figure 5 also shows the measured growth rates γ_{m-1} of the daughter wave amplitudes versus the normalized amplitudes of the parent wave, A_m, for $m = 2, 3$, and 4. Here, the growth rates γ are scaled by the central rotation time $\tau_R \equiv 2\pi/\omega_E(0) = 5.5 \ \mu$s. For comparison to theory estimates, the mode amplitudes A_m are calculated from the θ-components $\delta n_m(r)$ of the measured $n(r, \theta)$, by $A_m^2 \equiv \int d^2\mathbf{r} |\delta n_m(r)|^2 / \int d^2\mathbf{r} |n_0(r)|^2$. These measured density perturbations are directly proportional to the signals received on the wall sectors. The measured growth rates are all observed to depend on the vorticity profile.

This decay instability is apparently due to resonant damping of the beat wave resulting from nonlinear interaction of the pump and growing mode, shown diagrammatically in Fig. 4. The nonlinear couplings involved in beat-mode damping have been treated in general by Crawford and O'Neil (26), although no quantitative coupling rates Γ_{ij} have yet been calculated. Energy and angular momentum conservation between waves and particles predicts that the coupling between pump mode i and daughter mode j will occur at a

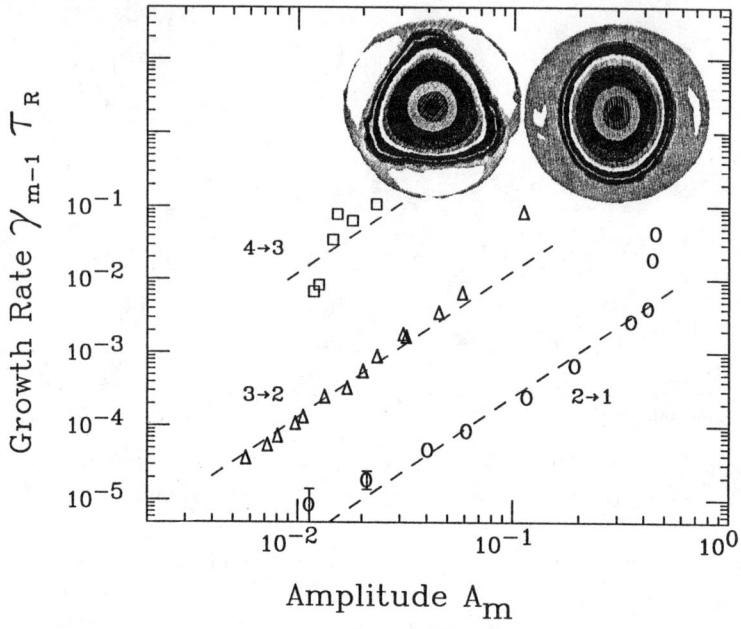

Figure 5. Measured decay instability growth rates $\gamma_{m-1}\tau_R$ versus mode amplitude A_m, for $m = 4$, 3 and 2. Insert shows $n(r, \theta)$ before and after the $3 \to 2$ instability.

radius r_s where

$$\omega_i - \omega_j = \omega_E(r_s^{i \to j})(i - j) \ . \tag{3}$$

The unstable mode j is predicted to grow as

$$\frac{d}{dt} A_j = (-\gamma_j + \Gamma_{ij}A_i^2)A_j \ , \tag{4}$$

where γ_j is the linear damping of mode j and the coupling rate Γ_{ij} is independent of amplitude A_i for small A_i. A similar nonlinear coupling to *external* field asymmetries A_i gives rise to an "induced scattering instability" which has been previously observed on these electron systems (27).

The predicted dependence on the square of the pump mode amplitude A_i in Eq. (5) agrees with the experimental data of Fig. 5. This allows us to estimate the coupling rates Γ_{ij}: the dashed lines in Fig. 4 give $\Gamma_{21} = 0.025/\tau_R$, $\Gamma_{32} = 1.3/\tau_R$ and $\Gamma_{43} = 120./\tau_R$. These measured rates are about 10 times less than estimated from first order perturbation theory (28). Presumably, this is because a nonlinear filamentary state forms more rapidly than the instability grows, so the true growth rate is determined by fine-scale diffusive or viscous processes smoothing out the filaments. This is similar to the situation with direct damping of large amplitude surface modes, where "bounce" oscillations and a less than expected damping rate have been observed (21).

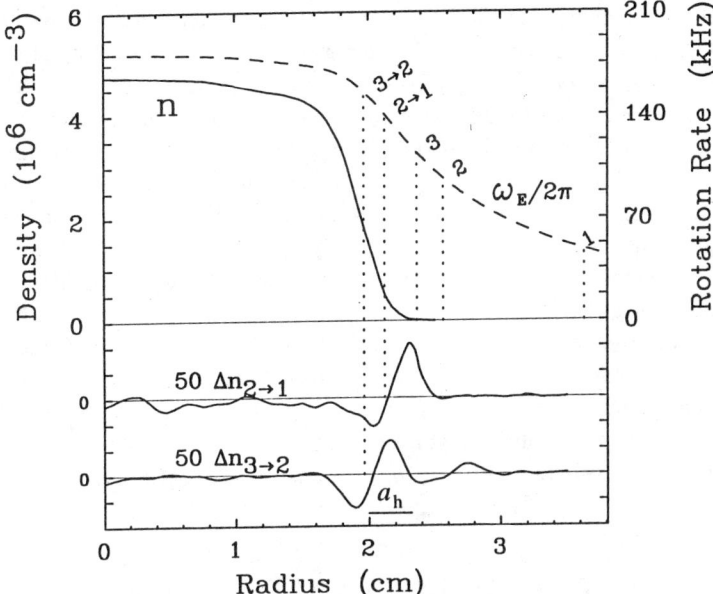

Figure 6. Measured density profile $n(r)$; calculated fluid rotation $\omega_E(r)/2\pi$, with $m = 1, 2, 3$ direct resonance radii, and $2 \to 1$, $3 \to 2$ beat resonance radii; and measured perturbations $\Delta n(r)$ from $2 \to 1$ and $3 \to 2$ decay instabilities.

Most significantly, the beat mode damping resonance signature of Eq. (4) is clearly seen in the measured perturbations in vorticity induced by the decay process (1). Figure 6 shows the measured z-averaged density profile $n(r)$ before any waves are excited, and the vortex rotation frequency $\omega_E(r)$ calculated from $n(r)$. The initial density profile was chosen so as to have no direct spatial resonances for modes $m \leq 3$, *i.e.* the measured mode frequencies projected onto $\omega_E(r)$ give resonant radii r_s at which there is essentially no vorticity. In contrast, the beat frequencies $\omega_2 - \omega_1$ and $\omega_3 - \omega_2$, have resonant radii $r_s^{2 \to 1}$, and $r_s^{3 \to 2}$ from Eq. (4) on the edge of the vorticity profile.

We obtain the vorticity perturbation induced by the decay instability of a small amplitude wave by subtracting $n(r)$ from the profile measured after mode m is excited and decays to mode $m - 1$, and mode $m - 1$ is damped by applied feedback. The measured $\Delta n_{2 \to 1}$ and $\Delta n_{3 \to 2}$ show particles transported from immediately inside to immediately outside $r_s^{2 \to 1}$ and $r_s^{3 \to 2}$. The true structure of the perturbations is probably narrower than the features in Fig. 6, due to averaging over the collimator area a_h; also, the effects of fine-scale diffusion or viscosity (7) are not resolved.

Our observations show that surface wave damping from direct and beat-wave resonances should contribute significantly to the relaxation of vortices in inviscid 2D flows. Indeed, the beat wave damping measured here suggests that when there are available longer-wavelength modes, surface waves of finite amplitude will always be damped, and fine-scale filamentation or cats-eyes within or outside the vortex will be generated. One consequence of this beat wave damping is that a long-wavelength mode (such as an $m = 2$ ellipse) would appear to actively damp or suppress shorter wavelength modes suchs as $m = 3$. These shape distortions are an inherent result of vortex-vortex interactions, including those leading to merger (29) or filamentation (30), so direct damping and beat-wave damping may affect these interaction processes.

Finally, we note that the near-linear perspective of the beat-wave decay instability shows qualitative correspondence with even the highly nonlinear evolution of Fig. 2. For example, the plasma at 40 τ_R has moderate $m = 3$ components δn_3 and $\delta\phi_3$, and this relaxes to predominantly $m = 2$ components by 100 τ_R. Of course, $\delta n_3(r)$ and $\delta n_2(r)$ do not necessarily correspond to the radial eigenfunctions for linear modes. From the nonlinear vortex perspective, this evolution is due to vortex merger. Both beat-wave damping and vortex merger result in energy flowing to long wavelengths, with concomitant formation of fine scale filamentation.

4. VORTICES STABLE IN SHEAR

From the nonlinear perspective, the rate of relaxation towards symmetry in Fig. 2 is largely determined by the longevity of coherent negative vortices, or "holes" (3). These holes are stable for hundreds of bulk rotation times, generally in a preferred configuration of two diametrically opposed holes on a distorted core (a "tripole"). On a longer time scale, these holes creep radially outward and are eventually destroyed. Because of these coherent holes, the observed fluctuation relaxation rate is about 50 times slower than expected from simple passive tracer mixing, and the measured fluctuations are strongly skewed from Gaussian.

The longevity of the vorticity holes in the images of Fig. 2 suggests that the holes can be regarded as negative vortices in equilibrium with the background shear flow. The inviscid equilibria of uniform vorticity patches in an imposed background flow are ellipses, as first derived by Moore and Saffman (31). In our case, the mean background flow is purely rotational, i.e. $<\mathbf{v}>= \omega_R(r)r\hat{\boldsymbol{\theta}}$, leading to a simple linear shear in the vicinity of any point. Therefore, the equilibrium depends only on the scaled shear

$$\sigma \equiv r_v \frac{d\omega_R}{dr}\big|_{r_v} / \Delta\zeta_v , \tag{5}$$

where $\Delta\zeta_v$ is the difference between the vorticity at the center of the vortex (at radius r_v) and the background vorticity $<\zeta>$ at r_v. Since the rotational shear considered here is everywhere negative, the equilibria are ellipses oriented along $\hat{\boldsymbol{\theta}}$ (for holes with $\sigma > 0$) or along $\hat{\mathbf{r}}$ (for clumps with $\sigma < 0$).

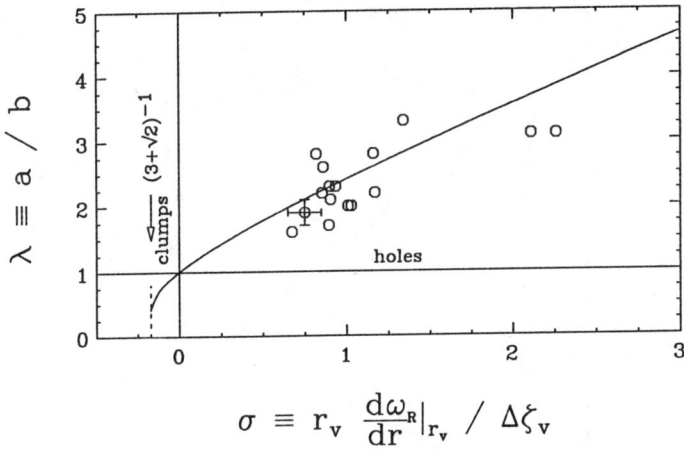

Figure 7. Measured hole aspect ratio λ versus normalized background shear σ and Moore-Saffman model (curve).

Figure 7 shows the predicted and measured hole aspect ratios $\lambda \equiv a/b$, where a and b are the elliptical axes along θ and r directions respectively. Here, λ and σ are measured directly from 16 images at times $40 \leq t/\tau_R \leq 400$. Reasonable agreement between the measurement and the model is found, validating the applicability of the model to individual holes in this system. The holes are robust because they are prograde with respect to the negative background shear, whereas clumps would be retrograde.

In the model, holes of any depth $\Delta\zeta_h$ can survive in the background flow with negative shear. Of course, a shallow hole would be very elongated, and the local model is no longer valid. In contrast, a clump must have relative vorticity

$$\Delta\zeta_c \geq (3 + 2\sqrt{2})\, r_c \left| \frac{d\omega_R}{dr} \right|_{r_c} \qquad (6)$$

to be in equilibrium. Such very intense clumps are not observed for evolutions from initially smooth distributions such as Fig. 2. If we approximate the background as $<\zeta> = \zeta_0[1 - (r/r_0)^2]$, giving $r d\omega_R/dr = -(\zeta_0/2)(r/r_0)^2$, then $\Delta\zeta_c$ cannot be large enough to satisfy Eq. (6) provided the clump vorticity is no greater than the central vorticity (*i.e.* $\zeta_c \leq \zeta_0$). However, long-lived clumps *have* been observed in this apparatus during evolutions of highly filamented initial conditions with low shear profiles, apparently satisfying the condition of Eq. (6) (32). Also, a similar limit to vortex stability has recently been measured in a system with a controlled background shear (33).

After about $40\, \tau_R$, the number of observed holes decreases with time, and the remaining holes tend to organize into symmetric configurations. After $t \simeq 80\, \tau_R$, virtually all the images are in the tripolar configuration of a de-

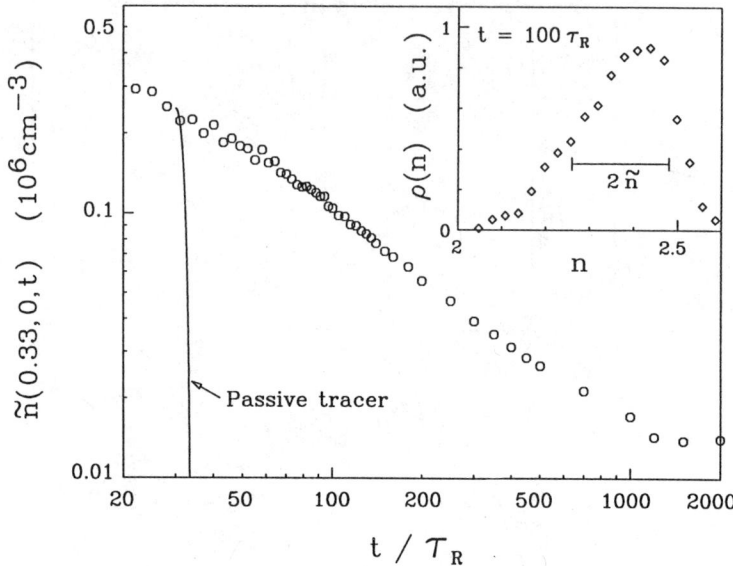

Figure 8. Measured fluctuation level $\tilde{n}$ versus time, and predicted passive tracer mixing. The insert shows the PDF $\rho(n)$.

formed central core and two elliptical holes opposite in θ, typified by Fig. 2 at $100\,\tau_R$. The decrease in the number of holes with time is apparently due to mutual advection and merger (34) of the holes. These merger events have been observed in a few camera images, but the shot-to-shot irreproducibility precludes imaging the dynamics. The "tripole" state is stable for hundreds of τ_R because of hole-induced distortion of the background flow: the elongated core in Fig. 2 at $100\,\tau_R$ maintains the θ-positions of the two holes.

In these evolutions, the quasi-stable tripolar configuration is eventually destroyed by a slow outward creep of the holes: the holes move from $r_h \simeq 0.35$ to $r_h \simeq 0.50$ in about $500\,\tau_R$. The cause of this outward motion is unknown at present. However, experiments varying L_p and B_z suggest that this outward creep is a 2D $\mathbf{E} \times \mathbf{B}$ drift effect rather than an axial end effect (3,35). When the holes approach the edge of the column at $r \simeq 0.57$, they apparently filament in the θ direction and are destroyed.

These long-lived coherent density holes cause the measured shot-to-shot fluctuations to decay much slower than expected from a simple passive tracer model (3). Figure 8 shows the RMS variation $\tilde{n}$ in the charge collected on C_r at $r = 0.33$, $\theta = 0$, from 1000 nominally identical evolutions at each time t. After growing during the initial instability period, $\tilde{n}$ relaxes with a time scale of $100-200$ rotations. A passive tracer relaxation model would postulate that fluctuations are distorted and filamented by the background shear (36), and then averaged over (coarse-grained) by the charge collector C_r. For fluctuations with $m = 2$, the passive tracer evolution is shown as the solid curve in Fig. 8, with an e-folding decay time $\simeq 3\,\tau_R$ determined from the background shear profile. The measured fluctuation decay rate is over 50 times slower than the passive tracer mixing.

The holes also produce a skewed non-Gaussian probability distribution function (PDF) for the density measurements. The insert in Fig. 8 shows the measured PDF of density $n(0.33, 0)$ at $t = 100\,\tau_R$. The localized density holes such as the ones shown in Fig. 2 give a skew at low densities, reflecting the probability that they are at the same (r, θ) as the collector C_r at the time of the measurement.

5. META-EQUILIBRIUM STATES

Eventually, the system relaxes to a low-noise meta-equilibrium state (MES). We observe that the MES is axisymmetric with a monotonically decreasing density profile, and lasts for about $10^4\,\tau_R$ before non-ideal effects on large scales cause it to evolve further (2). The total number of electrons, angular momentum, and electrostatic energy, given by

$$N \equiv R_w^2 \int d^2\mathbf{r}\, n \ ,$$

$$P_\theta \equiv \int d^2\mathbf{r}\, (1 - r^2) n/n_0 \ , \tag{7}$$

$$H_\phi \equiv -\tfrac{1}{2} \int d^2\mathbf{r}\, (n/n_0)(\phi/\phi_0) \ ,$$

are well conserved from the initial conditions to the MES. Here, $n_0 \equiv N/R_w^2$, and $\phi_0 \equiv eN$. However, less robust "ideal" invariants such as enstrophy Z_2 and mean-field entropy S, given by

$$Z_2 \equiv \tfrac{1}{2} \int d^2\mathbf{r}(n/n_0)^2 \ ,$$

$$S \equiv -\int d^2\mathbf{r}\, (n/n_0) \ln(n/n_0) \ , \tag{8}$$

vary significantly, due to measurement coarse-graining or dissipation of small spatial scales.

We have compared the measured MES density profiles to various theories based on maximization of entropy or minimization of enstrophy (37). The initial hollow and final monotonic density profiles for an evolution similar to Fig. 2 are shown in Fig. 9. Also shown are the profiles predicted by maximizing entropy (dashed) and by minimizing the enstrophy (solid). We find that

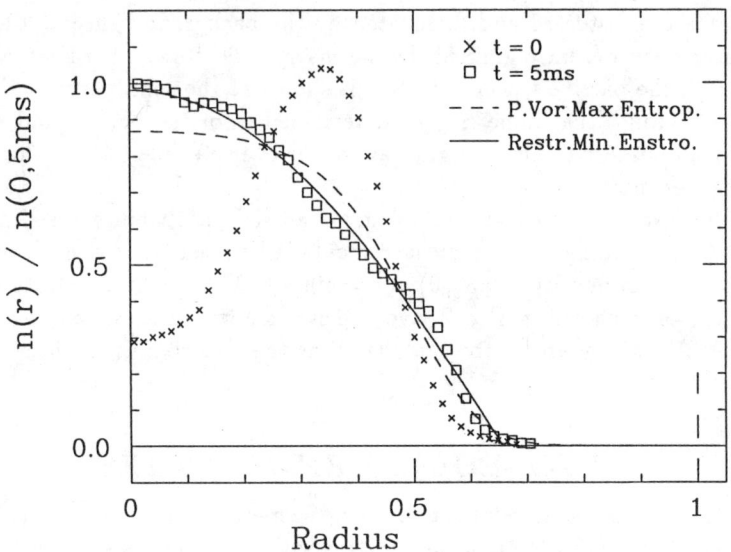

Figure 9. Measured radial density profiles of the initial and metaequilibrium state, and theoretical predictions from the four models discussed.

minimization of enstrophy accurately predicts the meta-equilibrium profiles for hollow initial conditions of moderate energy (2) such as shown in Fig. 2.

The enstrophy minimization gives

$$\frac{d^2 n_e}{dr^2} + \frac{1}{r}\frac{dn_e}{dr} + \beta^2(n_e + \gamma) = 0, \tag{9}$$

with the solution $n_e(r) = \alpha J_0(\beta r) - \gamma$, where J_0 is the zeroth order Bessel function. To keep $n \geq 0$, we take

$$n_e(r) = \begin{cases} \alpha[J_0(\beta r) - J_0(\beta r_0)] & 0 \leq r \leq r_0 \\ 0 & r_0 < r \leq 1, \end{cases} \tag{10}$$

where (α, β, r_0) are determined from the measured (N, P_θ, H_ϕ), with no adjustable parameters. This prediction, plotted as the solid curve in Fig. 9, is close to the experimentally measured meta-equilibrium density profile, supporting the "selective decay" hypothesis (38–41).

The predicted enstrophy $Z_2^{\min}$ of the minimum enstrophy state depends on the three robust invariants (N, P_θ, H_ϕ), but scaling allows us to remove the dependencies on N and P_θ (2). Actually, the dependence on N has already been removed by using n/n_0 in the definitions of P_θ and H_ϕ. The dependence on P_θ can be removed by defining $\hat{Z}_2^{\min} \equiv 4\pi(1 - P_\theta) Z_2^{\min}$; this rescaled minimum enstrophy is then only a function of excess energy $H_\phi^{\mathrm{exc}} \equiv H_\phi - H_\phi^{\min}$.

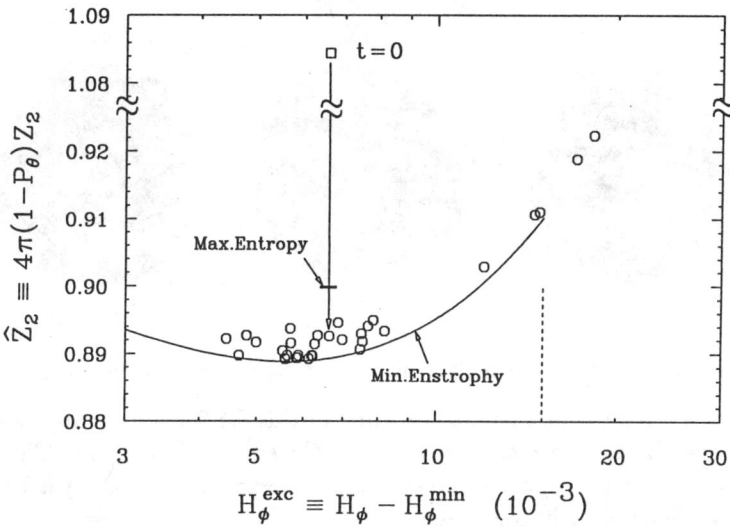

Figure 10. The rescaled enstrophy $\widehat{Z}_2$ of the metaequilibrium state versus excess energy H_ϕ^{exc}, from measurements (circles) and from the restricted minimum enstrophy model (curve).

Here, $H_\phi^{\text{min}} = \{1/2 - \ln[2(1 - P_\theta)]\}/2$ is the minimum energy possible for a column with given N and P_θ, which occurs for a uniform density profile.

We have experimentally measured the relaxed meta-equilibrium states resulting from initially hollow columns of various diameters, "hollowness" depths, and axial lengths, in several magnetic fields. The measured MES profiles give rescaled enstrophy $\widehat{Z}_2$ versus H_ϕ^{exc}, plotted as circles in Fig. 10. Here, the specific evolution of Fig. 9 is shown by the solid symbols: the curve shows the predicted $\widehat{Z}_2^{\text{min}}$. The measured enstrophy of the MES is close to the prediction of the restricted minimum enstrophy theory for each initial condition, typically within 1%–2% of the enstrophy available for dissipation. This reflects the fact that the measured $n(r)$ profiles agree well with Eq. (10), comparable or better than the MES shown in Fig. 9. For comparison, the $\widehat{Z}_2$ calculated from the maximum entropy solution of Fig. 9 is also shown. The experimental data of Figs. 9 and 10 is 3 to 4 times closer to the predictions of minimum enstrophy than to the predictions of maximum entropy.

For initial conditions with higher excess energy, the relaxation is less well understood both theoretically and experimentally. Figure 11 shows an evolution from a high excess energy ($H_\phi^{\text{exc}} \approx 45 \times 10^{-3}$), highly filamented initial condition. Here, the fine-scale vortices apparent at 1 τ_R relax to a sharply peaked ($n(0)/\bar{n} \sim 4$) meta-equilibrium, again with stable holes even at 600 τ_R.

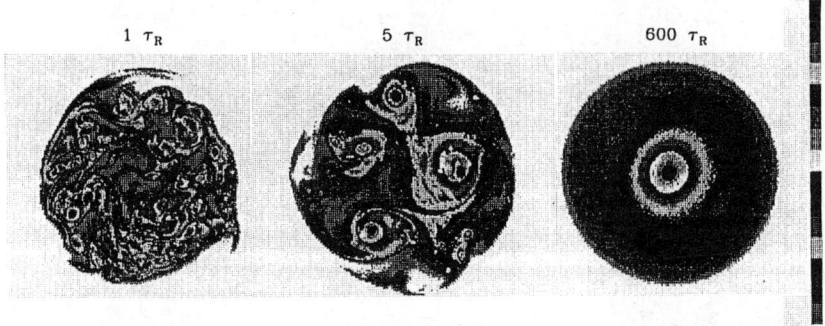

Figure 11. Evolution from a highly filamented, high excess energy initial condition.

However, the minimum enstrophy state for this H_ϕ^{exc} is not readily calculated, since off-axis states are expected to have the smallest enstrophy. Thus, quantitative comparisons between theory and experiment have not yet been done in this regime of high excess energy.

One striking feature which has been observed in this regime is the formation of long-lived geometric patterns of 10–15 intense vortices, *i.e.* "vortex crystals" (32). Here, the general process of energy flow to longer wavelengths is arrested when the vortices "anneal" into fixed positions rotating with the background. These vortex crystals are observed to be stable for $\sim 10^3 \tau_R$, until apparently non-ideal processes cause the intense vortices to decay in place. In one sense, these geometric patterns of intense positive vorticity are analogous to the long-lived "tripole" pattern of two holes symmetric about the high vorticity center. Future experiments and theory may clarify when these long-lived states occur, slowing or arresting the nominally inviscid relaxation to longer wavelengths, and what dissipative processes eventually allow further relaxation.

ACKNOWLEDGMENTS

This research is supported by Office of Naval Research grant N00014-89-J-1714, Department of Energy grant DE-FG03-85ER53199, and National Science Foundation grant PHY91-20240.

REFERENCES

1. T.B. Mitchell and C.F. Driscoll, *Phys. Rev. Lett.* **73**, 2196 (1994); T.B. Mitchell, Ph.D. Thesis, Univ. of Calif., San Diego (1993).
2. X.-P. Huang and C.F. Driscoll, *Phys. Rev. Lett.* **72**, 2187 (1994).
3. X.-P. Huang, Ph.D. Thesis, Univ. of Calif., San Diego (1993).
4. R.C. Davidson, *Physics of Nonneutral Plasmas* (Addison-Wesley, New York, 1990), Ch. 5,6.

5. R.J. Briggs, J.D. Daugherty, and R.H. Levy, *Phys. Fluids B* **13**, 421 (1970); R.H. Levy, *Phys. Fluids B* **8**, 1288 (1965).
6. C.F. Driscoll, and K.S. Fine, *Phys. Fluids B* **2**, 1359 (1990).
7. C.F. Driscoll, J.H. Malmberg, and K.S. Fine, *Phys. Rev. Lett.* **60**, 1290 (1988).
8. Lord Rayleigh, *Scientific Papers* (Cambridge Univ. Press, Cambridge, 1899), Vol. 1, p. 474.
9. V.I. Arnol'd, *J. Mec.* **5**, 29 (1966).
10. G. Rosenthal, G. DiMonte, and A.Y. Wong, *Phys. Fluids* **30**, 3257 (1987).
11. A.J. Peurrung and J. Fajans, *Phys. Fluids A* **5**, 493 (1993).
12. C.F. Driscoll, *Phys. Rev. Lett.* **64**, 645 (1990).
13. R.A. Smith and M.N. Rosenbluth, *Phys. Rev. Lett.* **64**, 649 (1990).
14. R.A. Smith, *Phys. Fluids B* **4**, 287 (1992).
15. S.N. Rasband, R.L. Spencer, and R.R. Vanfleet, *Phys. Fluids B* **5** 669 (1993); S.N. Rasband, *Bull. Am. Phys. Soc.* **39**, 1549 (1994).
16. W. Horton, J. Liu, J.D. Meiss, and J.E. Sedlak, *Phys. Fluids* **29**, 1004 (1986).
17. T. Huld, A.H. Nielsen, H.L. Pecseli, and J.J. Rasmussen, *Phys. Fluids B* **3**, 1609 (1991).
18. B. Song and A.K. Sen, Phys. Rev. Lett. **72**, 92 (1994).
19. W. Kelvin, *Nature* **23**, 45 (1880).
20. J.S. deGrassie and J.H. Malmberg, *Phys. Fluids* **23**, 421 (1980).
21. S. Pillai and R.W. Gould, *Bull. Am. Phys. Soc.* **38**, 1970 (1993).
22. H. Lamb, *Hydrodynamics*, 6th ed, New York: Dover Press, 1932, §158,159.
23. D.I. Pullin, *Ann. Rev. Fluid Mech.* **24**, 89 (1992).
24. M.V. Melander, J.C. McWilliams, and N.J. Zabusky, *J. Fluid Mech.* **178**, 137 (1987).
25. E.J. Hopfinger and G.J.F. van Heijst, *Ann. Rev. Fl. Mech.* **25**, 241 (1993).
26. J.D. Crawford and T.M. O'Neil, *Phys. Fluids* **30**, 2076 (1987).
27. D.L. Eggleston and J.H. Malmberg, *Phys. Rev. Lett.* **59**, 1675 (1987).
28. W.G. Flynn, Private communication.
29. T.B. Mitchell, C.F. Driscoll, and K.S. Fine, *Phys. Rev. Lett.* **71**, 1371 (1993); K.S. Fine, C.F. Driscoll, J.H. Malmberg, and T.B. Mitchell, *Phys. Rev. Lett.* **67**, 588 (1991).
30. D.G. Dritchell and D.W. Waugh, *Phys. Fluids A* **4**, 1737 (1992).
31. D.W. Moore and P.G. Saffman, *Aircraft Wake Turbulence and its Detection*, (Plenum, New York, 1971), p.339; S. Kida, *J. Phys. Jpn.* **50**, 3517 (1981); D.G. Dritchel, *J. Fluid Mech.* **210**, 223 (1990).
32. K.S. Fine, C.F. Driscoll, and A.C. Cass, *Bull. Am. Phys. Soc.* **39**, 1736 (1994).
33. D.L. Eggleston, *Phys. Plasmas*, to be published (1995).
34. P.S. Marcus, *Nature* **331**, 693 (1988); P.S. Marcus, *Ann. Rev. Astron. Astrophys.* **31**, 523 (1993).
35. A.J. Peurrung and J. Fajans, *Phys. Fluids B* **5**, 4295 (1993).
36. P.B. Rhines and W.R. Young, *J. Fluid Mech.* **133**, 133 (1983).
37. R.H. Kraichnan and D. Montgomery, *Rep. Prog. Phys.* **43**, 547 (1980).
38. F.P. Bretherton and D.B. Haidvogel, *J. Fluid Mech.* **78**, 129 (1976).
39. W.H. Matthaeus and D. Montgomery, *Ann. NY Acad. Sci.* **357**, 203 (1980).
40. C.E. Leith, *Phys. Fluids* **27**, 1388 (1984).
41. A. Hasegawa, *Adv. Phys.* **34**, 1 (1985).

Electron Vortex Dynamics in an Applied Shear Flow

D.L. Eggleston

Occidental College, Department of Physics
Los Angeles, California 90041

Abstract. An electron column in a modified Malmberg-Penning trap is used to study the behavior of a single two-dimensional vortex in an imposed irrotational shear flow. Phosphor screen images of the shearing process show a variety of phenonomena including the fission of the original vortex, the emission, stretching, and entrainment of filamentary arms, and turbulent diffusion. The vortex lifetime is measured as a function of applied shear, with vortex strength independently adjustable. These data are compared to the predictions of a fluid theory which correctly identifies the key dimensionless parameter (shear rate/vorticity) but not its critical value.

Introduction

A basic issue in vortex dynamics concerns the fate of a two-dimensional (2-D) vortex in a shear flow (plane strain). The interaction can be described as a competition between the shearing flow, which tries to disperse the vortex, and the vortical motion, which tries to maintain the vortex. Moore and Saffman[1] studied this problem analytically and found that an elliptical vortex patch (i.e. a region of constant vorticity bounded by an ellipse) could exist in an irrotational shear flow only if the strain rate was less than a critical value. Kida[2] later showed that when the strain rate exceeded this value the ellipse would undergo irreversible elongation and thus the vortex would be dispersed.

The work of Moore & Saffman and Kida is based on the classical theory of inviscid, constant density fluids as modeled by the Euler equations. As noted recently by Driscoll and Fine[3], such work can be applied directly to strongly magnetized pure electron plasmas since the two-dimensional drift-Poisson equations describing such plasmas are isomorphic to the Euler equations. In the plasma system the vorticity (a key fluid quantity) is proportional to the electron density, and is thus easily measured, while the viscosity and boundary drag are very small. These features make such systems ideal for testing 2-D vortex theory.

In this paper an electron column system is used to study the behavior of a single 2-D vortex in an imposed irrotational shear flow. The vortex evolution in a strong shear is characterized by the emission of filamentary arms and sometimes by the fission of the original vortex. The vortex lifetime is measured as a function of applied shear, with vortex strength independently adjustable, and compared to the prediction of Moore & Saffman. Consistent with their theory, the vortex lifetime increases significantly when the applied shear is less than a critical value. For subcritical shear values the vortex lifetime appears to be limited by a slow diffusion which gradually weakens the vortex.

Experimental Device

The apparatus used in these experiments, shown schematically in Fig. 1a, is a modified Malmberg-Penning trap. A long conducting cylinder is divided axially into eight electrically isolated parts labelled G1,S1,S2,S3,S4,S5,G2, and G3. Each of these cylinders is 6.00" long, has an I.D. of 3.05", and is separated from the others by a gap of 0.050". Two of the cylinders (G1 and G2) act as gates for the trap while the others are normally grounded and provide a well-defined boundary condition for the trapped electrons. Each of the cylinders S1 through S5 is divided azimuthally into eight sections which can be used to monitor the azimuthal motion of the electrons. The apparatus is placed in a vacuum of $< 10^{-9}$ Torr and is immersed in a uniform axial magnetic field produced by a large solenoid. For these experiments B = 500 G. At this magnetic field strength the Larmor radius is less than 0.1 mm and the electron dynamics are well described by the E×B drift motion.

The above description is fairly common for Malmberg-Penning traps. Some distinctive features of this apparatus are as follows: The electrons are produced by a small diameter (0.1") oxide-coated cathode which is placed off-axis (r = 1.55 cm.). This is used to inject a column of electrons into the system. The column's self-field causes it to spin around its axis (i.e., it forms a vortex). A long, thin (0.014" diam.), conducting wire which can be biased by the operator runs along the axis of the device. This is used to produce a radial electric field, and thus an azimuthal E×B drift velocity, which is a function of radius and has zero curl (i.e., an irrotational shear

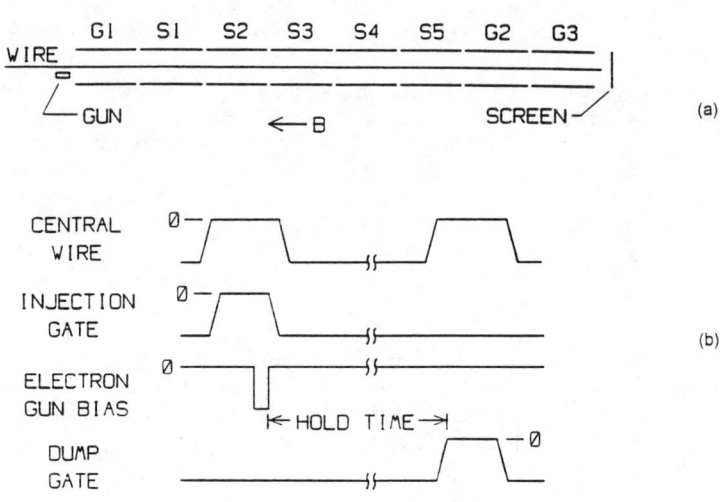

FIGURE 1. a) Schematic of experimental apparatus and b) timing diagram for one cycle of the experiment. Voltages applied to the central wire, injection gate (G1), electron gun, and dump gate (G2) are shown.

flow). The voltage ϕ_{cw} placed on the central wire of the device can be either negative or positive, and thus the applied shear flow can either oppose the vortex rotation (adverse shear) or favor it (favorable or prograde shear). Finally, a phosphor-coated screen is used to measure the total charge of the dumped electrons and to produce an image that shows the position and density of the electrons. The images are acquired with a CCD camera and stored and analyzed with a computer-controlled frame grabber.

The apparatus is run in cycles as shown in Fig. 1b. To start a cycle, the central wire bias and injection gate bias are switched to zero and a negative pulse is applied to the electron gun. This allows an off-axis column of electrons to fill the device. The density of the electrons can be adjusted by changing the gun bias. The injection gate is then returned to a negative bias which traps the electrons between G1 and G2. The central wire bias is then switched to a selected value, thus producing a shear flow. After a time selected by the operator, the central wire and the gate G2 are grounded, allowing the electrons to stream out of the device along magnetic field lines and be collected by the positively biased phosphor screen. The total charge in the machine at the time of the dump is thus obtained, as well as an axially integrated image of the electron positions. Typically, several cycles are required to produce enough light for a visible image. On subsequent cycles the hold time can be varied, so that a sequence of images representing the time evolution of the vortex is produced. The shot-to-shot variation in the injected vortex is small enough that each shot evolves identically for 20-80 μs after which shot-to-shot variations in the dump-time position of the vortex cause a smearing of the multi-cycle image. After this point one or more of the azimuthal sectors of cylinders S1 - S5 can be used to monitor the azimuthal motion of the vortex. This is done by attaching the sector to ground through a resistor and observing the voltage fluctuations across the resistor produced by the variations in the vortex image charges. Since only the amplitude of these signals is of interest, the shot-to-shot phase differences can be ignored.

Theoretical Model

The important quantities in the theoretical model of Moore & Saffman/Kida are the vorticity of the elliptical patch Ω and the strain rate e. The vorticity, which we define to be positive for our electron column, is given by $\Omega = nq/\varepsilon_o B$ where n is the electron density, q is the electron charge and B is the magnetic field strength. The theory assumes the vorticity is constant within the boundaries of the patch, whereas the density profile of our vortices is roughly Gaussian. To compare with theory, we assume that our vortex is comparable to a circular patch of radius R_v and density $N_L/\pi R_v^2$. Here $N_L = \int n dA/L$ is the number of electrons per unit length, L is the length of the column, $R_v = 1.5 \int n |r - R_{cm}| dA/N_L$, $R_{cm} = \int n r dA/N_L$ is the position of the vortex center of mass, $n = n(r,\theta)$ is the density which is determined from the pixel value at $r = (r,\theta)$, and the integral is over the cross sectional area of the device. The strain rate (or maximum rate of extension) e is defined in a frame where the vortex center is at rest. For our case, this will be a frame rotating with angular velocity $\omega(R_{cm})$, where $\omega(r) = v(r)/r$ and v is the E×B drift velocity produced by the biased center wire (plus a small component due to the vortex image charges). In this frame the applied flow is simple shear and $e = r(d\omega/dr)/2$ evaluated at $r = R_{cm}$. The strain rate is positive for positive center wire bias (i.e., the favorable shear case for the electron vortex). With these definitions, the key theoretical prediction can be simply stated: vortices with $e/\Omega < -(3-2\sqrt{2})/2 \cong -0.086$ will be dispersed, those with $e/\Omega > -0.086$ will

remain.   The theory assumes that in either case the
vortex will remain an elliptical patch, although the
axis ratio and orientation may change.

In order to make a valid comparison with theory, the
system parameters should be adjusted to satisfy the 2-D
approximation.   This requires that the axial transit
time of the electrons be small compared to the time for
motions in the $r-\theta$ plane.   Two characteristic $r-\theta$
motions are the vortex spin around its own axis and its
drift around the machine axis.   Let γ_s denote the axial
transit time over the vortex spin time and γ_d the
transit time over the drift time.   For the parameters
of these experiments $\gamma_s = 0.08 - 0.4$ and $\gamma_d = 0.003 -$
0.25 so the 2-D approximation is reasonable.   End
effects which can break the 2-D assumption are also
negligible.

Finally, we note that the theory approximates the
applied strain by employing a linear Taylor expansion
around the center of the vortex.   The next term in the
expansion goes like $2R_v/R_{cm}$.   In our experiments
$2R_v/R_{cm} = 0.25$ so the expansion is reasonable.

Qualitative Behavior of a Dispersing Vortex

Examples of vortex evolution are shown in the
phosphor screen images of Figs. 2-4.   Except as noted,
the screen bias is adjusted for each image so that full
exposure is obtained; if this were not done the later
images would be too dim to distinguish details.   Also,
we display negative images: darker parts of the image
correspond to brighter parts of the phosphor screen and
higher electron densities.

Fig. 2a shows the initial vortex.   For this case R_{cm}
$= 1.55$ cm and $R_v = 1.9$ mm.   The calculated density n is
2.08×10^7 cm^{-3}, $\phi_{cw} = -78.1$V, and $e/\Omega = -0.16$.   The
vortex center drifts clockwise around the central wire.
As shown in Figs. 2b and 2c, the vortex begins to
disperse by emitting filamentary arms (these two images
are overexposed so that these low density arms are
visible).   As time goes on, the arms continue to grow
longer and wrap around the central wire (Figs. 2d-f).
(Note: the vertical shadow in the images is produced by
a bar which supports the central wire).   The filaments
wind into a tighter and tighter spiral until the detail
can no longer be resolved (Fig. 2g).

Figure 3 shows the evolution for a slightly smaller
shear (n $= 2.08 \times 10^7$ cm^{-3}, $\phi_{cw} = -69.5$V, $e/\Omega = -0.14$).
The initial evolution is similar (Fig. 3a), but as the
inner filamentary arm completes its first wrap around
the central wire it re-joins the vortex (Fig. 3b; again
Figs. 3a-c are overexposed to show detail).   The

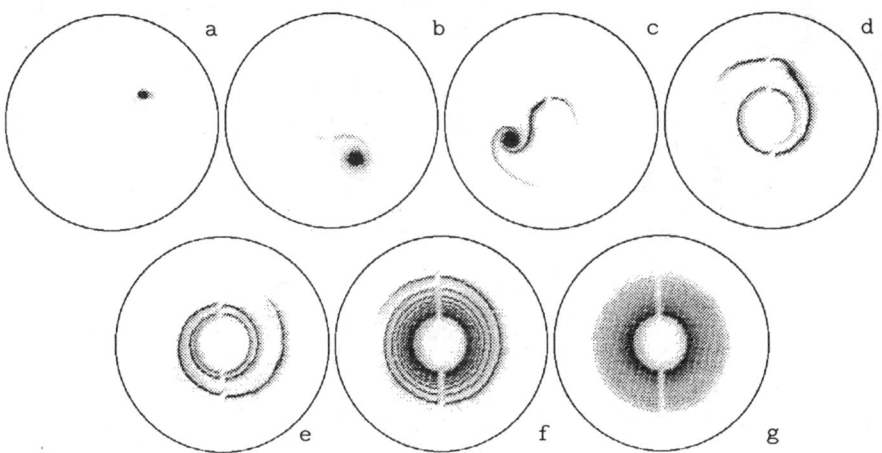

FIGURE 2. Phosphor screen images showing the emission of filamentary arms. Here $e/\Omega = -0.16$. Left to right, $t(\mu s) = 0$, 1.40, 2.70, 5.92, 9.02, 30.0, and 110.

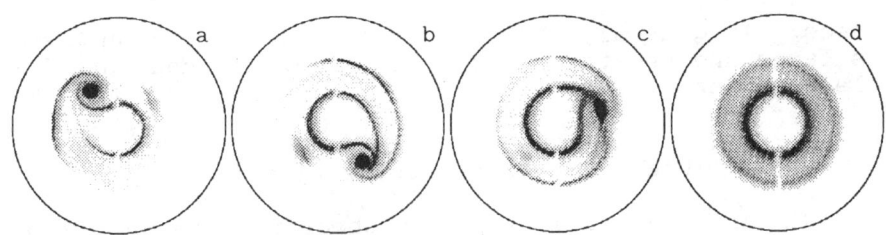

FIGURE 3. Phosphor screen images showing the entrainment of filamentary arm. Here $e/\Omega = -0.14$ and $t(\mu s) = 4.76$, 8.06, 20.1, and 135.

subsequent evolution shows turbulent variation from shot to shot; Fig. 3c is an average of eight shots. The turbulent dispersal of the vortex finally leads to a reproducable end state (Fig. 3d). Note that this end state is qualitatively different than the previous case (cf. Fig. 2g).

A final example of vortex evolution is shown in Fig. 4. Here the shear is stronger than in Fig. 2 and the density is lower: n = 9.33×10^6 cm^{-3}, $\phi_{cw} = -138.8$V, $e/\Omega = -0.64$. Rather than emitting filamentary arms the initial vortex undergoes fission, splitting into two and then three smaller vortices before being smeared into a long filament. The winding process then proceeds as before. The end state exhibits a more uniform distribution of electrons than the previous

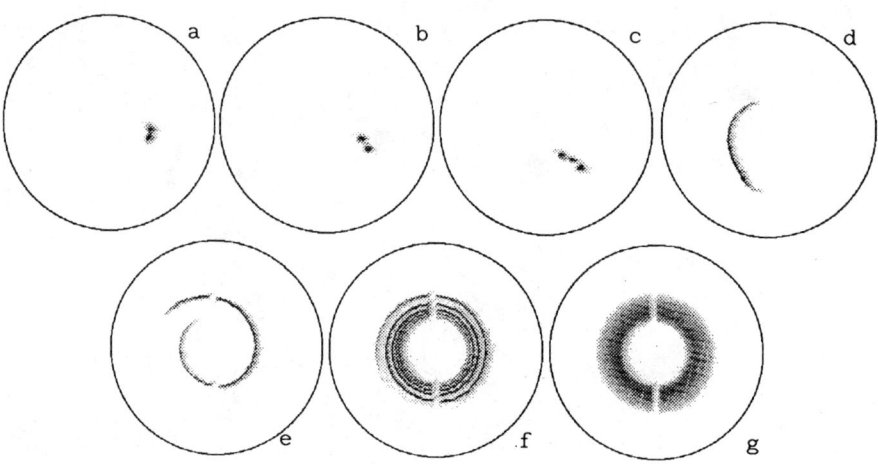

FIGURE 4. Phosphor screen images showing the breakup (fission) of initial vortex. Vortex at t = 0 looks the same as in Figure 2a. Here e/Ω = -0.64 and $t(\mu s)$ = 0.58, 0.68, 0.90, 1.70, 3.40, 13.6, and 77.0.

cases. In general, this vortex fission occurs when e/Ω is large and negative (i.e. large adverse shear and low electron density).

Measurements of Vortex Lifetime

In order to quantify the shearing process it is useful to define a quantity T_s, which measures the time it takes to shear (i.e., disperse) the vortex. The definition of T_s is chosen so that the two methods available for monitoring the vortex dynamics (i.e., screen images and wall probe signals) give comparable values. For images, the electron center of mass R_{cm} is calculated. Since the sector probe measures electrons within its 45° angular span, all pixel values within the span $\theta_{cm} \pm 22.5°$ are summed and divided by the sum at t = 0 (here θ_{cm} is the angle of the vector R_{cm}). The shear time T_s is defined as the time when this ratio drops to 0.5 (i.e. when half of the electrons have left the octant defined by $\theta_{cm} \pm 22.5°$). When the sector probe is used T_s is defined as the time when the amplitude of the probe signal drops to half of its initial value.

Data showing T_s versus the central wire bias ϕ_{cw} with column density (vorticity) as a parameter are presented in Fig. 5. The solid symbols are data points taken from screen images and the open symbols are data points acquired with the wall probe. The qualitative

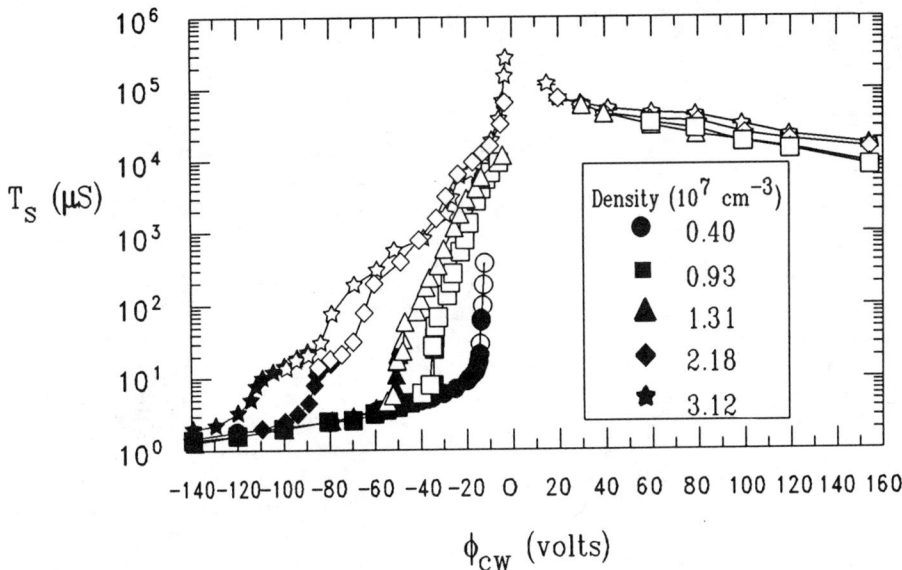

FIGURE 5. Vortex shear time T_s versus central wire bias ϕ_{cw} with column density as a parameter.

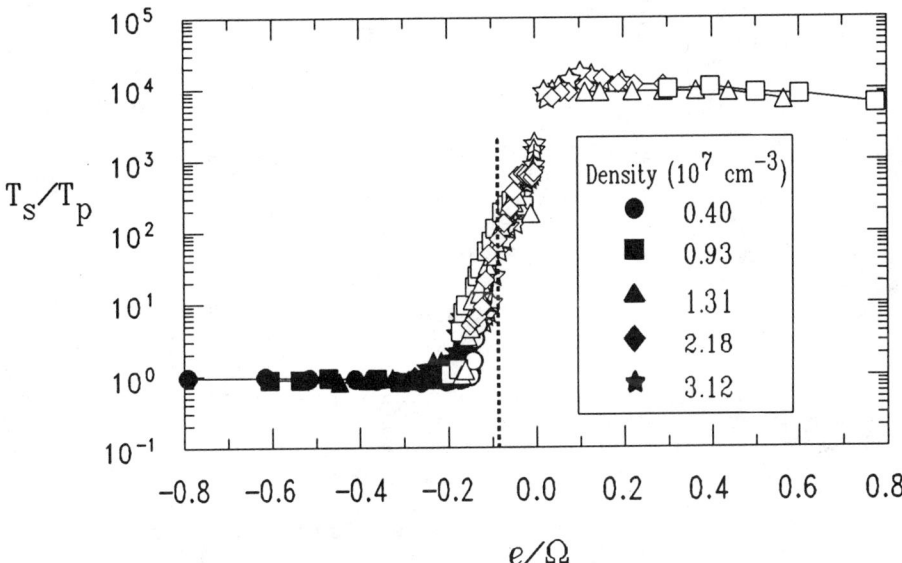

FIGURE 6. Data of Fig. 5 plotted using the normalizations suggested by theory. Dashed vertical line shows the theoretical critical value.

features are consistent with expectations. When the shear is adverse and large (large negative values of ϕ_{cw}), the vortex is quickly dispersed (small T_s). As the shear is reduced (ϕ_{cw} is made less negative), the vortex lifetime increases. The shear value at which this increase occurs depends on the column's density (vorticity); more shear is required to disperse a stronger vortex. We also note that for favorable shear ($\phi_{cw} > 0$) the vortex lifetime is long and roughly independent of the value of ϕ_{cw}.

In Fig. 6 this data is replotted using scaled quantities. The shear time T_s has been scaled to the passive shear time T_p, which is the time it would take to shear a column with zero vorticity. We take this time to be $R_{cm}\theta_s/2eR_v$ where θ_s is the angular spread defining vortex dispersal (here taken as $\pi/4$ rad). On the abscissa is plotted the shear strength e divided by the vorticity Ω, as suggested by theory.

The data now form one curve, showing that the scalings employed are the correct ones. The lifetime of a dispersing vortex is essentially T_p, and the parameter that determines the fate of the vortex is e/Ω. When e/Ω exceeds a critical value, the scaled vortex lifetime increases suddenly. The experimental critical value is slightly different for each of the cases considered, but the values do not exhibit any systematic dependence on density. The average experimental value is $(e/\Omega)_{crit} = -0.163 \pm 0.015$, in contrast with the theoretical value of -0.086. The vortex lifetime does not increase indefinitely at $(e/\Omega)_{crit}$. Rather, the lifetime jumps up abruptly by an order of magnitude and then follows a roughly exponential dependence on e/Ω, reaching a roughly constant maximum value of $10^4 T_p$ for $e/\Omega > 0$.

The dependence of T_s/T_p on e/Ω for $e/\Omega > (e/\Omega)_{crit}$ can be explained as follows. The wall probe signal is qualitatively different for $(e/\Omega)_{crit} < e/\Omega < 0$ and $e/\Omega > 0$. For the first case, the signal is roughly constant in time and then decreases abruptly, signaling that the vortex has been dispersed. Limited phosphor screen imaging corroborates this interpretation of the wall probe signal. For $e/\Omega > 0$, the probe signal decreases gradually until it is lost in the noise. The imaging data for this case shows that the vortex remains unsheared and that R_{cm} remains roughly constant. Both cases may be explained by the presence of a slow diffusive transport. When $(e/\Omega)_{crit} < e/\Omega < 0$ this transport causes the vortex density (and thus Ω) to decrease until e/Ω falls below $(e/\Omega)_{crit}$, at which point the vortex will be dispersed. For $e/\Omega > 0$, the shear is unable to disperse the vortex irregardless of the value of Ω, so the probe signal slowly decreases as electrons diffuse. The nature of this diffusion is unknown, but it is significantly faster than the transport that leads to particle loss. The time to lose half the electrons from the device is roughly ten seconds, whereas the maximum vortex lifetimes are on the order of 0.1 s.

We can offer no explaination for the factor-of-two discrepancy between experimental and theoretical values for the critical e/Ω. One can identify several areas where the experiment departs from the theoretical model (e.g., Gaussian vortex profile) but a quantitative analysis of such effects is, to our knowledge, not available.

Conclusions

A vortex in a strong shear exhibits a variety of behaviors: vortex fission, filament emission, stretching, and entrainment, and turbulent diffusion. While none of these behaviors is included in the theory of Moore & Saffman/Kida, their theory correctly identifies the key dimensionless parameter e/Ω. However, the predicted critical value is roughly half of the experimental value. When e/Ω is less than the critical value the vortex lifetime is the same as that of a zero vorticity patch. When e/Ω is above the critical value the vortex lifetime increases dramatically and appears to be limited by a slow diffusive process that gradually weakens the vortex.

Acknowledgments

The author gratefully acknowledges Dr. C.F. Driscoll's advice on the design of Malmberg-Penning traps; seminal discussions of vortex dynamics with Dr. K.S. Fine; and Dr. Pei Huang for pointing out an important error in the original manuscript. This work was supported by ONR grant N00014-89-J-1399.

References

1.   D.W. Moore and P.G. Saffman, in _Aircraft Wake Turbulence_, edited by J. Olsen, A. Goldberg, and N. Rogers, (Plenum, New York, 1971), pp. 339-354.

2.   S. Kida, J. Phys. Soc. Jpn. **50**, 3517 (1981).

3.   C.F. Driscoll and K.S. Fine, Phys. Fluids B **2**, 1359 (1990).

Stability of Highly Deformed, Asymmetric Single-Species Plasmas

J. Fajans

Physics Dept., U.C. Berkeley
Berkeley, CA 94720-7300

Abstract: Cylinders of pure electron plasma are routinely confined within cylindrically-symmetric Penning/Malmberg traps. When azimuthal asymmetries are imposed, the plasma deforms into appropriately asymmetric shapes. Such deformed plasmas have been observed experimentally, and are long-lived. The equilibrium and stability of these plasmas are analyzed in this paper. A very broad class of deformed plasmas are in equilibrium, but not all such plasmas are stable. Furthermore, the flux surface adiabatic invariant assures that the plasma will attain the appropriate equilibrium shape if the azimuthal boundary perturbations are applied slowly.

Equilibrium

When single-species plasmas are confined in a Penning/Malmberg trap with azimuthally asymmetric boundaries, the plasmas deform into non cylindrical shapes.[1, 2] These plasmas are often highly convoluted, and can last many seconds.[3] The azimuthal asymmetries are created by appropriately biasing the trap walls. Appropriate wall biases can be found which place *any* singly-connected, flat-top (constant density) plasma in equilibrium. For a flat-top plasma, the necessary and sufficient condition for equilibrium is that the boundary of the plasma must be an equipotential.

The appropriate wall biases can be found as follows. First, place a fictitious metallic boundary directly around the plasma. Inside this metal boundary, we have the well-posed problem of solving Poisson's equation $\nabla^2 \Phi_{\text{int}} = -4\pi e n$. Choosing some interior point as the origin, we can always construct the interior (of the plasma) potential Φ_{int} as an expansion

$$\Phi_{\text{int}}(r, \theta) = -\pi e n r^2 + -\pi e n \sum_{m=0}^{\infty} (a_m \sin m\theta + b_m \cos m\theta) r^m \qquad (1)$$

such that the plasma boundary is an equipotential. Next, place appropriate potentials on the wall boundary so that no surface charges are induced on the fictitious metal plasma boundary. Under these conditions, the fictitious metal boundary can be removed without perturbing the potential; consequently, the plasma boundary remains an equipotential and the plasma will be in equilibrium. Can this always be done? This problem is equivalent to finding an exterior (to the plasma) potential,

$$
\Phi_{\text{ext}}(r, \theta) = -\pi e n \sum_{m=0}^{\infty} \left[(A_m \sin m\theta + B_m \cos m\theta) r^m \right.
$$
$$
\left. + (C_m \sin m\theta + D_m \cos m\theta) r^{-m} \right] \tag{2}
$$

such that the potential and the electric fields are continuous across the boundary. The appropriate wall potentials are then given by Φ_{ext} evaluated at the wall, i.e. $V(\theta) = \Phi_{\text{ext}}(R_w, \theta)$. Note that terms proportional to r^m are permitted because Φ_{ext} is never evaluated at infinity. Specification of both the potential and the electric field on an irregular boundary is a highly unusual, but nonetheless well-posed problem which can always be solved. The freedom to match both the potential and the electric field comes from the existence of *two* complete sets of coefficients, (A_m, B_m) and (C_m, D_m). These coefficients can always be determined numerically. I have also developed a quasi-analytic formalism that yields the coefficients for simple geometries,[4] but it is too lengthy to be presented here.

Equilibrium and Stability of Circular Plasmas

Off-axis circular plasmas are always stable. The wall biases necessary to produce such circular plasmas are very easy to find. The potential generated by such plasmas is simply the standard potential from a cylindrical plasma, namely $\Phi_{\text{int}} = -\pi e n r^2$ inside the plasma and $\Phi_{\text{ext}} = -e n A_p [1 + 2 \ln(r/r_p)]$ outside the plasma. Here r_p is the plasma radius, $A_p = \pi r_p^2$ is the plasma area, and r is measured from the plasma center. The plasma boundary is clearly an equipotential, so the plasma will be in equilibrium. The required wall voltages $V(\theta)$ are found by evaluating Φ_{ext} along an appropriately off-center circle of radius R_w.

That these circular plasmas are stable can be proved using O'Neil and Smith's Maximal Energy Principle.[5] If the plasma motion is solely due to $\mathbf{E} \times \mathbf{B}$ drifts, the electrostatic energy of the plasma will be conserved if the wall voltages are not time dependent. If the electrostatic energy of a given plasma is extremal, there cannot be any nearby accessible states because such states would have different energy. Consequently, the plasma must be stable. Using this principle, Chu et al proved that slightly deformed plasmas are energy maxima and, consequently, are stable. This proof has been extended to any perfectly circular plasma.[4]

Equilibrium and Stability of Elliptical Plasmas

The equation

$$R_0(\theta) = \frac{(1 - b_2^2)^{1/4}}{(1 - b_2 \cos 2\theta)^{1/2}} R_c \tag{3}$$

defines a family of ellipses of area $A_p = \pi R_c^2$ and ellipticity $\lambda^2 = (1+b_2)/(1-b_2)$. The interior potential which produces these elliptical plasmas is

$$\Phi_{\text{int}} = -\pi e n r^2 (1 + b_2 \cos 2\theta). \tag{4}$$

With the aid of a symbolic manipulation program, I have found the exact expression for the exterior potential:

$$\Phi_{\text{ext}}(r, \theta) = -2\pi e n R_c^2 \ln r - \pi e n b_2 r^2 \left(1 - \frac{1 - \sqrt{1 - b_2^2}}{b_2^2} \right) \cos 2\theta \tag{5}$$

$$+ \pi e n \sum_{m=1}^{\infty} \frac{(-1)^m}{m(m+1)} R_c^{2m+2} (1 - b_2^2)^{\frac{(m+1)}{2}} r^{-2m} \cos 2m\theta \tag{6}$$

$$\times \sum_{p=0}^{\infty} \binom{2p + 2m}{m} \binom{2p + m}{p} \left(\frac{b_2}{2} \right)^{2p+m}$$

The appropriate wall voltages are readily obtained from this expression by evaluating it at the wall radius.

Chu et al[2] have derived a Hamiltonian expansion that accurately predicts the behavior of elliptical plasmas. The Hamiltonian $H(\theta_c, r_c, \phi, \lambda; b_2)$ is a function of the two conjugate pairs, (θ_c, r_c) and (ϕ, λ), where the first two variables describe the angle and radius of the center of charge, and the second two describe the angle of the major axis and the ellipticity of the plasma. The equilibrium values $\phi_e(b_2)$ and $\lambda_e(b_2)$ can readily be calculated.

The stability of elliptical plasmas was studied numerically by dynamically evolving an initially centered, cylindrical plasma while slowly increasing the wall perturbation b_2. Up to the onset of instability, the ellipticity of the plasma increases appropriately with the wall perturbation. When the wall perturbations are pushed beyond the stability boundary, the plasma moves off center exponentially along the line of its major axis. The specific direction along this axis is determined by the initial conditions.

The numerically observed instability threshold is well predicted by the condition

$$0 = \left. \frac{\partial^2 H(\theta, r_c, \phi_e, \lambda_e; b_2)}{\partial r_c^2} \right|_{r_c = 0}, \tag{7}$$

i.e. where the plasma is no longer in a maximal state relative to r_c. Equation (7) evaluates to the condition that $\beta = 1$, where

$$\beta = \frac{A_p}{4\pi R_w^2}\frac{1-\lambda^2}{\lambda} + \frac{\pi b_2 R_w^2}{A_p}$$ (8)

Since θ_c is conjugate to r_c, this condition implies that $\dot{\theta}_c = 0$ for small r_c, or equivalently, that the $\ell = 1$ diocotron motion stalls. The $\ell = 1$ diocotron period can readily be calculated from the Hamiltonian

$$T_D = \frac{2\pi}{\sqrt{1-\beta^2}},$$ (9)

and goes to infinity as the instability threshold is approached. The diocotron motion stalls because the radial electric field diminishes at ellipse vertices. At the instability threshold, this field is cancelled by the image field of an $\ell = 1$ perturbations, and there is no azimuthal drift. Note that the above analysis is linear; at high b_2 there are nonlinear, off-center equilibria.

Equilibrium and Stability of Large ℓ Plasmas

Consider the nearly square plasma generated by the interior potential

$$\Phi_{\text{int}} = -\pi e n r^2 - \pi e n r^4 b_4 \cos 4\theta.$$ (10)

A typical plasma produced by this field is drawn in Fig. 1. For large area plasmas, the value of b_4 is limited by the requirement that the vertices remain inside the wall boundary, and by the requirement that the normal electric field at the boundary never changes sign.

When the sides of the plasma are sufficiently concave, numeric simulations demonstrate that the plasma is unstable to $\ell = 2$ perturbations; the plasma stretches along one diagonal and compresses along the other. (The particular choice of diagonals is determined by initial conditions.) When the resulting deformation is still small, the perturbation can be represented by surface charges concentrated at the vertices (see Fig. 1). These surface charges induce electric fields E_{surf} with components parallel to the plasma surface which, in turn, attempt to further deform the plasma.

Consider an individual vertex with an excess of plasma, i.e. a vertex with a negative surface charge perturbation. The surface-charge-induced electric fields have opposite sense on the two sides of this vertex, and act to increase the surface charge on the side where the bulk rotation is sweeping the surface charge towards the vertex, and decrease the surface charge on the side where the bulk rotation is sweeping the surface charge away from the vertex. The bulk rotation itself stagnates at each vertex because the self electric field E_0 is partially canceled

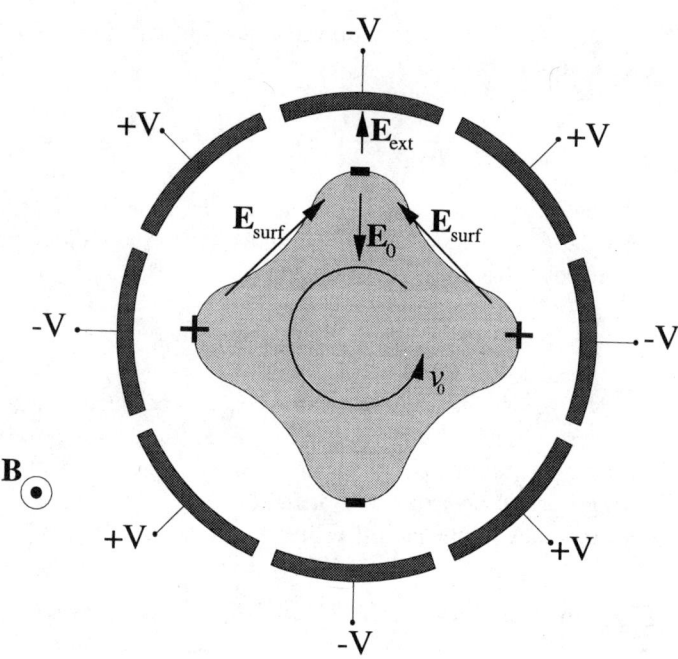

Figure 1: Negatively charged plasma deformed into a square. The quadrupole plus and minus signs indicate charge perturbations around the equilibrium.

by the exterior electric field from the wall biases $\mathbf{E}_{\mathrm{ext}}$. Instability occurs when more surface charge is generated than is swept through the stagnation point at each vertex. This criterion leads to the condition $v_0 < new/cB$, where v_0 is the bulk rotation velocity at each vertex, w is the path length between vertices, c is the speed of light, and B is the axial magnetic field. This criterion has been confirmed by numeric simulations.

Establishment of Equilibria

Asymmetric equilibria are normally established by applying the wall potentials to an initially centered, symmetric plasma. When the wall potentials are applied slowly enough, and when the particular applied potentials correspond to an actual equilibrium, the plasma will automatically deform into that equilibrium. Note that both the initial and final configurations are maximal energy states of their respective systems, but do not normally have the same energy. If the potentials are applied too rapidly, the plasma will oscillate around the equilibria and the energy will not be maximal.

That the plasma deforms correctly is somewhat remarkable, but follows from the conservation of the flux surface adiabatic invariant Φ_A described by Notte, et al.[6] This invariant corresponds to the area circumscribed by the orbit of the center of charge. For the initial, centered and undeformed plasma the center of charge is stationary, so the flux surface adiabatic invariant Φ_A is zero. If the final deformed plasma were not stationary, its center of charge would circumscribe some finite area, and Φ_A would be non-zero. Since such a change in the in Φ_A is forbidden, the final state must be stationary, and by definition, an equilibrium.

The time T_{ramp} required to establish the equilibria is the timescale of the lowest relevant modal period T_m. For instance, for a displacement of a circular plasma off the central axis, the relevant modal period is the $\ell = 1$ diocotron period, while for an elliptical deformation the relevant modal period is the $\ell = 2$ diocotron period; however, these modal periods may be significantly lengthened by the wall biases. As is typical with problems involving adiabatic invariants, the deviations from equilibria decrease proportional to $\exp(-\gamma T_{ramp}/T_m)$, where γ is a constant near unity.

I thank T. M. O'Neil, A. Portis, R. Smith and J. Wurtele for their insightful comments. This work was supported by the Office of Naval Research.

References

[1] J. Notte, A. J. Peurrung, J. Fajans, R. Chu, and J. Wurtele, Phys. Rev. Lett. **69**, 3056 (1992).

[2] R. Chu, J. S. Wurtele, J. Notte, A. J. Peurrung, and J. Fajans, Phys. Fluids B **5**, 2378 (1993).

[3] J. Notte and J. Fajans, Phys. Plasmas **1**, 1123 (1994).

[4] J. Fajans, 1995, to be published.

[5] T. M. O'Neil and R. A. Smith, Phys. Fluids B **4**, 2720 (1992).

[6] J. Notte, J. Fajans, R. Chu, and J. Wurtele, Phys. Rev. Lett. **70**, 3900 (1993).

Experiments with Trapped Positron Plasmas

R. G. Greaves, M. D. Tinkle, and C. M. Surko

Department of Physics, University of California, San Diego,

La Jolla, California 92093-0319, U.S.A.

Abstract. Cold positron plasmas are accumulated in a Penning trap by collisions with a nitrogen buffer gas. The trapped plasmas can be manipulated and diagnosed using nondestructive techniques, and they are being used for a range of physics experiments. One topic is the study of the physics of electron-positron plasmas, in which the masses of both charge species are equal. Another topic is the study of the interaction of slow positrons with various forms of ordinary matter. These experiments can address important issues in atomic and molecular physics, condensed matter physics and gamma-ray astrophysics. In this paper, we present an overview of the state of positron source and trapping technology. We then go on to describe recent progress in the accumulation and manipulation of cold positron plasmas and discuss experiments to study the electron-beam/positron plasma system, as well as the interaction of cold positrons with a variety of molecules.

1   INTRODUCTION

The confinement of low-energy positrons in Penning traps is of current interest in a number of areas of physics research. Low-energy positrons were first confined in traps by Schwinberg et al. (1). Subsequently, trapped positrons have been used or proposed for a variety of experiments. Plasma physics applications include the study of electron-positron plasmas (2–6) and transport studies in tokamaks (7) and in magnetic mirror devices (8). Other applications include the search for resonances in electron-positron collisions (9), the production of very cold positrons for a variety of purposes by sympathetic cooling with laser-cooled ions (10), astrophysical simulations (11,12), positron annihilation studies (13–15), positron ionization mass spectrometry (16–18), and antihydrogen production (19).

This paper describes recent progress in the accumulation and manipulation of low-energy antimatter in the form of positron gases and plasmas. We describe the utilization of these collections of antimatter for physics research in a number of areas. The recent focus of these experiments has been to refine the trapping techniques and to develop techniques to monitor, cool and manipulate collections of antimatter in a non-destructive manner. Experiments have now begun to use these collections of antiparticles to address outstanding physics issues.

This paper is organized in the following way. In Sec. 2, we give a brief review of slow positron technology and positron trapping techniques. In Sec. 3, we describe the operation of our trap in more detail. Section 4 deals with techniques to cool, heat and manipulate the trapped positron plasmas. In Sec. 5, we describe experiments that are now under way to use the positron plasmas for both electron-positron plasma studies as well as to study the interaction of positrons with ordinary matter. Section 6 concludes the paper.

2   SLOW POSITRON TECHNOLOGY

2.1   Positron Sources

At present, two types of positron sources are available, namely electron LINACs and radioactive sources.

Electron LINACS produce pulsed beams of electrons with energies up to about 100 MeV, which can be used to produce pulses of positrons by means of suitably designed converters (20,21). Electron LINACS are copious sources of positrons, and the positron pulses that they produce are relatively simple to trap by switching electrode voltages with the appropriate timing.

Radioactive sources. Many radioisotopes decay by positron emission, and several of these are available commercially. In early experiments, relativistic positrons were trapped directly into a magnetic mirror device from radioactive neon gas (8). Very intense, short-lived isotopes of copper have been used as positron sources, but they must be prepared on site in a high-flux nuclear reactor (22). The most commonly used positron emitter is ^{22}Na, which has a half-life $\tau_{1/2} \simeq 2.6$ yr and is available as sealed sources with activities up to 150 mCi. Other commercially available positron sources include ^{58}Co and ^{68}Ge. Because radioactive sources produce positrons continuously, the simple gate switching techniques cannot be used for trapping. However, various steady-state trapping techniques have been developed.

2.2   Positron Moderators

Positrons from either a radioactive source or a LINAC source have a broad energy spread of several hundred keV. For efficient trapping, the positrons must be slowed to energies of a few electron volts. This can be accomplished using a moderator. Several types of moderators are available including metal single crystals, insulators such as MgO, and rare gas solids.

Metal single crystals. Single crystals of various metals have been traditionally used for obtaining slow positrons. Tungsten (23), nickel (24) and copper (25) have been demonstrated, although tungsten is used most widely. These substances are used either in reflection geometries or as foils in transmission geometries, and they typically have moderation efficiencies, ϵ, of about 1–2×10^{-4}, defined as the ratio of the number of slow positrons emitted to the activity of the source. The energy spread of the emitted positrons can be as low as 65 meV for refrigerated moderators (26), but values of a few tenths of an eV are more typical at 300 K.

Rare gas solids. An important recent advance in positron moderator technology was the discovery that rare gas solids (RGS) can moderate positrons about an order of magnitude more efficiently than metal single crystals (27). The positron source is recessed into a conical or parabolic cup mounted on a cold head (28). Either a two-stage or three-stage closed cycle refrigerator is used to cool the source and a layer of rare gas is frozen onto the source at temperatures of about 7 K. The highest moderation efficiency is obtained using neon, although argon, krypton and xenon have also been found to have reasonable efficiencies (27, 29). Solid nitrogen has also been found to act as a moderator, although at a much lower efficiency (30). Recent studies suggest that krypton may be preferable to neon because its efficiency is only slightly lower than that of neon but it is much longer-lived (31, 32). Futhermore, it freezes at a higher temperature so that a less elaborate cold head can be used. Moderation efficiencies of about 0.003 are commonly obtained for krypton and neon.

2.3   Positron Trapping Techniques

Several techniques have been demonstrated for positron trapping, and a number of others have been proposed but not yet implemented.

Scattering from a buffer gas. Positrons can be trapped by inelastic collisions from a nitrogen buffer gas, which scatter positrons from a low-energy beam into a Penning trap. Single stage traps of this type have positron lifetimes of about 30 msec and have been utilized for simulations of positron annihilation in the interstellar medium (11, 33). We have developed a three-stage trap that uses differential pumping to increase the positron lifetime and improve the trapping efficiency (34). Trapping efficiencies in the range 0.25–0.4 have been obtained. The performance of this trapping scheme is described in detail in Sec. 3.

Magnetron drift. This technique injects positrons off-axis from either a bare source (1) or from a moderated source (35) into a Penning trap along a field line. Some positrons are reflected by the endcap and can magnetron-drift away from the load-line to become trapped. The trapped positrons are cooled by coupling their energy into a resonant circuit connected to the confining electrodes and by cyclotron emission. The efficiency of this scheme is very low ($\sim 10^{-5}$) but it has the advantage of not compromising the vacuum with a buffer gas. Recently, modifications to the scheme have been proposed which could lead to an increase in the trapping efficiency to about 10^{-3} (35).

Gate switching. Positrons from LINAC sources are produced in bunches, which can be trapped simply by switching the voltages on the confining electrodes at the appropriate time during the cycle. This technique has been implemented at several facilities, including the Oak Ridge National Laboratory (36), Lawrence Livermore National Laboratory (37) and Kyoto University (30).

Scattering from an ion plasma. Wineland *et al.* have proposed trapping positrons from a beam by scattering from a $^9Be^+$ plasma (10). Once trapped, the positrons would cool by cyclotron radiation and sympathetic cooling with the ions. Because the ions can be laser-cooled to very low temperatures, it should be possible to obtain very cold positrons as well. For large trapping potentials ($\sim$ 15 keV), trapping efficiencies as high as 0.91 have been predicted, while at lower trapping potentials (100–500 V), trapping efficiencies in the range 0.1–0.55 are calculated. An experimental demonstration is now in preparation.

Scattering from a stored positron plasma. Once a sufficiently dense plasma has been accumulated by one of the other techniques, it should be possible to trap incoming positrons by scattering from the trapped positron plasma, although fairly high densities are required. For example, to obtain a trapping efficiency of 0.1 for 1-eV positrons, a z-integrated positron density of about 10^{12} cm^{-2} is required.

Moderator ramping. Moderated positrons can be bunched in a Penning trap by applying a sawtooth voltage to the moderator (38). Like the magnetron drift technique, this method does not compromise vacuum quality. It has a high trapping efficiency of about 0.25, although the positrons are trapped only for a single cycle of the sawtooth (typically 1–10 ms), during which they must be utilized or ejected as a pulse. In combination with trapping by gate switching, this technique could potentially be exploited to accumulate large numbers of positrons in high vacuum.

Cyclotron heating. If positrons are injected into a magnetic mirror along a field line, they can be trapped by cyclotron heating. This involves the application of an rf signal at the cyclotron frequency, thereby increasing the transverse momentum of the positrons and shifting their velocity vectors out of the loss cone. This technique has been used to bunch positrons at frequencies of 1 kHz for use in a pulsed positronium beam, where trapping efficiencies of 0.63 were reported (39). It is also being developed as a means of producing dense steady-state relativistic positron plasmas for electron-positron plasma experiments (6). The technique could, in principle, be combined with a remoderator and the gate-switching technique to obtain low-temperature positron plasmas in high vacuum.

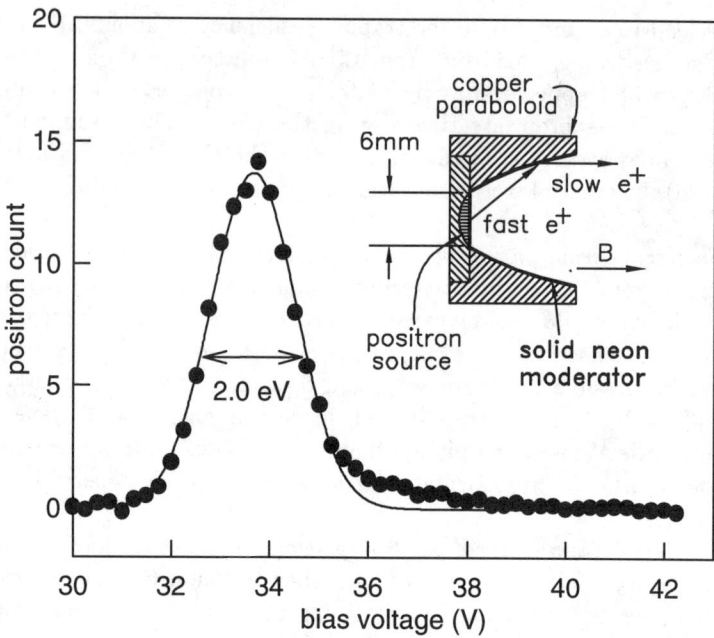

FIGURE 1: The energy spread of positrons from solid neon moderator. Inset: geometry for moderator.

3   DESCRIPTION OF THE EXPERIMENT

Here, we present a brief overview of the operation of our positron trap. A more detailed description can be found elsewhere (40). In our earlier experiments, positrons from a ^{22}Na source (typically 100 mCi) were moderated using a tungsten transmission moderator. We have recently installed a rare gas solid moderator, which has enabled us to increase the positron trapping rate by an order of magnitude. As shown in the inset to Fig. 1, the source is located near the focus of a copper paraboloid mounted on the second stage of a two-stage closed cycle refrigerator (APD model DE-204SLB) which cools the source to about 7 K. The moderator is formed by admitting neon at a pressure of $\sim 2 \times 10^{-4}$ torr for about 1 hour, after which a beam of about 8×10^6 slow positrons per second is obtained. In the presence of the buffer gas, the efficiency of the moderator decays with a half-life of about 24 hours. The energy spread is typically about 2 eV, as shown in Fig. 1. The maximum efficiency obtained for the moderator is about 0.003.

The moderated positrons are guided magnetically into the trap shown in Fig. 2, which has three stages at successively lower pressures and electrostatic potentials. Inelastic collisions with nitrogen gas molecules (denoted 'A', 'B', and 'C' in Fig. 2) cause incoming positrons to accumulate in stage III, where

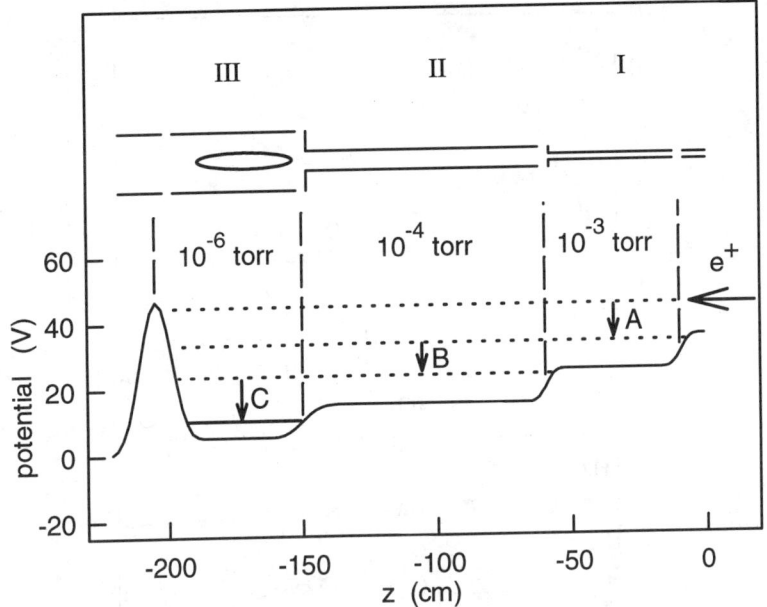

FIGURE 2: Schematic diagram of the three-stage positron trap.

they cool to room temperature in about 1 s (40). For positrons from a tungsten moderator, the efficiency of the trap was approximately 0.40, defined as the fraction of incident positrons trapped in stage III, per moderated positron entering the trap. For positrons from rare gas moderators, the larger energy spread results in a lower trapping efficiency, typically 0.25, but this is more than compensated by the larger moderation efficiency. We expect that the trapping efficiency can be further improved.

Confinement in stage III is limited by "direct" annihilation on the N_2 molecules, in which the positron annihilates with an electron on the molecule during an elastic collision. For a buffer gas pressure of 1×10^{-6} torr, the annihilation time is 40 s. Figure 3(a) shows positron accumulation and confinement from a 70 mCi ^{22}Na source with a solid neon moderator. The initial filling rate is more than 2.2×10^6 positrons per second. The apparent difference in the two lifetimes shown in Fig. 3(a) arises from changes to the filling rate caused by the positron space charge. Figure 3(b) compares positron confinement for the usual buffer gas pressure in stage III with that obtained when the buffer gas is pumped out after loading, and also for electrons in the same situation. The positron confinement is still limited by annihilation with residual gas molecules in the vacuum system, even at the base pressure of 6×10^{-10} torr. Our present vacuum system has Viton o-rings and is bakeable only to 60 °C, so we expect that the confinement can still be greatly improved.

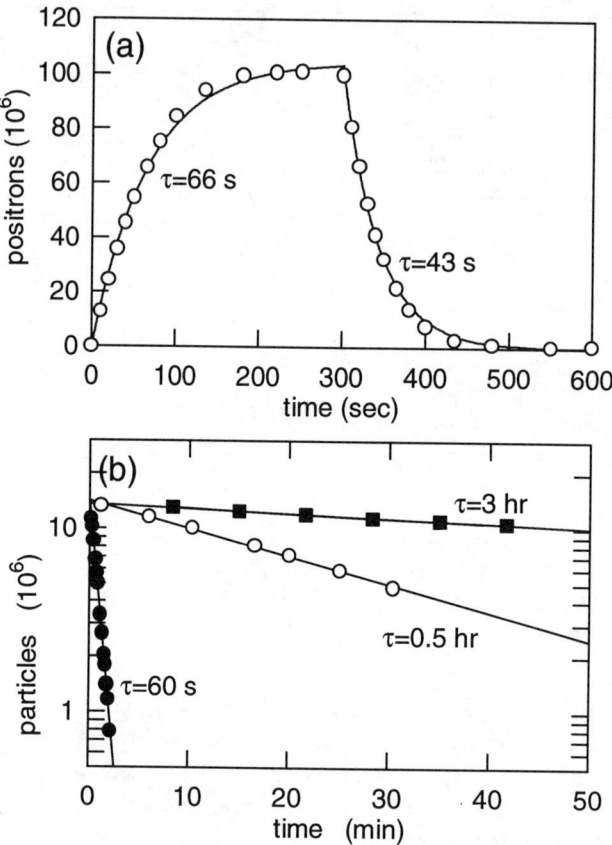

FIGURE 3: (a) Positron accumulation and storage; $p = 6 \times 10^{-7}$ torr. The source is switched on at $t = 0$ and switched off at $t = 300$ s. (b) Positron ($\circ$) and electron ($\blacksquare$) storage when the buffer gas feed is switched off after loading particles ($p = 6 \times 10^{-10}$ torr); positron ($\bullet$) storage in the presence of the buffer gas ($p = 6 \times 10^{-7}$ torr).

4   PLASMA CHARACTERIZATION

The density, temperature and radial profiles of the positron plasmas can be measured by dumping them (41). Line-integrated density at a given radius is measured by dumping the plasma onto a set of annular collectors. Temperature is measured by dumping the plasma onto a "magnetic beach" energy analyzer (42). Knowing the line-integrated density and the temperature, the plasma shape and the local density can be obtained by solving the Poisson equation. These techniques have the disadvantage that a plasma must be destroyed for each measurement. We have now developed non-destructive, remote sensing diagnostics of plasma parameters, by studying the collective modes (43) of these spheroidal plasmas (44, 45). These modes have also been studied in cryogenic pure electron plasmas (46) and in pure ion plasmas (47–49).

For convenience, the work on modes thus far has emphasized electron plasmas. Excitation and detection of the center-of-mass mode of plasma motion in the external potential well gives a measure of total particle number. This measurement can be calibrated by dumping the plasma once. The lowest-order compressional mode (the quadrupole mode) of the plasma in the direction along the magnetic field is strongly dependent on the temperature and aspect ratio, $\alpha = L/2r_p$, where L is the length and r_p is the plasma radius, and the frequency of this mode can be used to measure either of these quantities if the other is known. Measuring the frequency of a third mode provides enough information to obtain the plasma temperature, shape and total particle number, thereby characterizing completely these single component plasmas.

We have investigated the temperature and aspect ratio dependence of these modes in electron plasmas using an rf heating technique that we have developed (44). A typical result is presented in Fig. 4(a), which shows how the plasma temperature increases as a short rf pulse is applied, and then falls as the plasma cools on the buffer gas (41). The frequency shift of the quadrupole mode ($l = 2$) is shown in Fig. 4(b). We are also able to change the aspect ratio of the plasma by changing the magnetic field after the plasma is loaded, and so we were able to investigate the temperature dependence of the quadrupole mode for a range of aspect ratios (44,45).

5 POSITRON PLASMA EXPERIMENTS

5.1 Pure positron plasmas

The maximum density positron gas that we have achieved thus far is approximately 2.3×10^6 cm^{-3}, using a 70 mCi ^{22}Na source and a solid neon moderator. The positron temperature is 300 K, which gives a Debye length, λ_D, of 0.8 mm. The characteristic size of the spheroidal charge cloud is $L \simeq 20$ cm and $r_p \simeq 0.8$ cm. For these parameters, the number of positrons in a Debye sphere, N_D, is about 5×10^3, and $\omega_p \tau_{pn} \gg 1$, where τ_{pn} is the positron-neutral collision time. Thus, all the classical criteria for plasma behavior are comfortably satisfied, and we are able to create robust pure positron plasmas in which plasma modes can be easily excited. Figure 5 shows longitudinal plasma modes in a plasma of 7×10^7 positrons. To our knowledge, this is the first observation of collective plasma modes in an antimatter plasma.

5.2 Electron-positron plasmas

One of the goals of the experiment is to explore the physics of electron-positron plasmas. Such plasmas belong to the larger class of equal-mass plasmas, in which the dynamical symmetry between the electrons and positrons

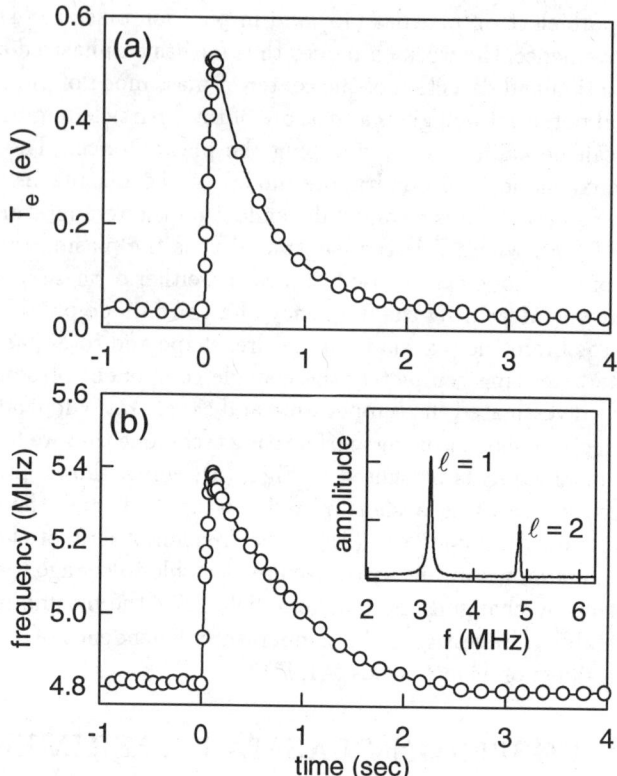

FIGURE 4: (a) Temperature of a pure-electron plasma during a cycle of heating by broadband rf noise and buffer gas cooling. Note that the temperature is close to 300 K before and after the heating cycle. (b) Quadrupole ($l = 2$) mode frequency, which is a measure of the plasma temperature.

leads to properties that differ from electron-ion plasmas (2–5). The electrostatic confinement schemes used currently for pure positron plasmas are not suitable for plasmas with oppositely charged species. However, magnetic mirror devices (2, 6) and Paul traps (50) may offer the possibility of studying equal-mass plasmas experimentally.

The simplest electron-positron plasma experiments to perform are those that study the beam-plasma system, because only one species needs to be confined. We have begun preliminary experiments of this type. The data shown in Fig. 6 were obtained by dumping the positron plasma after the transmission of an electron beam through the plasma. These data show a strong beam-plasma interaction which leads to rapid heating of the positrons. Figure 6(b) shows time-resolved heating occurring in the first few milliseconds after the electron beam is switched on. Such heating could arise from a beam-

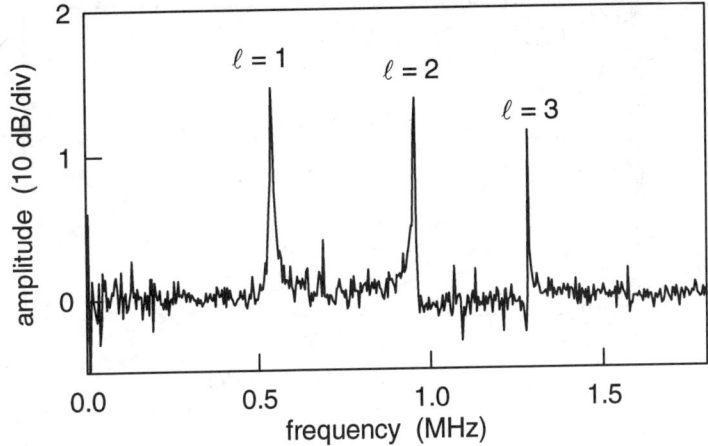

FIGURE 5: Longitudinal modes in a plasma of about 7×10^7 positrons at room temperature.

plasma instability (4).

We anticipate that other information on the nature of the interaction can be obtained by detecting modulation on the beam and signals excited on the confining electrodes. A number of physics issues are planned for investigation. One of the possibilities is the excitation of beam-plasma instabilities which have substantially larger growth rates than in conventional electron-ion plasmas (4). A second issue is the confinement properties of electron-positron plasmas, in which both species are equally and highly magnetized.

6 POSITRON-MATTER INTERACTIONS

6.1 Positron interactions with large molecules

The interaction of positrons with ordinary matter and the annihilation of positrons with electrons are understood in relatively few cases. Early in the development of our positron trap, we discovered that small numbers of impurity molecules, and in particular hydrocarbons, such as oils and greases, could severely limit the achievable lifetime in the trap (13). This effect had been previously recognized for the case of molecules such as butane (51, 52). We have been able to extend these measurements to much larger molecules and to study the effects of chemical composition, molecular symmetry, and the molecular vibrational modes on these interactions (53).

The positron trap provides a unique way of isolating two-body interactions between the molecules and positrons at a known temperature. The annihila-

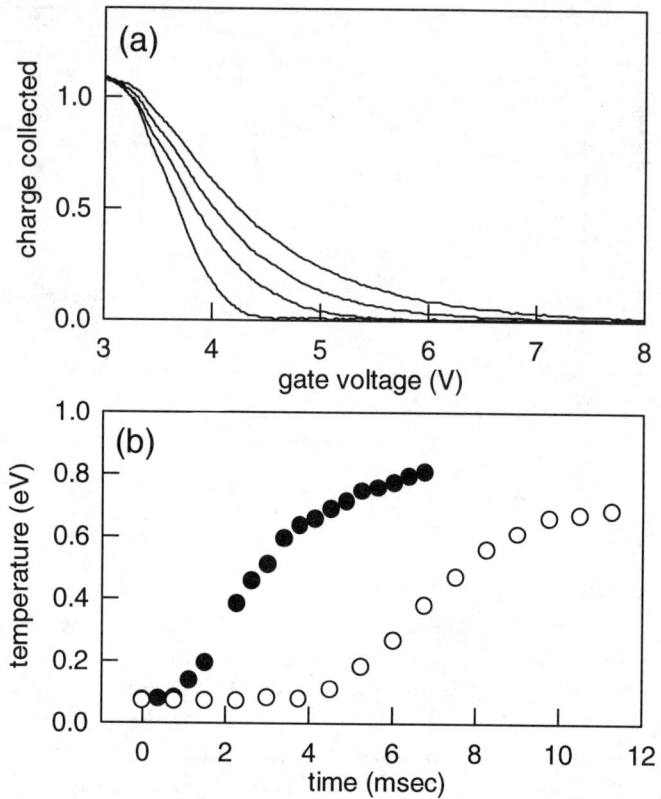

FIGURE 6: Electron-beam positron-plasma experiment. (a) Positron dump following transmission of electron beam for various times. The charge collected is plotted against gate voltage. (b) Time-resolved plasma heating after switch-on of electron beam: (•) $E_b = 4$ eV, (o) $E_b = 5$ eV.

tion rates are obtained by measuring the lifetime of positrons as a function of the pressure of the substance under test. The annihilation rate, Γ, is expressed in terms of $Z_{\text{eff}} = \Gamma/\Gamma_e$, where $\Gamma_e = \pi r_0^2 cn$ is the annihilation rate of a positron in an uncorrelated electron gas with a density, n, equal to the density of molecules, r_0 is the classical radius of the electron, and c is the speed of light (54). Figure 7 shows the values of Z_{eff} for a selection of molecules, illustrating the strong dependence of Z_{eff} on chemical structure (14,53). If the positron-molecule interaction occurs via elastic collisions, one would expect $Z_{\text{eff}} \lesssim Z$, where Z is the total charge on the molecule. The fact that Z_{eff} can be as much as four or five orders of magnitude larger than Z leads us to conclude that the interaction results in long-lived resonances between the positron and

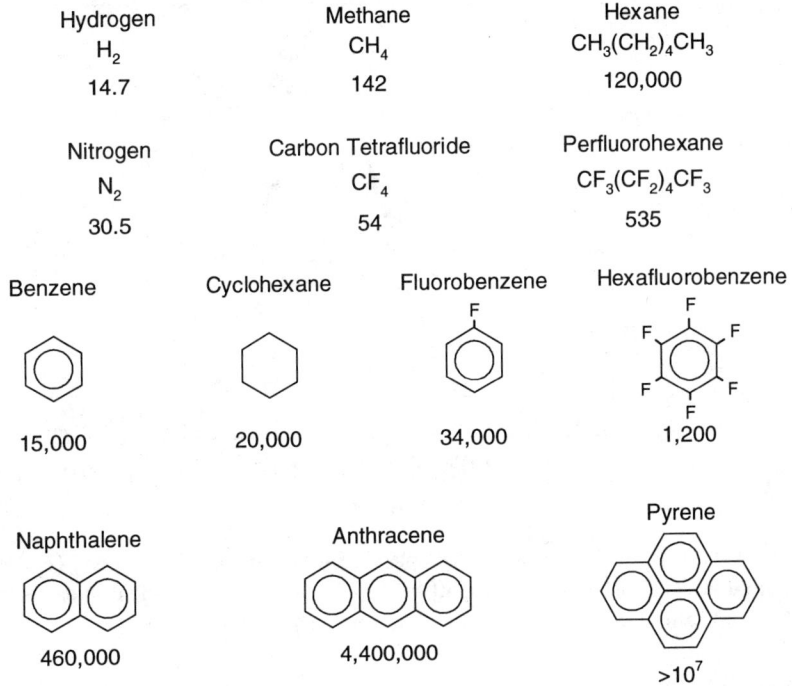

Hydrogen
H_2
14.7

Methane
CH_4
142

Hexane
$CH_3(CH_2)_4CH_3$
120,000

Nitrogen
N_2
30.5

Carbon Tetrafluoride
CF_4
54

Perfluorohexane
$CF_3(CF_2)_4CF_3$
535

Benzene
15,000

Cyclohexane
20,000

Fluorobenzene
34,000

Hexafluorobenzene
1,200

Naphthalene
460,000

Anthracene
4,400,000

Pyrene
$>10^7$

FIGURE 7: Positron annihilation rates, in units of Z_{eff}, for a selection of molecules.

molecule. In these resonances, the positron spends a much longer time in the vicinity of the molecule, and the probability of annihilation per collision is correspondingly larger.

We have found empirically that the annihilation rate is a function of the difference between the molecular ionization potential, E_I, and the binding energy of a positronium atom, E_{Ps}, which is 6.8 eV (14). We have speculated that this dependence might indicate a model of positron binding in which a highly correlated electron and positron pair (i.e., a "pseudo-positronium" atom) moves in the field of the positive molecular ion (40).

6.2   Annihilation gamma-ray spectra

An important feature of the interaction of positrons with matter is the energy spectrum of annihilation gamma rays. When a positron annihilates with an electron, two 511 keV-gamma rays with equal and opposite momenta are produced. If the center-of-mass of the electron and positron is moving, the gamma rays will be Doppler-shifted. In the case where a positron annihilates with a bound electron, there is a contribution to the center-of-mass momentum due to the momentum distribution of the bound electron (55). Thus, measurement of the gamma-ray spectra gives information about the site of the annihi-

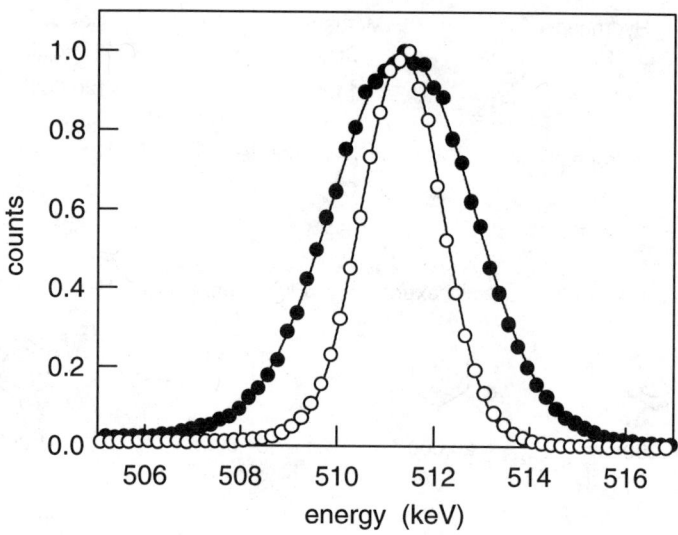

FIGURE 8: Annihilation gamma-ray spectra for hydrogen (o) and neon (•). When corrected for the instrumental resolution of 1.14 keV, the data yield energy widths of 1.62 keV and 3.28 keV for hydrogen and neon, respectively.

lation (15). The energy shift of the gamma rays will be $\Delta E \sim 511(v/c)$ keV, where v is the typical velocity of the electron in the atom. For annihilation with electrons in molecular bonds or in the outer shell of atoms, $\Delta E \sim 1$–3 keV. Shown in Fig. 8 are spectra for two molecular species. Comparison of the measured Doppler widths for hydrocarbons with calculations of the momentum distribution of electrons in the C–C and C–H bonds indicates that the positron annihilates predominantly with electrons in the C–H bonds (12). We expect that this fact will be useful in guiding theoretical calculations of E_A for these molecules. At present, progress in understanding the nature of positron interactions with large molecules is hindered significantly by the lack of such calculations.

6.3   Gamma-ray astrophysics

The positron annihilation line at 511 keV is the strongest gamma-ray line of astrophysical origin (56, 57). Large aromatic molecules, such as the four-ring compound pyrene, are estimated to be present in the interstellar medium at molecular abundances of about 10^{-7} that of hydrogen. On the other hand, our measurements indicate that the annihilation rate per molecule is approximately a factor of 10^{+7} greater for pyrene as compared with hydrogen. Thus, we have concluded that annihilation on large aromatics such as these could be a significant contribution to the gamma-ray radiation from the interstellar medium, and this will be reflected in the measured gamma-ray linewidths (12). Laboratory measurements of these effects are therefore expected to be important for the interpretation of data from the next generation of gamma-ray astrophysics missions. Other possible experiments include measurement of the annihilation rates and energy spectra when slow positrons annihilate on atomic clusters and dust grains, and the processes by which relativistic positrons slow down in partially-ionized gases (12).

7   POSITRONS FOR ANTIHYDROGEN PRODUCTION

For the production of antihydrogen (19), large numbers of positrons are required in a vacuum that is much better than the base pressure of our device. However, the presently available techniques for trapping positrons directly in UHV have very low efficiency (35). To accumulate sufficient positrons for even a single experiment might require weeks, if not months. Furthermore, such long-time confinement of large numbers of particles has not yet been demonstrated experimentally. The system that we have developed is six orders of magnitude more efficient than any other currently available using radioactive sources. For example, our trapping rate is more than 2.4×10^6 e$^+$/s from a 70 mCi source as compared with 0.14 e$^+$/s from a 12 mCi using the magnetron drift technique (35). It would therefore seem to be desirable to attempt to adapt the technique described here for accumulation in UHV.

We have now added a fourth stage to the trap, placed about 50 cm from stage III and operating at a pressure of 5×10^{-7} torr. Using it, we have shown that we can periodically (e.g., every 5 s or so) shuttle bunches of positrons from stage III to stage IV (41). Having demonstrated this stacking capability, we envision that it will not be difficult to arrange for stage IV to be in a much lower pressure region, separated from stage III by a lock consisting of a separately pumped vessel isolated by two gate valves. After accumulating for several minutes in stage III, the buffer gas feed would be switched off, and the buffer gas would be pumped out, taking about 30 seconds. The pressure in stage III would then be about 5×10^{-10} torr. The positrons could then be transferred to the lock and isolated, where further pumping would reduce the pressure to that of stage IV, at which time they could be stacked into the stage IV without compromising the vacuum quality. In this way, one could achieve high accumulation efficiencies and long accumulation times in an ultrahigh vacuum environment. With such improvements, we expect to be able to increase the total number of trapped positrons by another two orders of magnitude.

8   CONCLUDING REMARKS

The ability to efficiently accumulate and store large numbers of positrons and to heat, cool and otherwise manipulate the resulting positron gases and plasmas offers new and important research capabilities. In the near future, it is likely that such collections of antimatter will be used to pursue research opportunities in many areas, including astrophysics, plasma, atomic and molecular and condensed matter physics, as well as for a range of technological applications.

ACKNOWLEDGMENTS

We would like to acknowledge the collaboration of K. Iwata on the positron-molecule work presented here, and the technical assistance of E. A. Jerzewski. This work is supported by the Office of Naval Research and by the National Science Foundation under grant PHY 9221283.

REFERENCES

1. Schwinberg, P. B., Van Dyck, Jr., R. S., and Dehmelt, H. G., *Phys. Lett.* **81A**, 119 (1981).
2. Tsytovich, V. and Wharton, C. B., *Comments Plasma Phys. Contr. Fusion* **4**, 91–100 (1978).
3. Iwamoto, N., *Phys. Rev. E* **47**, 604–611 (1993).
4. Zank, G. P. and Greaves, R. G., "Linear and nonlinear modes in nonrelativistic electron-positron plasmas", submitted to *Phys. Rev. E*, (1994).
5. Stewart, G. A. and Laing, E. W., *J. Plasma Phys.* **47**, 295–319 (1992).
6. Boehmer, H., *AIP Conf. Proc.* **303**, 422–34 (1994).
7. Surko, C. M., Leventhal, M., Crane, W. S., Passner, A., and Wysocki, F., *Rev. Sci. Instrum.* **57**, 1862–7 (1986).
8. Gibson, G., Jordan, W. C., and Lauer, E. J., *Phys. Rev. Lett.* **5**, 141 (1960).
9. Cowan, T. E., Hartley, J., Howell, R. H., McDonald, J. L., Rohatgi, R. R., and Fajans, J., *Materials Sci. Forum* **105-110**, 529–32 (1992).
10. Wineland, D. J., Weimer, C. S., and Bollinger, J. J., *Hyper. Interact.* **76**, 115–25 (1993).
11. Brown, B. L. and Leventhal, M., *Phys. Rev. Lett.* **57**, 1651–4 (1986).
12. Surko, C. M., Greaves, R. G., and Leventhal, M., *Hyper. Interact.* **81**, 239–52 (1993).
13. Surko, C. M., Passner, A., Leventhal, M., and Wysocki, F. J., *Phys. Rev. Lett.* **61**, 1831–4 (1988).
14. Murphy, T. J. and Surko, C. M., *Phys. Rev. Lett.* **67**, 2954–7 (1991).
15. Tang, S., Tinkle, M. D., Greaves, R. G., and Surko, C. M., *Phys. Rev. Lett.* **68**, 3793–6 (1992).

16. Passner, A., Surko, C. M., Leventhal, M., and Mills, Jr., A. P., *Phys. Rev. A* **39**, 3706–9 (1989).
17. Donohue, D. L., Hulett, Jr., L. D., McLuckey, S. A., Glish, G. L., *et al.*, *Int. J. Mass Spectrom. Ion Proc.* **97**, 227–36 (1990).
18. Glish, G. L., Greaves, R. G., McLuckey, S. A., Hulett, L. D., Surko, C. M., Xu, J., and Donohue, D. L., *Phys. Rev. A* **49**, 2389–93 (1994).
19. Charlton, M., Eades, J., Horvath, D., Hughes, R. J., and Zimmermann, C., *Phys. Reports* **241**, 65–117 (1994).
20. Howell, R. H., Rosenberg, I. J., and Fluss, M. J., *Appl. Phys. A* **43**, 247–55 (1987).
21. Mohri, A., Mischishita, T., Yuyama, T., and Tanaka, H., *Jpn. J. Appl. Phys., Part 2 (Letters)* **30**, L936–9 (1991).
22. Lynn, K. G., Weber, M., Roellig, L. O., Mills, Jr., A. P., and Moodenbaugh, A. R., "A high intensity positron beam at the Brookhaven reactor", In *Atomic Physics with Positrons. Proceedings of a NATO Advanced Research Workshop*, edited by Humbertston, J. W. and Armour, E. A. G., Plenum, New York, NY, USA (1987), pages 161–74.
23. Gramsch, E., Throwe, J., and Lynn, K. G., *Appl. Phys. Lett.* **51**, 1862–4 (1987).
24. Zafar, N., Chevallier, J., Laricchia, G., and Charlton, M., *J. Phys. D* **22**, 868–70 (1989).
25. Mills, Jr., A. P., *Appl. Phys. Lett.* **37**, 667–8 (1980).
26. Brown, B. L., Crane, W. S., and Mills, Jr., A. P., *Appl. Phys. Lett.* **48**, 739–41 (1986).
27. Mills, Jr., A. P. and Gullikson, E. M., *Appl. Phys. Lett.* **49**, 1121–3 (1986).
28. Khatri, R., Charlton, M., Sferlazzo, P., Lynn, K. G., Mills, Jr., A. P., and Roellig, L. O., *Appl. Phys. Lett.* **57**, 2374–6 (1990).
29. Massoumi, G. R., Hozhabri, N., Lennard, W. N., Schultz, P. J., Baert, S. F., Jorch, H. H., and Weiss, A. H., *Rev. of Sci. Instrum.* **62**, 1460–3 (1991).
30. Mohri, A., Personal communication (1994).
31. Grund, T., Maier, K., and Seeger, A., *Materials Sci. Forum* **105-110**, 1879–82 (1992).
32. Mills, Jr., A. P., Voris, Jr., S. S., and Andrew, T. S., *J. Appl. Phys.* **76**, 2556–8 (1994).
33. Brown, B. L., Leventhal, M., and Mills, Jr., A. P., *Phys. Rev. A* **33**, 2281–3 (1986).
34. Surko, C. M., Leventhal, M., and Passner, A., *Phys. Rev. Lett.* **62**, 901–4 (1989).
35. Gabrielse, G., Haarsma, L., and Abdullah, K., *Hyper. Interact.* **89**, 371–379 (1994).
36. Xu, J., Hulett Jr., L. D., and Lewis, T. A., *AIP Conf. Proc.* **303**, 551–56 (1994).
37. Cowan, T. E., Beck, B. R., Hartley, J. H., Howell, R. H., Rohatgi, R. R., Fajans, J., and Gopalan, R., *Hyper. Interact.* **76**, 135–42 (1993).
38. Conti, R. S., Ghaffari, B., and Steiger, T. D., *Hyper. Interact.* **76**, 127–33 (1993).
39. Khatri, R., Lynn, K. G., Mills, Jr., A. P., and Roellig, L. O., *Materials Sci. Forum* **105-110**, 1915–18 (1992).
40. Murphy, T. J. and Surko, C. M., *Phys. Rev. A* **46**, 5696–705 (1992).

41. Greaves, R. G., Tinkle, M. D., and Surko, C. M., *Phys. Plasmas* **1**, 1439–1446 (1994).
42. Boyd, D., Carr, W., Jones, R., and Seidl, M., *Phys. Lett.* **45A**, 421 (1973).
43. Dubin, D. H. E., *Phys. Rev. Lett.* **66**, 2076–9 (1991).
44. Tinkle, M. D., Greaves, R. G., Surko, C. M., Spencer, R. L., and Mason, G. W., *Phys. Rev. Lett.* **72**, 352–5 (1994).
45. Tinkle, M. D., *Electrostatic oscillations of spheroidal single-component plasmas*, PhD thesis, University of California, San Diego (1994).
46. Weimer, C. S., Bollinger, J. J., Moore, F. L., and Wineland, D. J., *Phys. Rev. A* **49**, 3842 (1994).
47. Greaves, R. G., Tinkle, M. D., and Surko, C. M., "Modes in a pure ion plasma at the Brillouin limit", to be published in *Phys. Rev. Lett.*, Dec. 1994.
48. Bollinger, J. J., Wineland, D. J., and Dubin, D. H. E., *Phys. Plasmas* **1**, 1403 (1994).
49. Bollinger, J. J., Heinzen, D. J., Moore, F. L., Itano, W. M., Wineland, D. J., and Dubin, D. H. E., *Phys. Rev. A* **48**, 525–545 (1993).
50. Schermann, J. P. and Major, F. G., *Appl. Phys.* **16**, 225–230 (1978).
51. Paul, D. A. L. and Saint-Pierre, L., *Phys. Rev. Lett.* **11**, 493 (1963).
52. Heyland, G. R., Charlton, M., Griffith, T. C., and Wright, G. L., *Can. J. Phys.* **60**, 503 (1982).
53. Iwata, K., Greaves, R. G., Murphy, T. J., Tinkle, M. D., and Surko, C. M., "Measurements of positron annihilation rates on molecules", to be published in *Phys. Rev. A.*, Jan. 1995.
54. Dirac, P. A. M., *Proc. Cambridge Phil. Soc.* **26**, 361 (1930).
55. Brown, L. S. and Gabrielse, G., *Rev. Mod. Phys.* **58**, 233–313 (1986).
56. Schönfelder, V., *Adv. Space Res.* **10**, 243 (1990).
57. Leventhal, M., Barthelmy, S. D., Gehrels, N., Teegarden, B. J., Tueller, J., and Bartlett, L. M., *Ap. J., Lett.* **405**, L25–8 (1993).

Debye Shielding and the Dynamic Response of a Magnetized, Collisionless Plasma

C. Hansen and J. Fajans

Physics Department, University of California, Berkeley
Berkeley, CA 94720-7300

Abstract: The response of a collisionless plasma to an external perturbation does not necessarily obey the standard Debye result. In some cases, the plasma will behave in a manner opposite to the manner predicted by Debye theory; i.e. the plasma becomes more *positive* in the neighborhood of a positive test charge. In general, the plasma shields out an external perturbation, but this shielding results solely from electrons dynamically trapped in the neighborhood of the test charge. A new theory of dynamic shielding is in good agreement with experiments in pure-electron plasmas.

The response of a one dimensional (highly magnetized) plasma to an external perturbation is paradoxical. The insertion of a positive test charge into such a plasma locally accelerates the plasma electrons, causing them to move faster in the vicinity of the test charge. Since flux conservation requires that faster moving electrons have lower density, the density of the negatively charged electrons will *decrease* around the test charge.[1] The plasma *anti-shields* the test charge; instead of decreasing the net positive charge near the test particle, the plasma will *increase* the net charge. This phenomena has been sporadically recognized in the literature[2, 3, 4, 5, 6], but, to our knowledge, has never before been observed.

Though anti-shielding is observed when unusual initial conditions are employed, more commonly we observe the converse—shielding. We show here that this shielding results from the presence of electrons trapped in the potential well of the test charge. While several different mechanisms are observed to trap electrons, a ubiquitous, fast acting, transit-time mechanism always traps electrons when the test charge is introduced adiabatically.[7] That one dimensional (1-d), collisionless shielding requires trapping does not appear to have been previously recognized, and the explanation of shielding given in many textbooks and papers is incorrect or incomplete.[8, 2, 3, 4, 5, 6, 9, 10] Because the trapping results from dynamical processes, we call the resulting shielding "dynamic" shielding. Both the observed and calculated magnitude of dynamic shielding can be significantly smaller than

Debye shielding; eventually collisions transform the dynamic shielding to Debye shielding.

The difference between a situation which will produce shielding and one which will produce anti-shielding can be clarified with a simple phase space picture. As shown in Fig. 1, the phase space in the neighborhood of a positive potential well contains two distinct classes of orbits: free orbits on which electrons stream through the well, and trapped orbits that close within the well. Since the phase space area filled by any given set of free electrons lengthens in $\hat{z}$ as the set accelerates into the test well, Liouville's theorem requires that the $\hat{v}_z$ extent shortens concomitantly. In steady state, the phase space density distribution function $f(z, v_z)$ must be constant along any trajectory. Consequently, the electron density $n(z) = \int f(z, v_z)\, dv_z$ must decrease inside the test well if the trapped orbits are unpopulated, and the well will be anti-shielded. Shielding (density increase), as is found for adiabatically-created wells, can result only from the trapped orbits being populated.

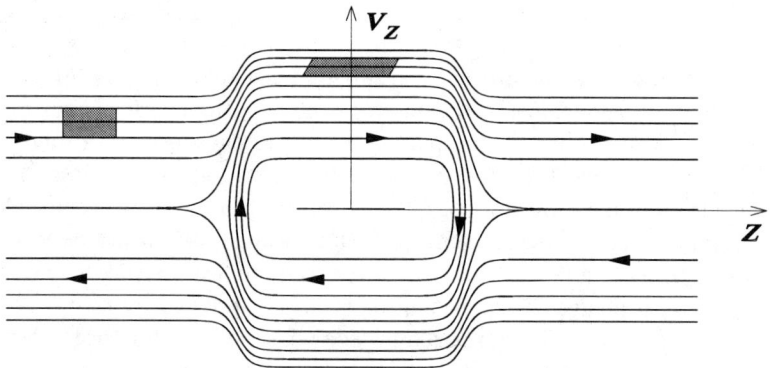

FIGURE 1. Electron orbit phase space in the vicinity of a positive test well. Note the change in aspect of the square set of electrons as they propagate into the well.

The results we report were observed in a pure electron plasma held in a Penning-Malmberg trap with cylinders of radius 1.905 cm and magnetic field 1800 G. More detailed descriptions of Penning-Malmberg traps can be found in the literature.[11]

We create the equivalent of a positive test charge by biasing a central electrode to create a square, secondary, positive electrostatic test well. The manner in which the test well is created determines the plasma response; when the test well is created nonadiabatically the plasma enhances (anti-shields) the test well depth, but when the test well is created adiabatically the plasma diminishes (shields) the test well depth. The final plasma radius is approximately 1 cm, the final density is $n_0 = 3 \times 10^7$ cm^{-3}, and the plasma temperature is 7.5 eV.

The total charge, and thus the number of electrons contained within the test well, equals the image charge on the test well electrode, and is measured by integrating the image current which flows onto the test well electrode. Figure 2 shows this total charge as a function of the test well depth. The total charge in the test well increases with test well depth for the adiabatically-created test well, but it decreases for the nonadiabatically-created test well. Consequently the test well is shielded in the first case and anti-shielded in the second.

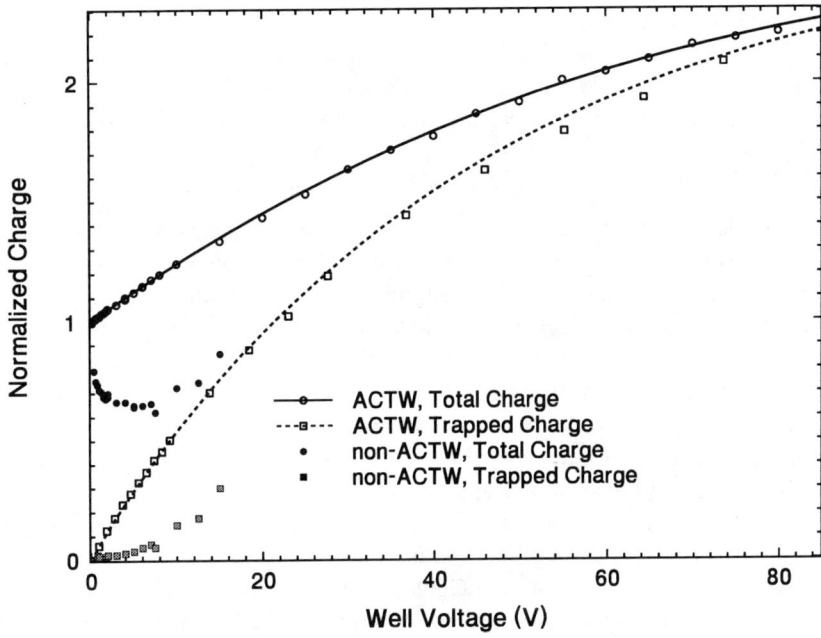

FIGURE 2. Charge measurements vs. well voltage (depth) for both types of test wells. The lines are calculated from the theory described in the text.

When the right wall of the overall confinement well is removed, electrons not trapped in the test well are no longer confined and stream out along the magnetic field lines. When the test well potential itself is removed, the trapped electrons are also released, and this trapped charge can be measured by monitoring the test well electrode image current. As shown in Fig. 2, the trapped charge increases linearly with test well depth for shallow, adiabatically-created test wells. In contrast, little charge is trapped by shallow nonadiabatically-created test wells. (Note that Fig. 2 shows the initial trapped charge; irregular instabilities are present in the nonadiabatic case that trap electrons. The onset of these instabilities occurs more rapidly as the test well depth is increased, and interferes with our nonadiabatically-created well measurements for well depths greater than ten volts.)

When the test wells are adiabatically-created, electrons are efficiently and automatically trapped by a transit time mechanism. The mechanism is discussed by Lifshitz and Pitaevskii:[7] Consider a slow-moving electron entering a slowly deepening test well. Although the electron gains kinetic energy while entering the test well and loses kinetic energy while climbing out of the test well, its kinetic energy remains constant inside the test well, even when the test well depth is varying. Since this mechanism is very general, it will occur regardless of whether the test charge is slowly increased in magnitude in place, or slowly introduced from outside the plasma.

The exact plasma response can be calculated using the bounce adiabatic invariant $J = \oint v_z dz$. We can show that when we include the plasma self consistent effects that for small V_c, the plasma density in the well region is given by:

$$n_w = n_0(1 + eV_c/kT). \tag{1}$$

This is identical to the density predicted by linearizing the Boltzmann relation, $n_w = n_0 \exp(eV_c/kT)$. Because of this numeric coincidence, most results, including derivations of Debye shielding, that mistakenly or implicitly use the Boltzmann relation will still be correct in the collisionless regime. However, the Boltzmann result increases exponentially with V_c, while our result remains roughly linear. Thus nonlinear Boltzmann-based, classic Debye shielding is stronger and more complete than transit-time trapping shielding.

In conclusion, we have shown that collisionless shielding in one-dimensional plasmas relies on transit-time trapping. Without trapping, the plasma anti-shields a positive test charge. Since the electrons in highly magnetized plasmas respond one-dimensionally, shielding due to transit time trapping is quite common.

We thank Drs. B.R. Beck, R. Gould, T.M. O'Neil, and J.S. Wurtele for their helpful comments. This work was supported by the Office of Naval Research.

References

[1] The paradox does not occur in three dimensions because the electron orbits bend.

[2] B. Abraham-Shrauner, Physica **43**, 95 (1969).

[3] D. Montgomery and F. Tappert, Phys. Fluids **15**, 683 (1972).

[4] E. W. Laing and A. L. Gibson, J. Plasma Phys. **14**, 433 (1975).

[5] P. Chenevier, J. M. Doloique, and H. Peres, J. Plasma Phys. **10**, 185 (1973).

[6] N. M. Meyer-Vermet, Am. J. Phys. **61**, 249 (1993).

[7] E. Lifshitz and L. Pitaevskii, *Physical Kinetics* (Pergamon Press Inc., New York, 1981), p. 146, the authors include an untraceable reference to A.V. Gurevich, 1967.

[8] Typically these papers either ignore trapping or implicitly assume that collisions maintain a Maxwellian distribution.

[9] R. C. Davidson, J. Plasma Phys. **6**, 229 (1971).

[10] F. F. Chen, *Plasma Physics and Controlled Fusion* (Plenum Press, New York, 1984), p. 8.

[11] J. H. Malmberg, C. F. Driscoll, B. Beck, D. L. Eggleston, J. Fajans, K. Fine, X. P. Huang, and A. W. Hyatt, in *Nonneutral Plasma Physics*, edited by C. Roberson and C. Driscoll (American Institute of Physics, New York, 1988), Vol. AIP 175, p. 28.

Ion Crystals in Paul Traps

J. A. Hoffnagle

I. B. M. Almaden Research Center
650 Harry Road
San Jose, CA 95120

Abstract

The dynamics of small ion crystals in a Paul trap has been studied by numerical solutions of deterministic equations of motion that include the time dependence of the trapping potential. For small trap voltage, the crystal shell structure agrees with earlier calculations based on a harmonic pseudopotential. Chaotic motion is also possible and, as in the 2-ion case, increasing the trap voltage gives rise to a transition from transient to stationary chaos that is consistent with boundary crisis theory. With increasing ion number, the critical trap voltage decreases, in accord with experiments on aluminum microspheres, but the lifetime scaling of the chaotic transients is independent of the number of ions.

INTRODUCTION

In one of the first experimental realizations of the Paul trap, Wuerker, Shelton, and Langmiur (1) observed that small ensembles of trapped, charged particles could assume symmetric, steady-state configurations. They also saw that these regular arrays could be made to "melt" by raising the trapping potential; the resulting, amorphous clouds then "crystallized" when the voltage was lowered again. Increasing the number of particles in the trap reduced the voltage at which melting occurred. These experiments were performed with electrostatically charged aluminum microspheres, but the dynamics are the same for particles of any size. After the introduction of laser cooling (2) experiments confirmed the existence of stable, regular arrays of trapped atomic ions (3, 4), sometimes called "ion crystals". They can be regarded as the few-body analogues of the crystalline phase of a one-component plasma (5), and have been the subject of considerable theoretical work. In this paper the dependence on ion number of the trapped ion dynamics is investigated through numerical simulations using the full, time-dependent equations of motion. Two issues are considered in detail: the shell structure of the crystals, and the way in which condensation from a disordered cloud to a crystal depends on the trapping potential and number of ions.

Most calculations of ion crystal structure have started by replacing the periodic potential of the Paul trap with a static, harmonic pseudopotential (6–11). The results agree well with observations for up to 16 ions (3, 4, 7, 12), and can be extended to much larger numbers of ions than have been observed to condense in experiments (9–11). However, recent calculations for the very simple system of two ions in a Paul trap indicate that for some trap parameters the exact solutions of the equations of motion can be qualitatively different than what one would calculate from the harmonic pseudopotential (13, 14). Consequently, it is of interest to systematically compare the ion crystal structures determined using the full trap potential with earlier results based on a harmonic pseudopotential.

While the equilibrium configurations of ion crystals have been the subject of several investigations, there has been relatively little work so far on the transitions between ordered and disordered motion, except for the special case of two ions. (See (15) for references to this system.) The dynamics of the two-ion system has some important features that also apply to larger ensembles, such as the identification of the crystal – cloud transitions with transitions between regular motion and deterministic chaos (16, 17), and the appreciation that transient chaos is at the heart of condensation (15, 18). However, many of the pioneering observations of (1) remain to be explained. Blümel et al. (17) interpreted numerical simulations of 5 ions in terms of competition between r.f. heating and laser cooling; in their model, transitions between order and chaos depended in a complicated way on the trapping and laser-cooling parameters, and they did not systematically study the ion-number dependence. Prestage et al. (19) simulated much larger ensembles, with several hundreds of ions. They made the important observation that crystal stability depends on the trap voltage, as seen in experiment (1), but found no dependence on ion number. The numerical simulations reported here systematically address the question of how condensation of ion clouds depends on the number of trapped ions. A purely deterministic model of the ion dynamics is used, in which the crystallization transition has an interpretation as a transition from stationary to transient chaos. The results of the simulations are consistent with the same boundary crisis scenario that is responsible for order-chaos transitions in the two-ion system, thus generalizing earlier work (15, 18) to larger ensembles of ions. The critical trap voltage decreases with increasing ion number, in qualitative agreement with (1).

EQUATIONS OF MOTION

The classical motion of N ions of charge e and mass m, with coordinates $\mathbf{r}^{(m)}$, $m = 1 \ldots N$, in the potential

$$V = [V_{DC} - V_{AC}(\cos \Omega t)] \frac{x^2 + y^2 - 2z^2}{2r_0^2} \tag{1}$$

is conventionally expressed in dimensionless form by measuring time in units of $2/\Omega$ and introducing scaled potentials $-a_3/2 = a_1 = a_2 = 4eV_{DC}/mr_0^2\Omega^2$ and $-q_3/2 = q_1 = q_2 = 2eV_{AC}/mr_0^2\Omega^2$. The trap potential can be described by the two parameters $a \equiv a_3$ and $q \equiv q_3$. Then the equations of motion for the Cartesian components $r_i^{(m)}$, $i = 1, 2, 3$, are a set of $3N$ coupled, second-order equations:

$$\ddot{r}_i^{(m)} + \alpha \dot{r}_i^{(m)} + (a_i - 2q_i \cos 2t) r_i^{(m)} = \sum_{n \neq m} \frac{r_i^{(m)} - r_i^{(n)}}{|\mathbf{r}^{(m)} - \mathbf{r}^{(n)}|^3}. \tag{2}$$

Here the unit of distance has been chosen to make the coefficient of the Coulomb term unity. The velocity-dependent term proportional to α accounts approximately for the energy dissipation due to laser cooling, and also accurately describes the aerodynamic drag in the experiment of Ref. (1). It is worth emphasizing that equations (2) are purely deterministic, with a single nonlinear term, due to the Coulomb interaction. Other studies of trapped-ion dynamics in a time-dependent trapping potential have either introduced stochastic terms to establish a well-defined ion temperature (19), or modeled in detail the fluctuating, nonlinear laser-ion interaction (12, 17, 20). The price to be paid for a more realistic description of the cooling process is that it is essentially impossible to disentangle the effects of multiple nonlinearities, or of stochastic and deterministic terms in the equations of motion. Since experiment (1) shows that the exact mechanism of laser cooling is not central to the existence of ion crystals, nor to the transitions between ordered and disordered motion, the model used in these calculations treats energy dissipation in the simplest possible way. Consequently, although numerical solutions often exhibit chaos, the dynamical system described by equations (2) has no well-defined temperature, and transitions between ordered and disordered motion are interpreted in terms of nonlinear dynamics, rather than statistical mechanics. Simulations of ion motion were carried out by choosing random initial conditions, $\mathbf{r}^{(m)}(t = 0)$ and $\dot{\mathbf{r}}^{(m)}(t = 0)$, corresponding to an initially disordered ion cloud, and numerically integrating the equations of motion with a Bulirsch-Stoer algorithm (21), thus modeling the crystallization of an initially hot cloud that is subject to energy dissipation.

SHELL STRUCTURE OF ION CRYSTALS

Before proceeding to the structure of ion crystals, it is necessary to consider a technical point that arises from the time dependence of the trap potential. If the Paul trap potential, equation (1), is replaced by a static pseudopotential, $\Psi = \left(\beta_1^2 x^2 + \beta_2^2 y^2 + \beta_3^2 z^2\right)/2$, where the secular frequencies $\beta_1 = \beta_2$ and β_3 are functions of a and q, then a crystal is an ionic configuration for which the total potential energy, with pseudopotential and electrostatic contributions, has a minimum (7). In general there may be multiple local minima and care is required to find the global energy minimum (9); still, each crystalline configuration has a well-defined energy. For a time-dependent potential the energy is not a constant of the motion, nor is the electrostatic potential energy alone. There are, however, stable solutions of the equations of motion for which the ions make micromotion oscillations at the trap frequency about their average positions $< \mathbf{r}^{(m)} >$. These solutions correspond to the crystals calculated using the pseudopotential, and it seems reasonable to define for them an average Coulomb energy (19),

$$< E_C >= \sum_{n \neq m} \left| < \mathbf{r}^{(m)} > - < \mathbf{r}^{(n)} > \right|^{-1}, \tag{3}$$

where the angular brackets indicate averaging over the micromotion. In this way it is possible to make a connection between the exact solutions of equations (2) and calculations of crystal structure which found the lowest-energy states of the pseudopotential. To make comparisons with the extensive calculations of crystals in spherically symmetric, harmonic potentials (9, 10), crystal structures were calculated for equations (2) with the parameters $a = -0.03963$ and $q = 0.39466$, giving secular frequencies $\beta_1 = \beta_2 = \beta_3 = 0.2$. The case of degenerate secular frequencies is interesting in its own right, since it is in this case that calculations for $N = 2$ show the largest discrepancy between solutions of the exact Paul trap potential and the harmonic pseudopotential (13, 14). Energy dissipation was chosen to be rather large, $\alpha = 1 \times 10^{-2}$, in order to obtain rapid crystallization and hence reduce the computation time. With this choice of parameters, the numerical simulations settled within a few hundred trap periods into ordered configurations that agree well with previously reported structures (9, 10). As in the harmonic trap, there are often several crystalline configurations, for which the differences in Coulomb energy can be very small, especially for large N. To try to find the lowest-energy configuration for a given N, the calculations were repeated for 20 different random sets of initial conditions. For some parameter values, it was also possible to find solutions of equations (2) that were frequency-locked to multiples of the trap

TABLE 1. Comparison of calculated crytals using the exact potential, equation (2), and a harmonic pseudopotential.

N	Paul trap[a]		Harmonic trap[b]	
	Structure	$< E_C >$	Structure	E_C
2	2	0.3930	2	0.3969
3	3	0.6877	3	0.6933
4	4	0.9380	4	0.9449
5	5	1.1775	5	1.1880
6	6	1.3933	6	1.4043
7	7	1.6094	7	1.6214
8	8	1.8082	8	1.8221
9	9	2.0003	9	2.0159
10	10	2.1866	10	2.2039
11	11	2.3696	11	2.3881
12	12	2.5401	12	2.5605
13	12 + 1	2.7123	12 + 1	2.7339
14	13 + 1	2.8802	13 + 1	2.9028
15	14 + 1	3.0410	14 + 1	3.0648
16	15 + 1	3.1989	15 + 1	3.2243
17	16 + 1	3.3538	16 + 1	3.3805
18	17 + 1	3.5065	17 + 1	3.5340
19	18 + 1	3.6558	18 + 1	3.6850
20	19 + 1	3.8049	19 + 1	3.8347
21	20 + 1	3.9489	20 + 1	3.9803
22	21 + 1	4.0926	21 + 1	4.1249
23	21 + 2	4.2332	21 + 2	4.2667
24	22 + 2	4.3718	22 + 2	4.4060
25	23 + 2	4.5087	23 + 2	4.5444
26	24 + 2	4.6428	24 + 2	4.6798
27	24 + 3	4.7758	24 + 3	4.8137
28	25 + 3	4.9073	24 + 4	4.9467
29	25 + 4	5.0375	26 + 3	5.0775
30	26 + 4	5.1655	27 + 3	5.2065
31	27 + 4	5.2915	28 + 3	5.3351
32	28 + 4	5.4170	28 + 4	5.4600
33	29 + 4	5.5418	29 + 4	5.5859
34	30 + 4	5.6646	30 + 4	5.7099
36	30 + 6	5.9068	30 + 6	5.9538
38	32 + 6	6.1437	32 + 6	6.1925
40	34 + 6	6.3785	34 + 6	6.4291
42	35 + 7	6.6090	35 + 7	6.6614
44	36 + 8	6.8352	37 + 7	6.8895
46	38 + 8	7.0580	38 + 8	7.1141
48	39 + 9	7.2779	39 + 9	7.3353
60	48 +12	8.5378	48 +12	8.6057
61	47 +13 +1	8.6393	48 +12 +1	8.7074
62	48 +13 +1	8.7399	48 +13 +1	8.8090

[a]Scaled by $\beta^{-2/3}$.

[b]From Hasse and Avilov (10).

period; in this, as in many other respects, the many-ion dynamics resembles that of the two-ion system (22, 23). The shell structures and average Coulomb energies for the crystals with the lowest values of $< E_C >$ are compared with pseudopotential calculations in Table 1. The units of energy are different in (10) than in equations (2), hence the values of $< E_C >$ for the Paul trap are scaled by $\beta^{-2/3}$ to make the comparison consistent. Except for a few instances near $N = 29$ the correspondence between the Paul trap and pseudopotential calculations is almost exact.

CRYSTALLIZATION OF COOLED IONS

For the calculations just described, the trap parameters were chosen to give radial and axial secular frequencies that were equal and relatively small, $\beta = 0.2$. Increasing β to 0.4 (corresponding to $a = -0.15308$, $q = 0.75820$) produces a drastic change: initially disordered ions fail to crystallize, even if the simulations are extended to time intervals orders of magnitude longer than are required to see condensation at $\beta = 0.2$. In their study of the transition between ordered and disordered states of Coulomb clusters, Wuerker, Shelton, and Langmuir (1) chose to operate the Paul trap with line $V_{DC} = 0$, and extensive experimental and theoretical work on the two-ion system has also focused on this special case, therefore the simulations to be described in this section were all done with $a = 0$. (Note that the shell structure is different for $a = 0$ than for the spherically symmetric crystals described in the previous section: in particular the second shell already begins to form at $N = 6$.) The process of condensation has a natural interpretation in terms of nonlinear dynamics, if the disordered "cloud" is understood as a state in which the ion motion exhibits deterministic chaos. For low q, the crystal is the only stable state of motion – it corresponds to an attractor to which all trajectories converge in the limit $t \rightarrow \infty$. Persistent chaotic motion is also possible, but only if $q > q_c$, where the critical parameter q_c depends on the number of ions. Ions prepared in the chaotic state will be observed to condense if the trap voltage is reduced so the q becomes smaller than q_c.

The system of two ions in a Paul trap exhibits a similar transition between regular and chaotic motion, which has been shown experimentally (18) and theoretically (14) to follow the boundary crisis scenario of Grebogi, Ott, Yorke, and colleagues (24). For q slightly smaller than q_c, this model predicts chaotic transients, intervals of chaotic motion which are very long, but ultimately result in abrupt crystallization. The duration of an individual transient is unpredictable, depending sensitively on initial conditions, but for given q the lifetimes of an ensemble of transients are exponentially distributed, with an

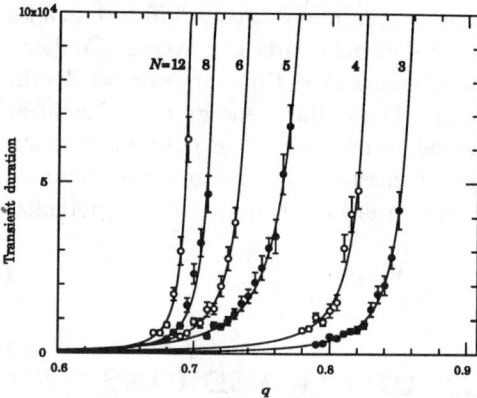

FIGURE 1. Simulated transient lifetimes of multi-ion crystals in the Paul trap. Open and closed circles distinguish simulations with different values of N. The error bars are statistical and the solid curves are best fits to equation (4).

average value $T(q)$. Near the critical parameter the average lifetime diverges according to a power law:

$$T(q) \propto (q_c - q)^{-\gamma}. \tag{4}$$

The parameters q_c and γ can in principal be calculated from more fundamental phase-space constructions; this calculation is feasible for $N = 2$, but probably not for larger numbers of ions, due to the large dimensionality of phase space. However, the existence of chaotic transients, with lifetimes scaling as in equation (4), is a prediction of crisis theory which is easily testable for moderate values of N.

Numerical solutions of equations (2) were indeed seen to exhibit transient chaos for suitable values of q. In a typical transient, initially disordered ions swarm in a featureless cloud for a period that is long compared to the time scale for energy dissipation, before suddenly falling in into place in a crystalline array. Average lifetimes computed for a range of it q-values and for ion numbers $3 \leq N \leq 12$ are shown in Figure 1. Each point represents an average of 50 simulations in which randomly chosen initial conditions were integrated for up to 5×10^4 trap periods. As before α was chosen to be 1×10^{-2}. For each value of N the parameters of equation (4) were adjusted to minimize χ^2 (21), resulting in the solid curves, all of which are in satisfactory agreement with the numerical data.

The shift of the order-chaos transition to lower q with increasing N, first seen experimentally (1), is evident in the simulations. This can be seen more clearly by plotting the fitted values of q_c against the number of ions (Fig. 2a). It is noteworthy that q_c shows no sign of a discontinuity at $N = 6$, where the second shell begins to fill. The points at the largest values of N appear to

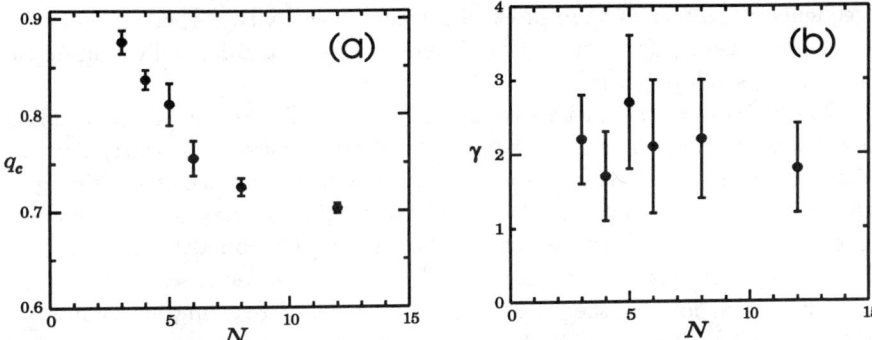

FIGURE 2. Dependence on ion number of the parameters in equation (4): (a) critical q-parameter; (b) critical exponent. The error bars are the estimated uncertainties derived from the covariance matrix of the fit.

level off near $q_c = 0.65$ which was the melting point of the largest crystals ($N \sim 100$) in the experiment of Ref. (1). If q_c does approach an asymptotic value with only ~ 10 ions, this would also be consistent with calculations that do not show any N-dependence in the stability properties of much larger crystals (19).

Whereas q_c shows a clear dependence on the ion number, the critical exponent γ is the same, to the accuracy of these computations, for all values of N (Fig. 2b). The average of all the simulations, $\bar{\gamma} = 2.1 \pm 0.3$, is also close to the value of 2 that was predicted (15) from boundary crisis theory for the two-ion system, in the limiting case $\alpha \to 0$. However, there is no general theoretical requirement for γ to be independent of N. In fact, the similarity of the transient behavior for all values of N (figure 1) is an unexpectedly simple result of these calculations, suggesting that the order-chaos transition arises in a similar way for ion crystals of all sizes.

CONCLUSIONS

Small ion crystals in a Paul trap have been studies by numerical simulations, with emphasis on the systematic variations in crystal structure and dynamics as the number of ions increases. The model used was purely deterministic, treating the oscillatory trap potential and inter-ion Coulomb interaction exactly, and approximating ion cooling by simple, linear energy dissipation. Shell structures of ion crystals were carried out for $N \leq 62$ with trap potentials chosen to give degenerate radial and axial secular frequencies. For small secular frequencies the results were in excellent agreement with earlier, more extensive, calculations of the structure of ion crystals in a spherically symmetric, harmonic pseudopotential.

The process by which ions condense from a diffuse cloud to a crystalline array was interpreted as a transition from deterministic chaos to regular motion. As in the case of two ions in a Paul trap, the order-chaos transition goes through a stage of transient chaos and the lifetime scaling of the transients is consistent with the theoretical predictions of the boundary crisis model. For the special case $a = 0$, the critical q-parameter decreases with increasing number of ions in the trap, and shows signs of reaching an asymptotic value $q_c \approx 0.65$ for large N. This is in qualitative agreement with the first observations of crystallization of charged particles, made 35 years ago. To the accuracy of these calculations, the dependence of q_c on N is not correlated with the shell structure of the ion crystals. Remarkably, the critical exponent that describes the scaling of transient lifetimes near the onset of chaos is independent of N over the range $N \leq 12$ investigated here. This suggests that there may be an underlying, as yet unidentified, similarity in the condensation of clouds containing any number of ions.

ACKNOWLEDGEMENT

This work arose in the course of a long standing collaboration with R. G. Brewer, to whom I am grateful for many valuable comments.

REFERENCES

1. Wuerker, R. F., Shelton, H., and Langmuir, R. V., *J. Appl. Phys.* **30**, 342–349 (1959).

2. Neuhauser, W., Hohenstatt, M., Toschek, P., and Dehmelt, H., *Phys. Rev. Lett.* **41**, 233–236 (1978).

3. Diedrich, F., Peik, E., Chen, J. M., Quint, W., and Walther, H., *Phys. Rev. Lett.* **59**, 2931–2934 (1987).

4. Wineland, D. J., Bergquist, J. C., Itano, W. M., Bollinger, J. J., and Manney, C. H., *Phys. Rev. Lett.* **59**, 2935–2938 (1987).

5. Ichimaru, S., *Rev. Mod. Phys.* **54**, 1017–1059 (1982).

7. Itano, W. M., Bergquist, J. C., and Wineland, D. J., in *Proc. Workshop on Crystalline Ion Beams*, Wertheim, ed. Hasse, R. W., Hofmann, I., and Liese, D., 1989, pp. 241–254.

8. Lozovik, Yu. E. and Mandelshtam, V. A., *Phys. Lett. A* **145**, 269–271 (1990).

9. Rafac, R., Schiffer, J. P., Hangst, J. S., Dubin, D. H. E., and Wales, D. J., *Proc. Nat. Acad. Sci. USA* **88**, 483–486 (1991).

10. Hasse, R. W. and Avilov, V. V., *Phys. Rev. A* **44**, 4506–4515 (1991).

11. Tsuruta, K. and Ichimaru, S., *Phys. Rev. A* **48**, 1339–1344 (1993).

12. Blümel, R. *et al.*, *Nature* **334**, 309–313 (1988).

13. Emmert, J. W., Moore, M., and Blümel, R., *Phys. Rev. A* **48**, 1757-1760 (1993).

14. Hoffnagle, J. and Brewer, R. G., *Appl. Phys. B* (in press).

15. Hoffnagle, J. and Brewer, R. G., *Phys. Rev. A* (in press).

16. Hoffnagle, J., DeVoe, R. G., Reyna, L., and Brewer, R. G., *Phys. Rev. Lett.* **61**, 255–258 (1988).

17. Blümel, R., Kappler, C., Quint, W., and Walther, H., *Phys. Rev. A* **40**, 808–829 (1989).

18. Brewer, R. G., Hoffnagle, J., and DeVoe, R. G., *Phys. Rev. Lett.* **65**, 2619–2622 (1990).

19. Prestage, J. D., Williams, A., Maleki, L., Djomehri, M. J., and Harabetian, E., *Phys. Rev. Lett.* **66**, 2964–2967 (1991).

20. Casdorff, R. and Blatt, R., *Appl. Phys. B* **45**, 175–182 (1988).

21. Press, N H., Flannery, B. P., Teukolsky, S. A., and Vetterling, W. T., *Numerical Recipes*, Cambridge, Cambridge Univ. Press, 1987.

22. Hoffnagle, J. and Brewer, R. G., *Phys. Rev. Lett.* **71**, 1828–1831 (1993).

23. Hoffnagle, J. and Brewer, R. G., *Science* **265**, 213–215 (1994).

24. Grebogi, C., Ott, E., Romeiras, F., and Yorke, J. A., *Phys. Rev. A* **36**, 5365-5380 (1987), and references therein.

Charged Vortex Structures and Electron Transport in a Non Neutral Electron Plasma

N. A. Kervalishvili

Institute of Physics, Georgian Academy of Sciences
6 Tamarashvily Street,380077, Tbilisi, Georgia

Abstract: The properties and dynamics of charged vortex structures in a nonneutral electron plasma in crossed electric
and magnetic fields and their role in the processes of electron transport along and across the magnetic field lines have been considered.

Introduction

The steady-state nonneutral electron plasma can be obtained and confined in the devices with crossed electric and magnetic fields. Nonneutral electron plasma is a unique plasma medium in which the electric fields from external and internal objects are not screened. Such property imposes certain restrictions on experimental and theoretical methods of investigation. At the same time due to this property some physical phenomena may take place in a nonneutral electron plasma having no analogy in neutral plasma. One of such phenomena namely charged vortex structures are discussed below.

Charged Vortex Structures

The vortex structures in a nonneutral electron plasma of high-voltage discharge in crossed electric and magnetic fields were observed in three geometries of the discharge devices: magnetron,[1] Penning cell,[2] and inverted magnetron.[3] The structures are local formations with high density of electrons stretched along the magnetic field lines. Life time of vortex structures are much higher than that of electron-neutral collisions[4] In decaying (absence of ionization) nonneutral electron plasma of Penning traps[5, 6, 7] the vortex structures are formed, as a rule, by external effect and in the case of stable distribution of electron density are conserved practically until a complete decay of plasma. Being electrically nonneutral the structures have their electric fields and therefore rotate about their own axis as well as about discharge device axis. Just this fact allows us to call them vortices. Besides due to the exist of own electric fields the vortices interact with each other, the electrodes of discharge device and with electrons located around vortices at sufficiently great distances. It can be said that charged vortex structures have the long-range properties in contrast to dipole and neutral vortices.

It is assumed that the reason of vortex structures format in the nonneutral electron plasma is the diocotron instability great number of theoretical investigations is devoted to the study of diocotron instability in collisionless pure electron plasma[8, 9, 10, 11, 12, 13, 14] and rather precise pattern of linear and weakly nonlinear phases of development of this instability are obtained. The only weak point of this pattern is that the initial distribution of unperturbed electron density is taken arbitrarily while in collisional nonneutral electron plasma of gas discharge the profile of such distribution is formed independently and it is necessary to find it from the equations describing the electron plasma. This was carried out in [15], where the analytical expressions for the equilibrium radial distribution of electron density as a function of time were obtained. As for the highly nonlinear phase of development of diocotron instability, just at which the vortex structures are generated, it has not been as yet studied theoretically. However, recently in[16] the solution was found for the equilibrium state with the already existing highly nonlinear, as the authors call them "coherent", structures.

Evolution of electron plasma and dynamics of vortices. As follows from the experimental results[1, 2, 3, 4] the general pattern of the evolution of gas discharge nonneutral electron plasma is as following. From the moment of discharge striking the formation of equilibrium stable profile of the plasma is initiated as a result of electron-neutral collisions. The slow evolution of the profile continues until the conditions of diocotron instability are originated. It is not very important exactly which mode of diocotron instability is excited at first. It develops quickly and leads, as a rule, to the formation of several vortex structures. However, one cannot consider structures to be the azimuthal modes of instability. They move at different drift radii and with different angular velocity and therefore, are independent formations. The vortices interact intensively with each other (approach each other, merge, sometimes split) until a single stable vortex remains. The whole process of formation of such stable vortex takes place for the periods of time much less than that of electron-neutral collisions and is accompanied by ejection of a part of electrons of plasma layer onto the endplate electrodes (cathodes) along the magnetic field lines.

Further, the evolution of plasma proceeds again slowly. Life time of a stable vortex is much greater, than the electron-neutral collision time and its behavior depends substantially on the geometry of discharge device.

a) In the geometry of the inverted magnetron[3, 4] a stable vortex decays slowly

It is not blurred but just decays. The density of electron background is increased simultaneously. Both processes lead to the fact that after a certain time interval the vortex completely disappears and for some period of time the plasma remains quiet. At the same time the mean electron density continues to increase. At last comes the moment when the diocotron instability occurs and everything is repeated again.

b) In geometries of magnetron and Penning cell a stable vortex doesn't decay. Simply its orbit is slowly displaced to the anode.[1, 2] At some instant of time the orbit becomes unstable and the vortex begins to oscillate radially. The movement of vortex is spiral - it alternately removes from and approaches the anode. These movements are rather rapid, they take place for the period of time much less than the time of electron-neutral collision and are accompanied by electron ejection along the magnetic field lines . As a result of such movements the electron density decreases and the vortex orbit is stabilized in the position removed from the anode. Then the whole process is repeated again.

Thus, a nonneutral electron plasma of the discharge in crossed electric and magnetic fields is characterized by periodicity of the processes of vortex structure behavior and the periodicity itself is determined by the collision time.

The pattern of the behavior of a single vortex in a nonneutral electron plasma described above takes place at the pressures of the order of or less than 10 microtorr. At the pressures of the order of or more than 100 microtorr there exist simultaneously several vortices and the main mechanism of electron ejection along the magnetic field lines is related to the periodic approaches of the vortices.[1] At these pressures the periods of vortex approach are of the same order of magnitude as the time of electron-neutral collisions.

Electron Transport by Vortices

Now consider in more detail the role of vortices in the processes of electron transport along and across the magnetic field lines. a) The electron current along the magnetic field lines to the endplate electrodes being under the cathode potential is already well known as "anomalously high energy electron" current. The mean value of this current is sufficiently high and is 30-50% of the value of discharge current. However, the mechanism of its formation was unknown and only now after the revelation of vortex structures it is possible to explain its origination.[1, 2, 3] As follows from the experimental results, the ejection of electrons onto endplate electrodes (cathodes) along the magnetic fields lines takes place in three cases: when the vortices are formed;[3] when two or more vortices approach each other[1] and when the vortex is moved away from the anode surface.[1, 2] At the same time, the electron ejection is strictly localized. It is located in the places of formation, existence or approach of the vortices. If the electron ejection is continued for a sufficiently long time, the area of ejection drifts together with the vortex about the discharge device axis. Thus the electron ejection occurs when and where the local increase of electron density takes place that, in its turn, leads to the decrease of retarding potential between the electron cluster and endplate electrode. In other case the electron ejection takes place when the electron cluster itself displaces along a radius into a region with less retarding potential. However, the decrease of potential barrier in itself is not sufficient for the ejection of electrons. It is necessary for a part of electrons in plasma to have sufficient "longitudinal energy" at the moment in order to overcome this decreased retarding potential. As it is known the electrons in discharge in crossed electric and magnetic fields acquire the transverse velocity at the expense of electric field and "longitudinal energy" is accumulated at the expense of electron scattering at electron-neutral collisions. "Longitudinal energy" increases as the electrons are displaced to the anode. However, this energy will be always less than the retarding potential barrier on the corresponding discharge radius. From the above it follows that, first the time of "longitudinal energy" accumulation must take place during the collision time, and second, the electron ejection may occur not for that distribution of the potential at which the accumulation took place, but at the other more favorable one.

In fact, the electron ejection observed experimentally occur; after the time intervals defined by collision time, and the moment and the place of the ejection always correspond to rapid (for the time much less than that of electron-neutral collision) decrease of retarding potential barrier.

b) Now consider the equilibrium stable layer of the magnetized nonneutral electron plasma located between two cylindrical electrodes with a single stable vortex the drift trajectory of which remains constant . Both the vortex and the electron plasma drift about discharge device axis. However, the plasma layer has a radial shear of velocities and the vortices drifts as an unit. The vortex still rotates about its own axis for the presence of its own electric field. But it is not all. The electric field of the vortex propagates far beyond its boundary (long-range forces). As a result the electrons of plasma Layer located round the vortex will rotate (drift) not only about the discharge device axis, but also about the vortex itself. As there is a shear of velocity in the plasma layer, then the all electrons of plasma layer will be periodically involved in the rotation about the vortex . Thus, the existence even of a single stable vortex with a stable orbit leads to the global radial displacement and mixing of electrons in the plasma layer. For the radial velocity of the electrons this is not of a great importance as it is defined by the value of radial electric field at the given radius of the discharge device. However, the "longitudinal energy" is accumulated as a result of electron- neutral collisions and has a maximum value near the anode surface. The radial mixing of the electrons results in the fact, that the electrons with large "longitudinal energy" located at the anode surface might appear near the cathode and, therefore can easily overcome a small retarding potential barrier along the magnetic field lines. Thus, in the equilibrium electron layer even with a single stable vortex having a stable orbit a continuous electron flow must be observed onto the endplate electrodes along the magnetic field lines.

Actually, a single stable vortex with a stable orbit is observed experimentally in the magnetron at the pressures of 10-20 microtorr.[1] Though, at the same time, periodical electron ejections are practically absent, the electron current on the endplate electrodes (cathodes) has the same 30-50% of the value of discharge current as in the case of periodical ejections.

In summary, we have shown that the vortex structures in a nonneutral electron plasma in crossed electric and magnetic fields fulfill various functions. The charged vortex structures regulate the mean electron density in the layer of non-neutral electron plasma . Although the specific mechanisms of electron ejection along the magnetic field lines are various, the result is the same- the mean electron density is maintained at the definite level independently of gas type. The other result of the vortex structures "activity" is the radial mixing of electrons, This leads to the fact , that the "longitudinal energy" of electrons must: not depend on the discharge radius. And finally, the vortex structures maintain the electron layer of the discharge in the state of continuous periodical variation that can be characterized as the state of dynamical equilibrium.

References

[1] N. A. Kervalishvili, Sov. J. Plasma Phys. **15**, 98 (1989).

[2] N. A. Kervalishvili, Sov. J. Plasma Phys. **15**, 211 (1989).

[3] N. A. Kervalishvili, Sov. J. Plasma Phys. **15**, 436 (1989).

[4] N. A. Kervalishvili, Phys. Lett A **57**, 391 (1994).

[5] J. S. DeGrassie and J. H. Malmberg, Phys. Fluids **23**, 63 (1980).

[6] C. F. Driscoll and K. S. Fine, Phys. Fluids B **2**, 1359 (1990).

[7] K. S. Fine, C. F. Driscoll, J. H. Malmberg, and T. B. Mitchell, Phys. Rev. Lett. **91**, 588 (1991).

[8] R. H. Levy, Phys. Fluids **8**, 1288 (1965).

[9] R. H. Levy and R. W. Hockney, Phys. Fluids **11**, 766 (1968).

[10] R. J. Briggs, J. D. Daugherty, and R. H. Levy, Phys. Fluids **13**, 421 (1970).

[11] R. C. Davidson and K. Tsang, Phys. Rev. A **30**, 488 (1984).

[12] R. C. Davidson, Phys. Fluids **38**, 1937 (1985).

[13] S. A. Prasad and J. H. Malmberg, Phys. Fluids **29**, 2196 (1986).

[14] R. A. Smith, Phys. Rev. Lett. **64**, 649 (1990).

[15] N. A. Kervalishvili, Phys. Lett A **188**, 170 (1994).

[16] S. M. Lund and R. C. Davidson, Phys. Fluids B **5**, 1421 (1993).

Production and Storage of Ground State O^{2+} and Charge Transfer of O^{2+} with Neutrals at eV Energies

Victor H. S. Kwong and Z. Fang

Department of Physics, University of Nevada, Las Vegas, Las Vegas, NV 89154

Abstract. The $2p^2\,^3P$ ground state O^{2+} ions were produced from solid iron oxide targets by laser ablation and stored in a rf ion trap. By using this technique, we measured charge transfer rate coefficients between the ground state O^{2+} ion (at mean energy of about 2.5 eV or 2×10^4 K) and neutrals (He, H_2, N_2 and CO)

INTRODUCTION

In the past thirty years, ion traps have been used extensively in precision spectroscopy, in ion cooling, in the measurement of mean lives of metastable atomic states, and in the study of collisional processes between ions and neutrals (1-5). Traditionally, ions were produced from atomic or molecular gases by electron impact ionization. The interaction between the stored ions and the parent gas may severely limit its application. Recently, Kwong (6) introduced a novel approach that combined laser ablation ion source and ion trap to produce and store a wide range of low energy multiply charged ions. With this technique, ions such as Mo^{6+}, W^{2+}, and O^{2+} have been produced and stored in the trap for various studies (6-10). We will use this combined approach to measure the charge transfer rate coefficient of the ground state O^{2+} and atomic and molecular target gases.

The charge transfer reaction $O^{2+} + He \rightarrow O^+ + He^+$ has generated considerable interest as a potential mechanism for the destruction of O^{2+} ions, and the source for O^+ and He^+ ions in the terrestrial ionosphere, in supernovae, and in the upper atmospheres of Mars and Venus. There are, however, significant discrepancies between the results of calculations and measurements on this charge transfer reaction. We will report the results of our recent measurements of the charge transfer rate coefficients for $2p^2\,^3P$ ground state O^{2+} ion and He. In addition, we will also report the charge transfer rate coefficients of O^{2+} and H_2, N_2, and CO at eV energies.

LASER ABLATION ION SOURCE AND ION STORAGE

The experimental method used in the measurement has been reported earlier (6-8).   We used the second harmonic of a Nd:YAG laser with output energy of about 0.3 mJ and 20 ns pulse duration as the ablation source. O^{2+} ions are produced by laser ablation of two 99.9% pure solid iron oxide targets inside an UHV reaction chamber.   The O^{2+} ions were cooled by colliding two beams of laser induced plasmas inside the ion trap at 90°. We have discussed the crossed-beams cooling mechanism in previous publications (6,7). The unique properties of the expanding laser induced plasmas in vacuum can result in the production of ions and neutral atoms mainly in their ground states (11).   Those low energy O^{2+} ions were stored in a rf ion trap of which the trapping parameters were optimized for storage of O^{2+} ions (rf frequency f = 1.4 MHz, amplitude V_0 = 380 V, and DC bias U_0 = 35 V, were chosen).   The number of O^{2+} ions stored in the trap was about 10^3, the energy of the ions was estimated to be about 2.5 eV, and the storage time was about 10 seconds. Ferric oxide was chosen as target material because of the distinct mass difference between low charge states of iron ions (m/q = 56 for Fe^+, m/q = 27 for Fe^{2+}) and O^{2+} ions (m/q = 8) and we want to demonstrate that gaseous ions can be produced from their solid compound. To measure the number of ions stored in the trap, we extract the ions from the trap by applying a synchronous push/pull voltage pulse on the trap's end caps. The ejected ions were mass analyzed by a 0.3-m time-of-flight mass spectrometer equipped with an electrostatic lens and a dual microchannel plate.

ELECTRONIC STATE OF O^{2+} IONS

The electronic state of the stored O^{2+} ions was verified experimentally (10) by comparing our measured charge transfer rate coefficients of H_2 with that for the $2p^2\ ^3P$ ground state O^{2+} (1.71×10^{-9} cm^3 s^{-1}) and the long-lived (37 s) $2p^2\ ^1D$ metastable state O^{2+} (9.6×10^{-9} cm^3 s^{-1}) measured previously by Church and Holzschieter (12). Figure 1 shows the time dependent ion signal can best be fitted by a single exponential function indicating that ions exist in one state. The rate coefficient obtained from the slope of the decay rates at five different densities is $2.36(0.22) \times 10^{-9}$ cm^3 s^{-1} and is within 30% of the result for the $2p^2\ ^3P$ ground state reported by Church and Holzschieter (12). We concluded that the stored O^{2+} ions are in their ground state.

CHARGE TRANSFER EXPERIMENT

Charge transfer rate measurements are performed by measuring the relative number of O^{2+} ions remaining in the trap as a function of the time after their

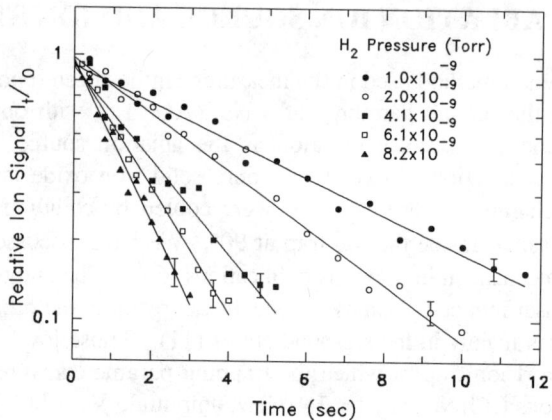

FIGURE 1. The decay curves of O^{2+} ion signal vs storage time for different hydrogen pressures.

production. The ion signals are measured alternately at delay time, t, and at time, t_0 = 0.4 s. The intensity ratio, I_t/I_0, is computed to obtain a normalized intensity. More than ten such measurements are obtained for each time t. By scanning the delay time t, we obtain the ion decay curve. The use of the intensity ratio enables us to minimize both the short term and the long term ion signal fluctuation and drift, caused by the instability of laser power and the changing surface condition of the target.

Typically, the ion signal scans are made at four to five different gas densities. Figure 2(a) shows the O^{2+} intensities as a function of storage time for $O^{2+} + N_2$ at different gas pressures. The solid lines represent the least-squares-fits of the data to a one-exponential function.

The charge transfer rate coefficient (K) is derived from the slope of the ion decay rate ($1/\tau$) vs gas density (ρ): $1/\tau = 1/\tau_0 + K\rho$, where $1/\tau_0$ is the interaction rate of O^{2+} with the residual gas in the vacuum chamber. The plots of the decay rates of the stored O^{2+} ions versus N_2 gas pressure are shown in Figure 2(b). The slope in the figure is obtained by the weighted least-squares-fits to a linear function.

SUMMARY OF RESULTS

The measured rate coefficients of the charge transfer between the $2p^2\ {}^3P$ ground state O^{2+} ion and He, H_2, N_2, and CO are:

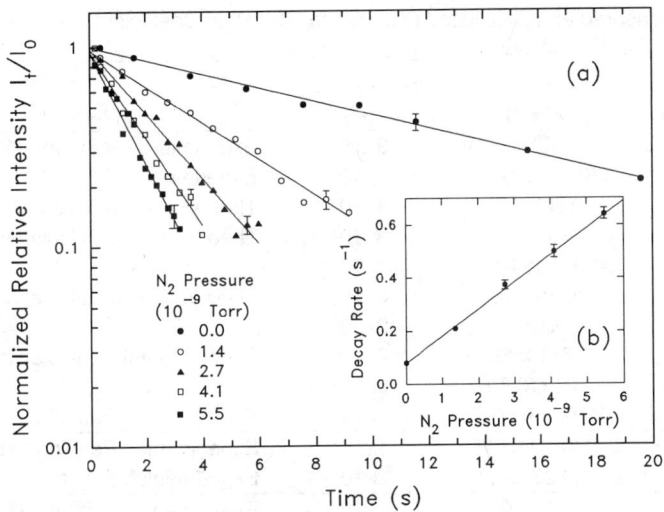

FIGURE 2. (a). The decay curves of O^{2+} ion signal vs storage time for different nitrogen pressures. (b). O^{2+} decay rate vs nitrogen pressure. The slope of the straight line fit gives the charge transfer rate coefficient of the O^{2+} ion with N_2.

$$K\,(O^{2+} + He) = 1.12(0.15){\times}10^{-12}\ cm^3\,s^{-1},$$
$$K\,(O^{2+} + H_2) = 2.36(0.22){\times}10^{-9}\ cm^3\,s^{-1},$$
$$K\,(O^{2+} + N_2) = 3.15(0.26){\times}10^{-9}\ cm^3\,s^{-1},$$
$$K\,(O^{2+} + CO) = 3.40(0.29){\times}10^{-9}\ cm^3\,s^{-1}.$$

The estimated uncertainty of the result is mainly due to the uncertainty in the statistical fluctuation of the ion signal and the uncertainty of the absolute measurement of the gas density.

Table 1 summarizes the results of various calculations and measurements. While our results for $O^{2+} + H_2$ and for $O^{2+} + N_2$ are consistent with the published values, our measured value for $O^{2+} + He$ disagrees with all the available theoretical and experimental values by as much as three orders of magnitude. No theoretical and experimental values for $O^{2+} + CO$ are available for comparison.

ACKNOWLEDGMENTS

This work is supported by FY 93 EPSCoR fund from the State of Nevada.

Table 1. Measured and Calculated Charge Transfer Rate Coefficients for Ground State O^{2+} with He, H_2, N_2 and CO.

Reaction	K (cm^3 s^{-1})	Temperature (K)	Method
O^{2+} + He	$\leq 10^{-14}$	3×10^2	Experimental, drift tube (13)
	$3.5(1.5)\times10^{-11}$	4×10^2	Experimental, Drift tube (14)
	2×10^{-11}	4×10^2	Theory, distorted wave ((15)
	10^{-14}	3×10^2	Theory, quantal calculation (16)
	2×10^{-11}-1×10^{-10}	1×10^4	
	8×10^{-12}	3×10^2	Theory, quantal calculation (17)
	3.9×10^{-10}	2×10^4	
	0.7×10^{-10}	2×10^2	Theory, quantal calculation (18)
	1.01×10^{-9}	2×10^4	
	$1.12(0.15)\times10^{-12}$	2×10^4	Experimental, Paul trap (9)
O^{2+} + H_2	$1.71(0.15)\times10^{-9}$	1×10^4	Experimental, Penning trap (12)
	$2.36(0.22)\times10^{-9}$	2×10^4	Experimental, Paul trap (10)
O^{2+} + N_2	$1.3(0.3)\times10^{-9}$	3×10^2	Experimental, drift tube (19)
	$3.15(0.26)\times10^{-9}$	2×10^4	Experimental, Paul Trap (this work)
O^{2+} + CO	$3.40(0.29)\times10^{-9}$	2×10^4	Experimental, Paul Trap (this work)

REFERENCES

1. Dehmelt, H. G., *Adv. At. Mol. Phys.* **3**, 53 (1967); **5**, 109 (1969).
2. Wineland, D. J., Itano, W. M., and Van Dyck, R. S., jr., *Adv. At. Mol. Phys.* **19**, 135 (1983).
3. Brown, L. S., and Gabrielse, G., *Rev. Mod. Phys.* **58**, 233 (1986).
4. Church, D. A., *Physics Report* **228**, Nos. 5 & 6, 253 (1993).
5. Kwong, V. H. S., Fang, Z., Gibbons, T. T., Parkinson, W. H., and Smith, P. L., *Astrophys. J.* **411**, 431 (1993).
6. Kwong, V. H. S., *Phys.Rev. A* **39**, 4451 (1989).
7. Kwong, V. H. S., Gibbons, T. T., Fang, Z., Jiang, J., Knoche, H., Jiang, Y., Rugar, B., Huang, S., Braganza, E., and Clark, W., *Rev. Sci. Instrum.* **61**, 1931 (1990).
8. Kwong, V. H. S., Fang, Z., Jiang, Y., and Gibbons, T. T., *Phys. Rev. A* **46**, 201 (1992).
9. Kwong, V. H. S., and Fang, Z., *Phys. Rev. Lett.* **71**, 4127 (1993).
10. Fang, Z., and Kwong, V. H. S., *Rev. Sci. Instrum.* **65**, 2143 (1994).
11. Drewell, N., *Studies of Laser Selective Excitation of Atoms*, University of Toronto, UTIAS Report No. 229, 1979, ch. 6, pp. 45-53.
12. Church, D. A., and Holzscheiter, H. M., *Phys. Rev. A* **40**, 54 (1989).
13. Howorka, F., Viggiano, A. A., Albritton, D. L., Ferguson, E. E., and Fehsenfeld, F. C., *J. Geophys. Res.* **84**, 5941 (1979).
14. Johnsen, R., and Biondi, M. A., *J. Chem. Phys.* **74**, 305 (1981).
15. Bienstock, S., Heil, T. G., and Dalgarno, A., *Phys. Rev. A* **29**, 503 (1980).
16. Dalgarno, A., Butler, S. E., and Heil, T. G., *J. Geophys. Res.* **85**, 6047 (1980).
17. Butler, S. E., Heil, T. G., and Dalgarno, A., *Astrophys. J.* **241**, 442 (1980).
18. Gargaud, M., Bacchus-Montabonel, M. C., and McCarroll, R., *J. Chem. Phys.* **99**, 4495 (1993).
19. Johnson, R., and Biondi, M. A., *Geophys. Res. Lett.* **5**, 847 (1978).

PFX - The Penning Fusion Experiment

T. B. Mitchell, M. H. Holzscheiter, M. M. Schauer,
D. W. Scudder and D. C. Barnes

Los Alamos National Laboratory

Abstract. An experiment is being constructed at Los Alamos to investigate the possibility of using a Penning trap as a fusion confinement device. The goal is to demonstrate electron densities greater than the Brillouin density at the center of the trap by filling it with particles occupying a very restricted region of phase space. The trap under construction has a containment region diameter of 6 mm, standoff lengths of 25 mm, and gaps between electrodes of 1.5 mm. High voltage tests have demonstrated standoff breakdown voltages of >100 kV, and established that titanium electrodes offer better performance than stainless steel or copper ones. The main diagnostic of the degree of density focusing will be transmitted beam deflection measurements. The electron source will be biased to inject a probe beam through the trap center, and this will then be imaged on a phosphor screen outside the trap.

PFX CONCEPT

The PFX trap is designed for the production and study of dense spherical plasmas. The project is based on recent theoretical work[1] on charge-nonneutral plasmas which has suggested how to produce dense, well confined, thermonuclear plasmas in a Penning trap. Access to such operation is possible if plasma density varies significantly in either space or time. In the trap, a large spatial variation will be produced in a spherical system by maintaining a nonthermal velocity distribution of the confined particles. Cold electrons injected along the magnetic field axis should fall into a quadratic and spherical (in the rigid-rotor frame) potential well, and form a beam distribution oscillating through the center.

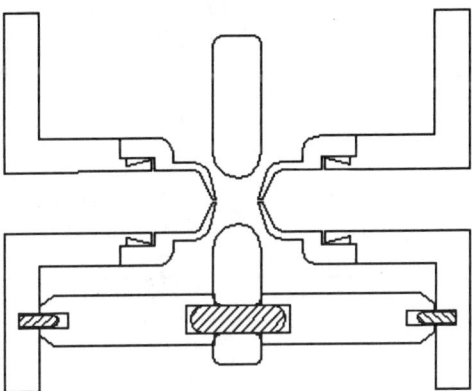

Figure 1. Schematic of the PFX trap electrodes, actual scale.

A schematic of the trap under construction is shown in Figure 1. In order to achieve large electron densities, the trapping potential must be as high as possible. The limit to the trap voltages is set by the danger of surface flashover along the standoffs supporting the structure, and of vacuum breakdown between the narrow (1.5 mm) gaps between the anode and endcaps. A literature search was conducted, and high voltage experiments performed, to assist in the design of PFX.

HIGH VOLTAGE EXPERIMENTS

Initiation of a surface flashover occurs from the emission of electrons from the cathode triple junction (where the insulator, cathode and vacuum are in close proximity). The design of the standoff at this point is thus crucial. Also known to be important is the choice of standoff material, and the preparation of its surface; for a recent review on surface flashover of high voltage standoffs, see Miller[2].

The geometry of the standoffs designed for PFX can be seen in Figure 1; their length is 25 mm. The chamfer was added to relieve the electric field at the cathode, through self-generation of negative charge on the chamfer surface. This feature has been shown to increase the breakdown voltage of Lucite standoffs[3]. A second feature which reduces the field is the use of pins, protruding into the standoff, to join the surfaces. Wesgo Al-300 alumina, polished after grinding, was used as the standoff material since it had been previously found to provide superior performance as well as to benefit from a Cr_2O_3 coating[4].

The breakdown voltage (V_b) of the PFX standoffs has been measured, in a test stand with high vacuum (5×10^{-7} Torr). Two different base materials were used, titanium and 303 stainless steel. Voltages up to 100 kV could be generated, and both positive and negative polarities were used to examine the effect of the chamfer (which would be expected to increase V_b only if it were at the cathode). Table 1 shows the experimental results. The chamfer appears to greatly improve the performance of the standoff, when it is located at the cathode.

TABLE 1. Standoff Performance

Base	Coating	Chamfer	V_b (kV)
Ti	no	anode	-57
Ti	no	cathode	95
Ti	Cr_2O_3	cathode	>100
SS 303	no	cathode	>100
SS 303	no	anode	-58

High voltage tests were done on flat electrodes (area: 400 mm^2) of OFHC copper, titanium, and various types of 300-series stainless steel. The finish and cleanliness of electrode surfaces has been found to affect breakdown[5]; the test electrodes were lathe-turned to give each a similar surface finish, and then ultrasonically cleaned with methanol. The tests were conducted in high vacuum, and the separation between the electrodes could be varied. The different questions which were addressed, and the conclusions reached, are as follows:

Spark Conditioning: The first breakdown of 'virgin' electrodes occurs at low voltages, and it is necessary to continue increasing the voltage to 'spark condition' them[5]. The duration of the spark is limited by having a protection resistor (2 MΩ) between the power supply and electrode gap. One concern about spark conditioning is that excessive conditioning might result in a net reduction in the holdoff capability of the electrodes. Experimentally, it was found this is not a concern with 1.5 mm separations, but that with gaps of less than 1.0 mm a net reduction in holdoff can indeed occur.

Gas Conditioning: This is a very common conditioning method, and consists of introducing gas while the high voltage is on. A quenching of the prebreakdown emission current is then observed. The efficacy of the treatment has, however, been found to be reduced by subsequent high temperatures, and also to be very dependent on the gas species and electric field[6]. Experimentally, we find that nitrogen (introduced at 5 x 10^{-4} Torr) results in effective conditioning for gaps of 1.5 mm, and that titanium and stainless steel condition as well as copper.

However, we also found that the conditioning benefits can be negated by subsequent spark conditioning. Typically, the measured current from gas conditioned electrodes was about a factor of 10 lower than it had been, and the electrodes could be safely brought to somewhat higher ($\sim$ 15%) voltages. If the electrodes were exposed to further spark conditioning, however, their voltage holdoff characteristics tended to return to their pre-conditioned values.

Titanium performance: Titanium is well known to have good high voltage standoff properties. Compared to copper, aluminum and stainless steel, however, there have been few HV experiments on this metal reported in the literature. Experimentally, we find that titanium exhibits superior performance in the 1.0 to 2.5 mm range. Figure 2 shows some measured breakdown voltages versus separation, for electrodes of copper, stainless steel and titanium. The electrodes have been both spark and gas conditioned. Such data has scatter of $\pm$3 kV, but in the full data set titanium always performed as well or better than stainless steels, and was clearly superior to copper. As a consequence of this high voltage performance, plus its non-magnetic properties, titanium was chosen as the material for the PFX electrodes.

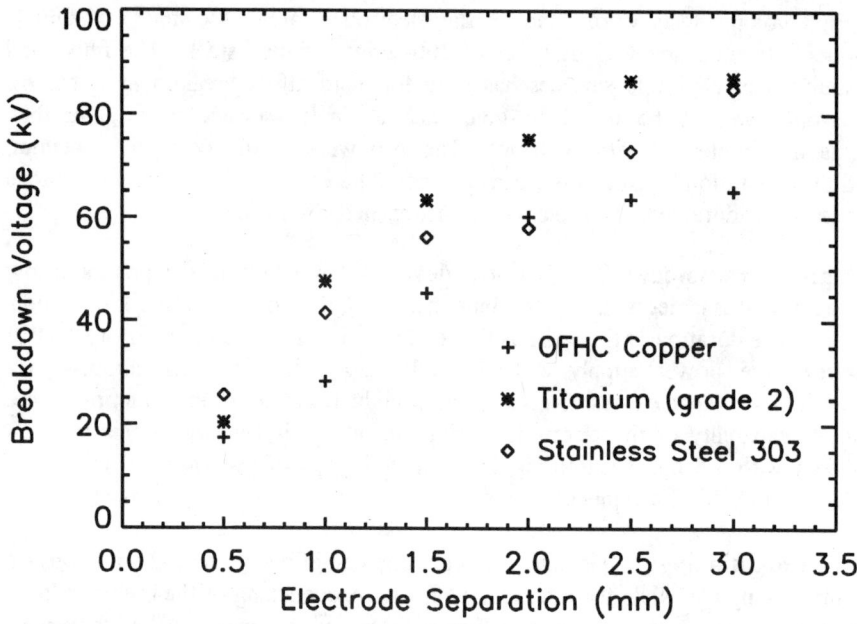

Figure 2. High voltage performance of copper, titanium and stainless steel electrodes after spark and gas conditioning.

DENSITY FOCUS DIAGNOSTIC

The most common diagnostics used in Penning trap experiments are measurements of electrostatic modes and charge collection onto a Faraday cup. However, these can't directly measure the quantity of most interest, the degree of density focusing. We therefore plan to diagnose the density focus by biasing the electron source to inject a probe beam through the trap. The strong electric fields of a focused core will deflect the gyrocenters of the beam's electrons, and increase their gyroradius. The beam will then be imaged on a phosphor screen located outside the trap and opposite the beam source. We have modeled this diagnostic using the beam optics program EGUN, and find that even modest focusing should cause measurable beam deflections.

In Figure 3, we show the predicted probe beam images resulting from cores, consisting of a conserved number of electrons, whose radius is progressively decreased. The core densities which resulted in the six traces shown are 0, 0.2, 1.6, 3.8, 12.8 and 59 (in units of the Brillouin density). The basic effect is that as the core density increases, the width of the probe beam also increases. Here, the phosphor screen is located where the magnetic field is 0.02 its value at the trap.

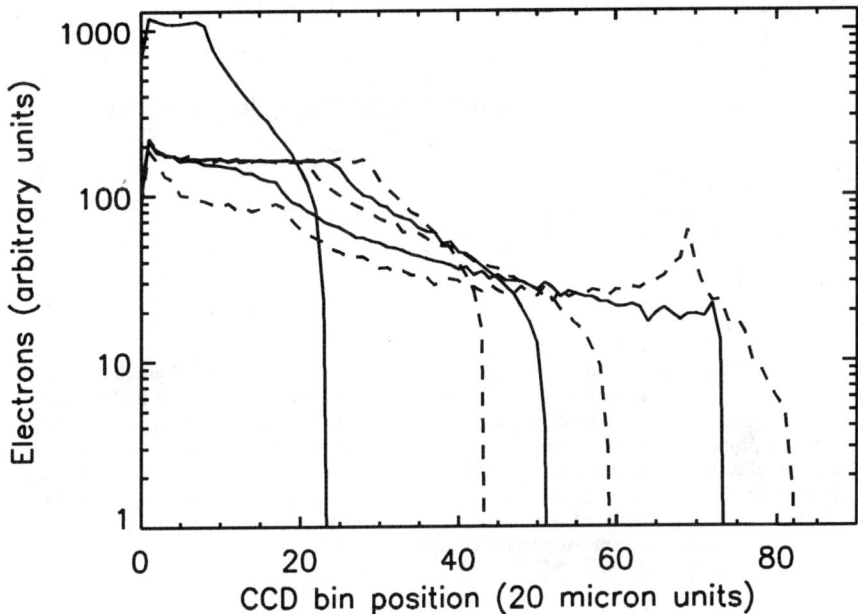

Figure 3. CCD image (from EGUN simulations) predicted to result from a probe beam passing through a density focus core with conserved particle number but progressively larger densities. The core densities, in units of the Brillouin density, are 0, 0.2, 1.6, 3.8, 12.8 and 59.

There are several nice features about this diagnostic. Since the beam deflections scale with Brillouin density, the density focusing effect can be studied while operating the trap at low voltages, if desired. Because we will use a microchannel plate in front of the phosphor screen, the probe beam can be a low density, fast burst of electrons; this allows the diagnostic to be relatively non-perturbing, and gives good time resolution. Finally, because we will be imaging the probe beam with a phosphor screen/CCD camera, we will be able to observe 2D behavior of the dense core, such as might arise from non-azimuthally symmetric modes.

REFERENCES

1. D. C. Barnes, R. A. Nebel, and L. Turner, *Phys. Fluids B* **5**, 3651 (1993).
2. H. C. Miller, *IEEE Trans. Electr. Insul.* **24**, 512-527 (1993).
3. N. C. Jaitly and T. S. Sudarshan, *IEEE Trans. Electr. Insul.* **22**, 801-810 (1987).
4. N. C. Jaitly and T. S. Sudarshan, *IEEE Trans. Electr. Insul.* **23**, 231-242 (1988).
5. R.V. Latham, *High Voltage Vacuum Insulation*, New York: Academic Press, 1981.
6. S. Bajic, A. M. Abbot and R. V. Latham, *IEEE Trans. Electr. Insul.* **24**, 891-897 (1989).

Pressure Measurement using
a Pure Electron Plasma

D. A. Moore, R. C. Davidson, S. M. Kaye and S. F. Paul

Princeton Plasma Physics Laboratory,
P.O. Box 451, Princeton, NJ 08543

Abstract. The Electron Diffusion Gauge (EDG) is a Malmberg trap configuration used to investigate the application of pure electron plasmas to the measurement of background gas pressure. To form a useful gauge, the rates of different types of transport in EDG must be properly ordered and understood. The dependence of the asymmetry transport rate on plasma parameters has been determined experimentally. The relaxation of the plasma profile to a meta-equilibrium state during asymmetry transport has also been observed. A local model of the asymmetry flux consistent with the observations is presented. Implications for pressure measurements are summarized.

INTRODUCTION

A pure electron plasma can theoretically be confined indefinitely (1), during which time it relaxes to a thermal equilibrium state described by the rigid-rotor distribution (2). An imperfect vacuum, imperfect confining fields, and finite wall resistivity limit the confinement in actual experiments. Previous experiments have shown that plasma density profiles relax toward quasi-equilibrium profiles on a time scale much faster than that predicted by simple collisional models (3,4). Other measurements on pure electron plasmas have demonstrated approximately linear diffusion rates over five orders-of-magnitude of the background gas pressure (5).

The Electron Diffusion Gauge (EDG) experiment was designed to test the feasibility of using a pure electron plasma as a pressure-sensing medium, and to determine if a primary pressure standard could be developed. In particular, a primary pressure standard could be achieved if the plasma remained near rigid-rotor thermal equilibrium while expanding due to elastic collisions with the background gas, because the expansion rate could be predicted theoretically.

The possibility of indefinite confinement of pure electron plasmas is a result of conservation of angular momentum. The dominant component of the canonical angular momentum for the low-density plasmas in the EDG is

$$L_z(t) = \frac{eBZ}{2c} \int_0^R 2\pi r n(r,t) \left(R^2 - r^2\right) dr \tag{1}$$

where $-e$ is the charge on the electron, B is the axial magnetic field strength, Z is the plasma length, c is the speed of light *in vacuo*, R is the wall radius, and $n(r,t)$ is the plasma density profile, which is assumed to be cylindrically symmetric. For constant angular momentum, the mean squared radius of the plasma,

$$\langle r^2 \rangle \equiv \frac{1}{N} \int_0^R 2\pi r^3 n(r,t) dr \tag{2}$$

where N is the total number of electrons, remains constant. Only effects which change the total angular momentum of the plasma cause any net expansion.

The range of operation of a primary pressure standard can be ,determined by comparing transport rates due to three mechanisms. First, electron-electron collisions and collective effects act to produce a rigid-rotor thermal equilibrium state. Second, electron collisions with the background gas change the plasma angular momentum, causing expansion and temperature changes. The upper pressure limit occurs when this transport is so large that quasi-equilibrium cannot be maintained by the electron-electron transport. The third mechanism is anomalous transport, which also changes the plasma angular momentum. The lower pressure limit occurs when the transport due to the background gas cannot be distinguished from noise in the rate of anomalous transport.

Experiments have been performed to determine both the upper and lower pressure limits. Reported here are the results of experiments with the expansion dominated by anomalous transport, and their implications for the lower pressure limit of a primary vacuum standard based on EDG.

RELAXATION OF PROFILES

Plasmas dominated by anomalous transport relax to characteristic radial density profiles during their expansion. This effect has been applied previously to the production of uniform plasma "initial conditions," using pulsed magnetic field tilts to increase the rate of anomalous transport (6). To understand the characteristic profiles, which we have termed Meta-Equilibrium States from Asymmetries

(MESA's), it is necessary to separate relaxation from expansion. This can be done by defining coordinates ρ and η that expand with the plasma, as measured in terms of the $\langle r^2 \rangle$ moment, *i.e.*,

$$r_0 = \sqrt{\langle r^2 \rangle}$$
$$\rho \equiv \frac{r}{r_0} \tag{3}$$
$$\eta(\rho,t) \equiv r_0^2\, n(\rho r_0,t).$$

The profile $\eta(\rho,t)$ has constant zeroth and second radial moments, similar to a non-expanding plasma.

Figure 1 shows a typical relaxation. Averaging over multiple shots has been used because random diocotron modes present in the initial condition cause shot-to-shot variation. In the first 0.1 sec, the diocotron modes damp to low levels, and the characteristic outer edge of the plasma forms. The scaled profile remains approximately constant from 0.2 sec to 1.3 sec. Wall effects alter the profile after 1.3 sec, with 10% of the total trapped charge being lost by 2.67 sec.

The MESA profiles can be fit very accurately to rigid-rotor equilibria, as long as the plasma temperature is treated as a free parameter. The temperatures derived from the fits to the MESA profiles are consistently lower than the directly measured plasma temperatures. For the data shown in Fig. 1, the fit gives a temperature of 0.5 eV, and direct measurements give ~2 eV.

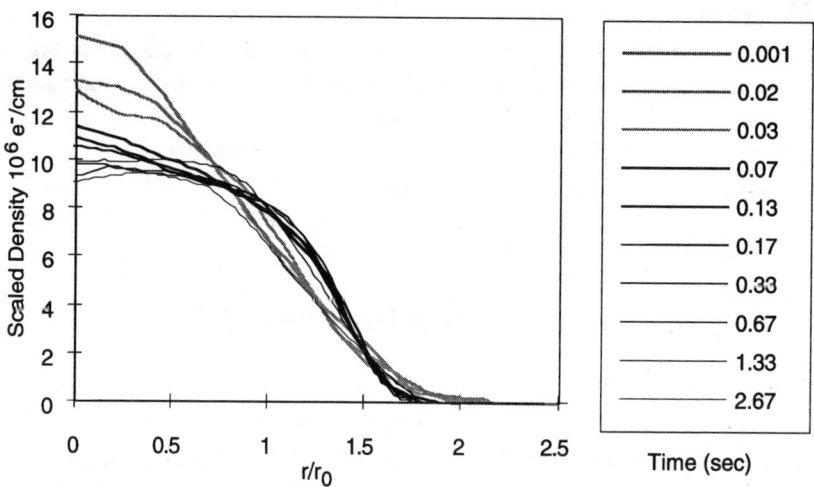

FIGURE 1. Relaxation of average density profiles caused by anomalous transport.

RADIAL EXPANSION

The expansion of plasmas due to anomalous transport has been studied by observing the rate of change of canonical angular momentum. Figure 2 shows the time evolution of the angular momentum derived from the profiles shown in Fig. 1. The rate of change of angular momentum is faster while diocotron modes are present, but approaches a constant value as the relaxation proceeds. This constant rate, occurring while the plasma has a MESA profile, is unaffected by changes in plasma density and temperature caused by the expansion, and even remains approximately constant as the first 10% of the plasma is lost to the wall.

The linear dependence of angular momentum on time indicates two important features of the rate of anomalous transport. First, since the temperature changes during the evolution but the expansion rate does not, the anomalous transport rate must be independent of plasma temperature. Second, the anomalous transport rate is constant while N and the plasma scaled shape are constant, regardless of the plasma size r_0.

For the data shown in Figs. 1 and 2, the plasma has had minimal contact with the wall so that any related effects can be neglected. As the plasma continues to expand, wall interaction increases, and the total charge in the trap will begin to decay along with the angular momentum. During the decay, the total remaining charge is approximately proportional to $1/t$. This scaling has been reported previously in experiments dominated by neutral gas transport (7). This is the behavior expected theoretically when the mobility term is dominant in the neutral gas

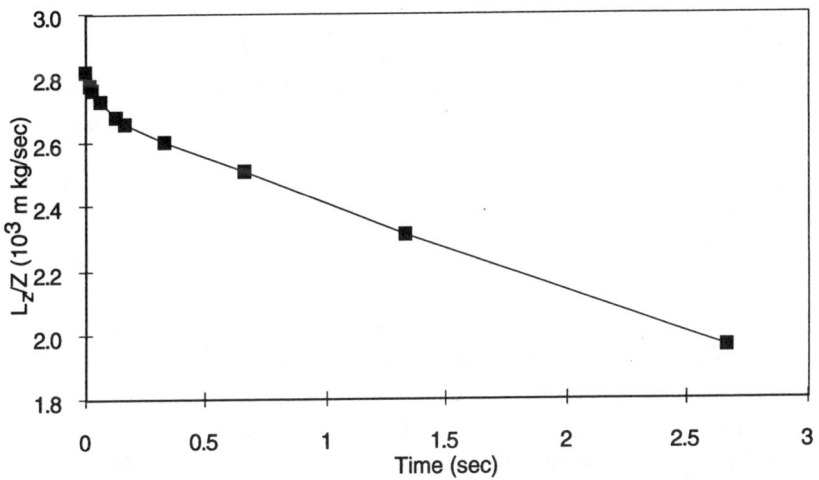

FIGURE 2. Decay of angular momentum caused by anomalous transport.

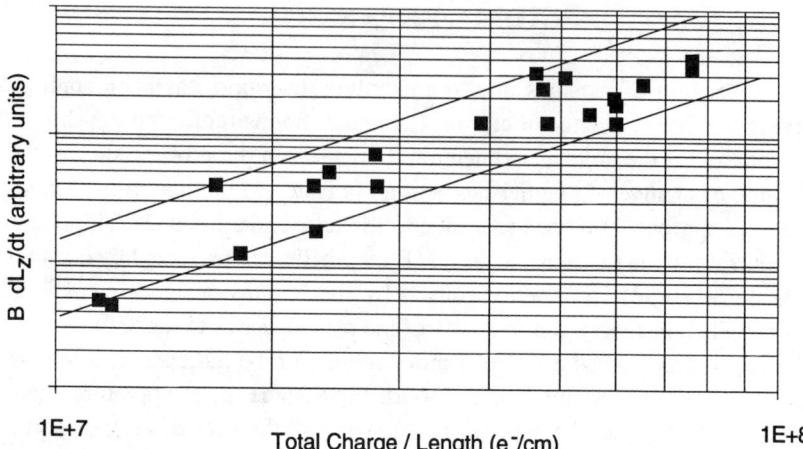

FIGURE 3. Dependence of anomalous transport rate on total charge. N^2 indicated.

transport and the plasma temperature remains approximately constant.

For anomalous transport, $N(t) \propto 1/t$ means that

$$\frac{1}{Z}\frac{dL_z}{dt} \propto \left(\frac{N}{Z}\right)^2 T^\varepsilon \qquad (4)$$

where T^ε is a weak temperature dependence. The upper limit for ε depends on the accuracy of the $N(t)$ measurements and the amount of variation in T that occurs. Equation (4) can be checked by comparison to the early expansion, before wall losses, of plasmas with different amounts of total charge. That data is shown in Fig. 3, The previously known dependence on axial magnetic field strength has also been incorporated (8), so that data with B in the range 50–990 G could be plotted. The data shows a factor of 5 spread if the N^2 scaling and limits indicated on the graph are correct. Each data point comes from the slope of a large linear region like that seen in Fig. 2.

IMPLICATIONS FOR PRESSURE MEASUREMENT

The empirical scaling relations for anomalous transport have been determined. Both in terms of plasma density and magnetic field, the scaling for pressure transport and anomalous transport are the same, and no direct benefit is gained by either raising or lowering either parameter. Since anomalous transport is only weakly dependent on temperature, if at all dependent, using slightly higher temperature plasmas allows lower pressures to be measured. As previously determined, shorter plasmas have lower rates of anomalous transport, and are therefore better suited to pressure measurement.

It is reasonable to expect that a trap with anomalous transport equivalent to a pressure of 10^{-8} Pa could be constructed. The lower limit of pressure measurement would depend on the accuracy desired, the elimination of residual gas, and the ability to regulate extremely low gas pressures. The upper limit of pressure measurement would depend on the rate of electron-electron collisional transport, and the characteristics of that transport have not yet been determined.

CONCLUSIONS

Anomalous transport in pure electron plasmas leads to the formation of MESA's and expansion of the electron cloud. The expansion rate, as described by the rate of change of canonical angular momentum, is proportional to N^2, but is only weakly dependent on temperature. The relative rates of anomalous transport and background gas driven transport cannot be changed by adjusting either density or magnetic field. Therefore, the expected minimum equivalent pressure is 10^{-8} Pa. The maximum pressure must still be determined.

ACKNOWLEDGMENTS

This work has been supported by the Office of Naval Research, the National Institute of Standards and Technology, and the SMC Corporation.

REFERENCES

1. T. M. O'Neil, *Phys. Fluids* **23**, 2216–2218 (1980).

2. R. C. Davidson and N. A. Krall, *Phys. Rev. Lett.* **22**, 833 (1969).

3. D. H. E. Dubin, T. M. O'Neil and C. F. Driscoll, "Transport Toward Thermal Equilibrium in a Pure Electron Plasma," in *Proc. of Workshop of US-Japan Joint Institute for Fusion Theory Program*, Nagoya University, 265-279 (1986).

4. C. F. Driscoll, J. H. Malmberg and K. S. Fine, *Phys. Rev. Lett.* **60**, 1290 (1988).

5. J. H. Malmberg, *et. al.*, "Experiments with Pure Electron Plasmas," in *Non-neutral Plasma Physics* (C. W. Roberson and C. F. Driscoll, editors), New York: American Institute of Physics, 28 (1988).

6. K. S. Fine, Ph.D. thesis, University of California, San Diego, 1988 (unpublished).

7. J. S. deGrassie and J. H. Malmberg, Phys. Fluids 23, 63–81 (1980).

8. C. F. Driscoll and J. H. Malmberg, *Phys. Rev. Lett.* **50**, 167–170 (1988).

COMPUTER SIMULATION OF THE DIOCOTRON INSTABILITY

S. C. Neu and G. J. Morales
Physics Department, University of California, Los Angeles,
Los Angeles, CA 90024

OVERVIEW

A particle simulation study demonstrates that an initial-value analysis is essential in understanding the early-time growth of diocotron modes. It is also demonstrated that the equilibrium properties of a rigid rotor (including rotation on both the high and the low frequency branches) are well described by particle simulation.

NONNEUTRAL PARTICLE SIMULATION CODE

An electrostatic, particle-in-cell code[1,2] is used to model the behavior of a strongly magnetized, nonneutral system in the plane (x,y) perpendicular to the confining magnetic field (along z). The system is bounded by two planar walls (located at $x = \pm a$) on which the value of the potential is zero (ground). The code is 2-1/2 dimensional, i.e., it follows 3 velocity components (v_x, v_y, v_z) and two spatial coordinates (x,y) on a time scale smaller than a gyroperiod. Time is scaled to the spatially averaged $< \omega_{pe}^2 >^{-1/2}$. The y coordinate is periodic (with length L) and the z-variation is ignorable. In this study we consider two different initial configurations: a nonneutral plasma slab parallel to the walls, and a small rotating cylinder far from the walls.

EVOLUTION OF A NONNEUTRAL PLASMA SLAB

A pure electron plasma slab having uniform density and width $2r_p$ is centered between the walls at $t = 0$. The slab is initialized by giving each electron an initial velocity $v_y(t=0) = \omega_{pe}^2 x/\Omega_e$, corresponding to the expected $\underline{E} \times \underline{B}$ drift. A small random velocity is added to yield a low temperature Maxwellian in the drift frame. The system is divided into 32 grids along x and 1024 grids along y. More than 32,000 particles are typically followed, and the system is well below the Brillouin limit; i.e., $\Omega_e/\omega_{pe} = 6$.

Figure 1a shows the configuration space during the early stage of the diocotron instability and Figure 1b exhibits the formation of vortices at a later time. The time evolution of each Fourier mode (having $k_n = 2\pi n/L$) is monitored using the potential along $x = 0$. It is found that there is an inherent mixture of transient and beat behavior which can be confused with an instability or a nonlinear mode-coupling process. This phenomenon is illustrated in Fig. 2, where the solid curves correspond to the simulation results. Figure 2a shows that mode

20 exhibits a monotonic exponential growth, while Fig. 2b shows that mode 26 also grows initially but is followed by an abrupt drop ($t > 15$) in amplitude. The initial growth in Fig. 2b arises both because of transients and because of the beat between two surface modes (one on the top interface and the other on the bottom) moving past each other. A proper description of the phenomenon requires an initial-value analysis too lengthy to report here. The method uses a Laplace transform to generate a differential equation in x which is solved via a Green's function using the method of Case[3]. The inverse transform is obtained by evaluating the residues, but one remaining integration over x must be done numerically. The result of this analysis is indicated by the dashed lines in Fig. 2. It is evident that mode 20 corresponds to a true instability, while the unusual evolution of mode 26 is a transient feature not contained in the steady-state stability analysis.

The growth rates of the unstable modes observed in the simulation are shown in Fig. 3. These growth rates are measured by recording the potential energy of each simulation mode over a finite time interval. The solid curve is the well-known[4] expression for the growth rate γ of an unstable mode (of wave number k) of a slab of width $2r_p$ in the limit $a \to \infty$:

$$\frac{\gamma}{\omega_{pe}} = \frac{\omega_{pe}}{2\,\Omega_e}\sqrt{e^{-4kr_p} - (1 - 2kr_p)^2}. \tag{1}$$

Excellent agreement between the asymptotic theory and the simulation results is obtained over a wide range. The possible discrepancies associated with the high-k modes (18-21) are likely to be due to the sampling of the growth rates at early times. The initial-value theory shows that early in time these growth rates are larger than their asymptotic values. We have also simulated systems with different ratios of r_p/a and found excellent agreement with the theoretical predictions. As $r_p/a \to 1$, the system remains stable, as expected.

RIGID ROTOR EQUILIBRIUM

We have investigated the behavior of a small rigid rotor of radius r_0 bounded by planar walls ($r_0/a \simeq 0.16$) as sketched in Fig. 4. The magnetic field strength B_0 has been varied over the range $1.5 \leq \Omega_e/\omega_{pe} \leq 3$ in order to measure the equilibrium rotation frequencies and the radial oscillation frequencies. For each choice of B_0 the system is initialized by assigning to each particle an azimuthal velocity $r\omega_{re}$, to which a small random component is added to represent a small thermal spread, and where

$$\frac{\omega_{re}}{\Omega_e} = \frac{1}{2} \pm \frac{1}{2}\sqrt{1 - \frac{2\omega_{pe}^2}{\Omega_e^2}}, \tag{2}$$

corresponds to the two possible rotation frequencies for a rigid rotor in vacuum. The average radius and average rotation frequency of the cylinder are monitored as a function of time. It is found that the rotor exhibits long-term stable oscillations in both the mean radius and the mean rotation frequency. The top of Fig. 4 shows that the simulated system follows the classical rotation diagram, while the bottom shows that the frequencies of the radial oscillations follow the solid curve predicted by the magnetron frequency

$$\frac{\omega_{mag}}{\Omega_e} = \sqrt{1 - \frac{\omega_{pe}^2}{\Omega_e^2}}. \tag{3}$$

ACKNOWLEDGMENTS

This work is sponsored by the Office of Naval Research. Drs. V. K. Decyk and H. Ramachandran played a key role in the development of the computer code.

REFERENCES

1. H. Ramachandran, G. J. Morales, and V. K. Decyk, Phys. Fluids B **5**, 2733 (1993).

2. C. K. Birdsall and A. B. Langdon, *Plasma Physics via Computer Simulation* (McGraw-Hill, New York, 1985).

3. K. M. Case, Phys. Fluids **3**, 143 (1960).

4. For example, O. Buneman, R. H. Levy, and L. M. Linson, J. Appl. Phys. **37**, 3203 (1966).

5. R. C. Davidson, *Physics of Nonneutral Plasmas* (Addison-Wesley, Redwood City, CA, 1990), p. 41.

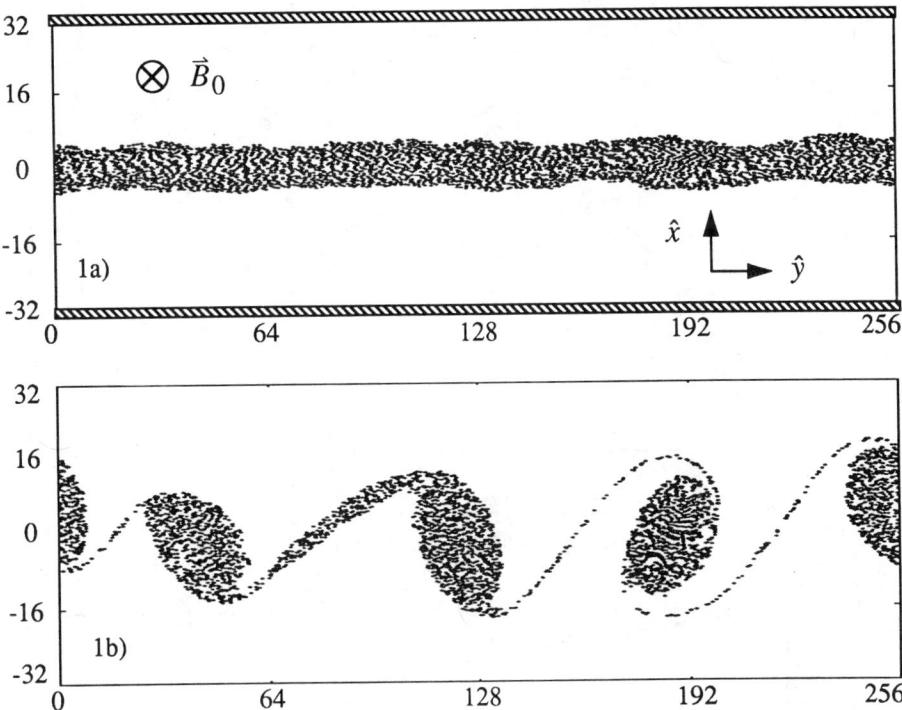

Fig 1: Nonneutral plasma slab simulation at 1a) time $t = 30$, and 1b) time $t = 70$. Spatial coordinates are scaled to Debye lengths, and $r_p/a = 0.15$.

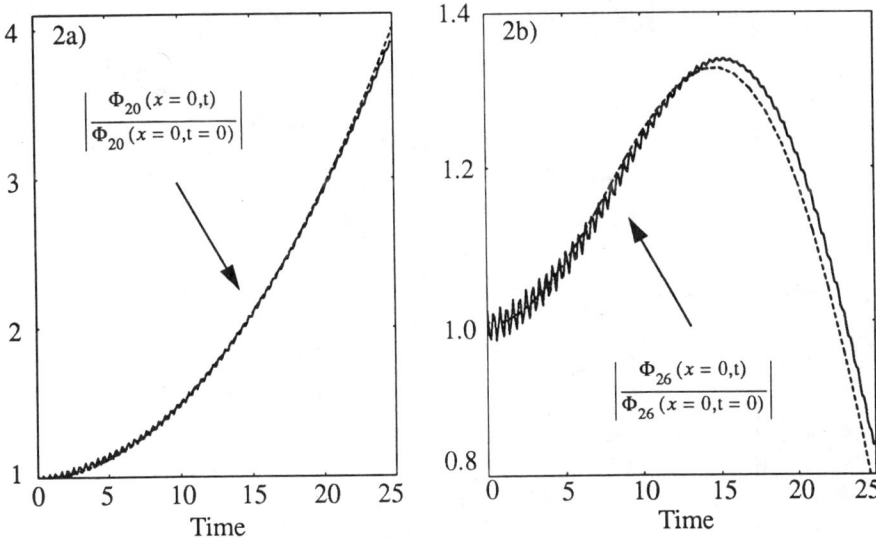

Fig 2: Initial-value theory (dashed line) and simulation data (solid line) along $x = 0$ for 2a) mode 20, and 2b) mode 26.

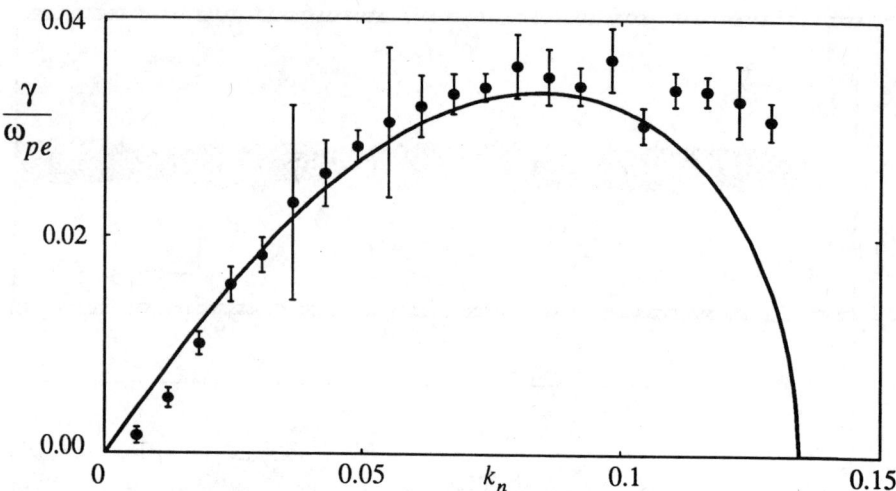

Fig. 3: Measured growth rates of the unstable modes of
the nonneutral plasma slab in Figure 1.

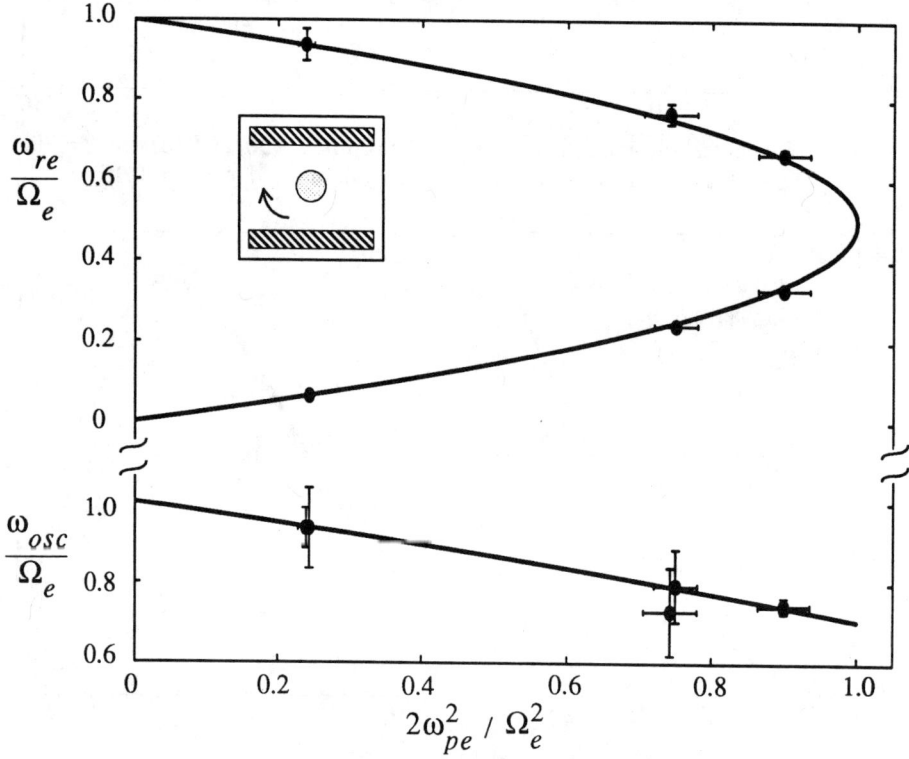

Fig 4: Measured equilibrium rotation frequencies and radial
oscillation frequencies of the nonneutral rigid rotor.

Coupled gyrators: a paradigm for space-charge effects in cyclotron mass spectrometry

A.J. Peurrung, S.E. Barlow, and R.T. Kouzes

Pacific Northwest Laboratory, Richland, Washington 99352

Abstract. Space-charge effects are a known limitation to the accuracy and precision of an entire family of mass spectrometers. We propose a new model for the study of space-charge effects in which charged ion clouds each containing ions of a particular species interact as coupled gyrators. This model predicts a range of phenomena with implications for mass spectrometry including resonant energy exchange, synchronization, modulation, and chaotic motion.

INTRODUCTION

Ion cyclotron resonance (ICR) mass spectrometers have become a common tool for atomic and molecular mass measurements. Mass accuracies in excess of one part in 10^8 have been reported over a wide range of masses [1,2]. Fourier transform ICR mass spectrometry in particular is able to measure the charge-to-mass ratios of many different ions simultaneously, allowing the rapid determination of molecular structure [3]. These devices work by first storing an ion sample in a Penning ion trap; that is, trapping is accomplished through the combined action of static electric and magnetic electric fields [4]. A mass measurement begins when a radio-frequency electric excitation pulse causes the ions to gyrate around the magnetic field lines at approximately the cyclotron frequency. If the trap's magnetic field strength is known, an ion's mass depends only on this gyration frequency, which is accurately measured by long-term detection of the motion of image charges induced in the trap walls.

When two or more species are simultaneously weighed, space-charge effects arise from the Coulombic force between gyrating ion clouds, each of which contains ions of a particular mass. These effects limit the dynamic range of a mass spectrometer, since there is a point beyond which the addition of more ions to the trap is accompanied by a deterioration in the mass measurement resolution, accuracy, and precision. Ion motion during and after the excitation pulse is very complicated, and previous analyses have neglected the detailed dynamics of interacting charge clouds [5–7]. An improved understanding of the causes and consequences of space-charge effects would greatly aid further developments in the field of cyclotron mass spectrometry.

We propose a new model for the study of space-charge effects in cyclotron mass spectrometry and numerically investigate its predictions. This model holds that the dynamics of Coulombically interacting ion clouds are like those of a system of coupled gyrators, and that the detailed form for the interaction is relatively unimportant. Using this model, we predict a range of phenomena that are experimentally manifested as space-charge effects. As in any system of coupled oscillators, a substantial exchange of energy between ion clouds can occur when the gyration frequencies are approximately commensurate (a common case in mass spectrometry). If the coupling is sufficiently strong, separate gyrators may synchronize, causing failure of the mass spectrometer to resolve different masses. When three or more gyrators are mutually commensurate, ion cloud motion becomes chaotic, losing all periodicity and exhibiting extreme sensitivity to initial conditions. We provide a scaling law that allows prediction of the severity of space-charge effects and test this law for a simulated xenon mass measurement.

Historically, interesting dynamical systems have arisen from consideration of problems in applied physics. A notable example is the Henon and Heiles problem [8,9] used to model the gravitational three-body problem. The dynamics described in this paper may bear some similarity to stellar dynamics in galaxies [10], planetary ring dynamics [11], and planetary dynamics in solar systems [12]. In all of these systems, the interaction of oscillators with nearly commensurate frequencies has a substantial long-term dynamical effect.

Figure 1 shows the geometry of a typical cyclotron resonance mass spectrometer. The axial magnetic field $B\hat{z}$ and the bias voltage V_t applied to the trap's end plates provide effective ion trapping. Collisions with neutral gas molecules are routinely used to cool the ion population, allowing the trap to operate at very low electric field strengths while still confining ions near the trap center. After the excitation pulse, ions with the same mass and charge generally move together as a single ion cloud. Throughout this paper we denote the gyrofrequency and the gyroradius of the various ion clouds by Ω_A, r_A, Ω_D, r_D, etc.

A central point of this paper is that the essential features of ion cloud dynamics do not depend strongly on the detailed form of the interaction. A variety of assumed forms are possible for the Coulombic coupling between ion clouds. The long range force between ion clouds can be computed by modeling the ion clouds shape as a sphere or an infinitely long cylinder [5]. One may also choose to include the effect of image charges in the trap walls thereby improving overall accuracy at the expense of simplicity [13]. Finally, there are a number of ways to eliminate the infinity that occurs for two point-like ion clouds whose positions momentarily overlap [5,7]. Unless otherwise stated, our numerical computations do not account for image charges and assume that ion clouds are point-like objects. We remove the singularity in short-range interactions by assuming that whenever two ion clouds with cloud radius ρ_c are separated by a distance $s < \rho_c$, their overlap causes a reduction in the Coulombic force by a factor $(s/\rho_c)^2$. Our assumed interaction is most appropriate for compact, spherical ion clouds that never closely approach the trap's outer walls.

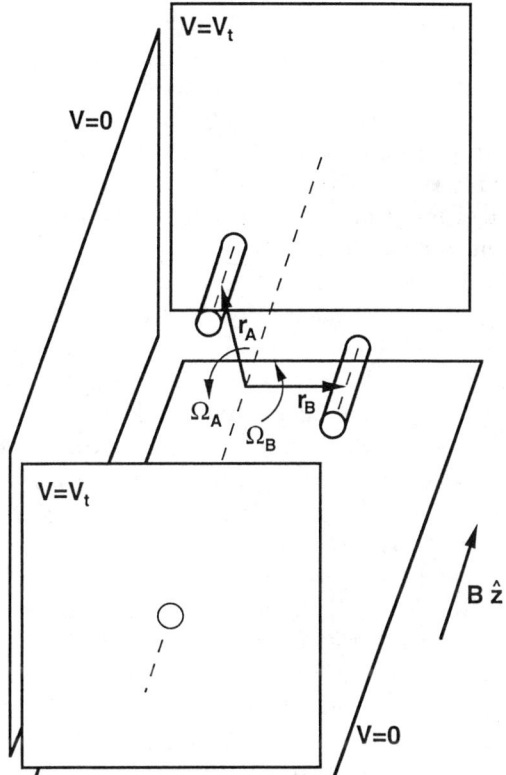

FIG. 1. Typical cyclotron mass spectrometer cell geometry. Two of the side-plates are not shown.

Unlike ordinary oscillators, each gyrator has two degrees of freedom. Since we choose here to ignore the effect of the trap walls, a system of coupled gyrators exhibits both translational and rotational invariance. Three easily recognizable constants of the motion are the total energy and the two components of the canonical angular momentum. Three or more coupled gyrators therefore have three or more degrees of freedom, leading us to expect chaotic motion for this case [9].

DYNAMICAL BEHAVIORS

Even the dynamical system consisting of only two coupled gyrators exhibits a rich variety of behavior. When the coupling is sufficiently strong and the gyration frequencies are sufficiently close, it is possible for the two gyrators to effectively synchronize (mode lock). In this case a mass spectrometer would observe the coalescence of the spectral peaks for the masses in a doublet. The remaining interacting systems are called commensurate if the gyration frequency ratio is approximately an integer fraction and called incommensurate otherwise. The dynamical analysis of incommensurate gyrators can be greatly simplified as a result of phase averaging; the positions where ion clouds "collide" are uniformly distributed with respect to the phase of either gyrator. The behavior of gyrators with commensurate orbits is much more complicated and is the subject of much of the remainder of this letter. When only two gyrators are present, a new, slower time scale appears on which the gyrators periodically exchange some amount of energy and angular momentum. Three or more gyrators often behave chaotically, and the system may require a statistical treatment when $N \to \infty$. All of the above behaviors are important for mass spectrometry where they may cause peak shifting, peak broadening, or even the appearance of artificial, modulation peaks [7].

Previous work has shown that a single gyrator may phase lock to the phase of a nearly resonant electrostatic wave [14]. We here derive a rough criterion for the synchronization of two similar ($r_A \approx r_B$, $\Omega_A \approx \Omega_B$) gyrators via their mutual interaction. The relative speed between two ion clouds is $v_{rel} \approx r_g \Delta\Omega$, where r_g is approximately the gyration radius for both gyrators ($r_g \approx r_A, r_B$) and $\Delta\Omega = |\Omega_A - \Omega_B|$ is the gyration frequency difference. Since synchronization only occurs when $\Delta\Omega \ll \Omega$, two such ion clouds interact sufficiently slowly that gyration-averaged $\mathbf{E} \times \mathbf{B}$ drift dynamics must occur in a frame of reference rotating at the mean gyration frequency. In such a frame, each ion cloud acquires a drift velocity $v_D = Q/4\pi\epsilon_0 s^2 B$, where Q is the total charge in the other cloud and s is the separation distance. If each ion cloud contains roughly N_i ions of mass approximately m, synchronization should occur when $v_{rel} < v_D$, or

$$\frac{\Delta\Omega}{\Omega} < \frac{mN_i}{4\pi\epsilon_0 r_g s^2 B^2}.$$

(1)

Two ion clouds containing 10^5 singly ionized $m \approx 100\,\mathrm{u}$ ions each should synchronize when $\Delta\Omega/\Omega < 4 \times 10^{-4}$. Although the above analysis is only a rough estimate, it serves a valuable purpose; the frequency shifting and modulation that occur for nonsynchronized ion clouds should become increasingly severe as the gyration frequency difference approaches the limit given in Eq.(1).

Incommensurate gyrators generally experience small but important frequency shifts possibly accompanied by a slow frequency modulation. Two weakly coupled gyrators with the same gyration radius and guiding center position will on average exert a mutually outward force on each other, thereby lowering both gyration frequencies. However, the magnitude and even the sign of a particular gyrator's frequency shift depends on its position relative to all other gyrators. A slow modulation of the gyrator frequency occurs when the time for a "collision" between two ion clouds is short compared to the period of either gyrator, i.e., when

$$\frac{s_{min}}{r_g \Delta \Omega} \gg \frac{2\pi}{\Omega}, \tag{2}$$

where s_{min} is the distance of closest approach between ion clouds. When modulation occurs, false sideband peaks spaced at intervals of the modulation frequency ($\Delta \Omega$) appear in the mass spectrum. These peaks often appear very close to real peaks representing different isotopes or molecules with differing protonation [7].

The strong, short-range interactions between commensurate gyrators occur at phases that, to lowest order in the motion, do not vary with time. Repeated "collisions" between gyrators can therefore have a relatively large effect. Commensurate gyrators are the standard case rather than the exception in mass spectrometry because molecular weights often differ by nearly an integer multiple of 1 u. In addition, differing ionic charge states have precisely commensurate gyration frequencies due to the constancy of the electron's charge. The dimensionless number

$$\Lambda = \frac{4\pi \epsilon_0 r_g^3 B^2}{N_i m} \tag{3}$$

quantifies the potential severity of energy and angular momentum exchange between coupled gyrators. One interpretation of Λ is that it represents the angle through which the typical coupling force would need to act in order to accelerate the ion cloud from rest to its gyration velocity. Equation (3) allows us to predict, for example, that if the magnetic field strength were doubled and the number of ions in an ion cloud quadrupled, the general behavior of the system would remain the same. All of our numerical computations have lent support to Eq. (3).

NUMERICAL RESULTS

Figure 2a shows the maximum change in the gyration radius of one of two coupled gyrators as a function of the ratio of the gyration frequencies. This

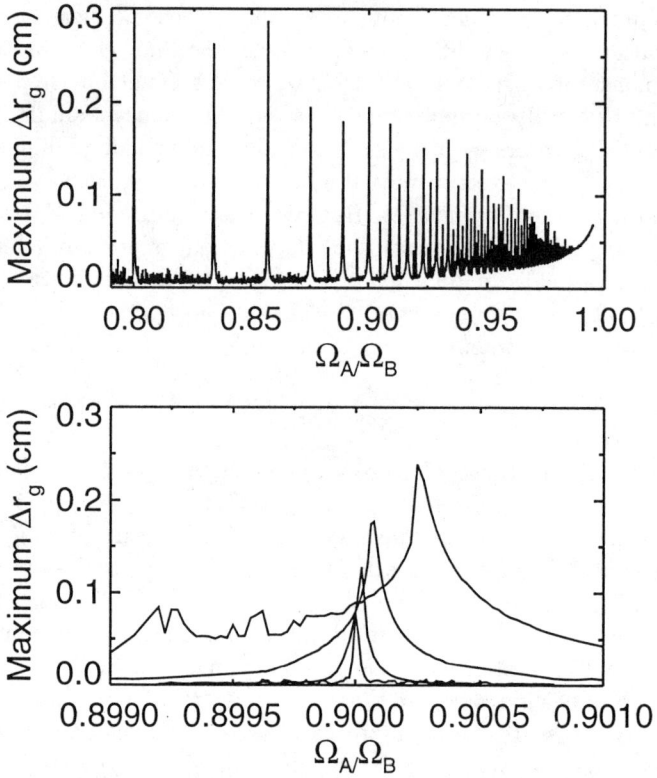

FIG. 2. Maximum change in the gyroradius Δr_g of one of two interacting ion clouds vs. gyration frequency ratio Ω_A/Ω_B. The mass of the lighter ion is varied while the mass of the heavier ion is fixed at $m = 10^4$ u. The ion clouds start 180 degrees apart with $r_g = 1.0$ cm, $B = 3$ T, and charge $Z = 6e$. The ion clouds contain 3000 ions each in 2a. For 2b, the two ion clouds each contain 94, 375, 1500, and 6000 ions for successively increasing peak heights.

figure is obtained by numerical integration of the equations of motion for two coupled gyrators. The two gyrating ion clouds are here assumed to start 180 degrees apart in phase and with the same gyration radius. The largest energy exchanges occur for commensurate frequency ratios of the form $(i-1)/i$, because in these cases the closest approaches between gyrators always occur at the same phase. There is, of course, a symmetry between gyrators A and B so that a plot similar to Fig. 2a, but with $\Omega_B/\Omega_A < 1.0$ would appear identical. The exact form assumed for ion cloud coupling affects only some details in Fig. 2a, such as the presence of energy exchange for a small number of other frequency ratios $(i-2)/i$. The amount of energy exchange near a commensurate frequency ratio depends on the strength of the interaction.

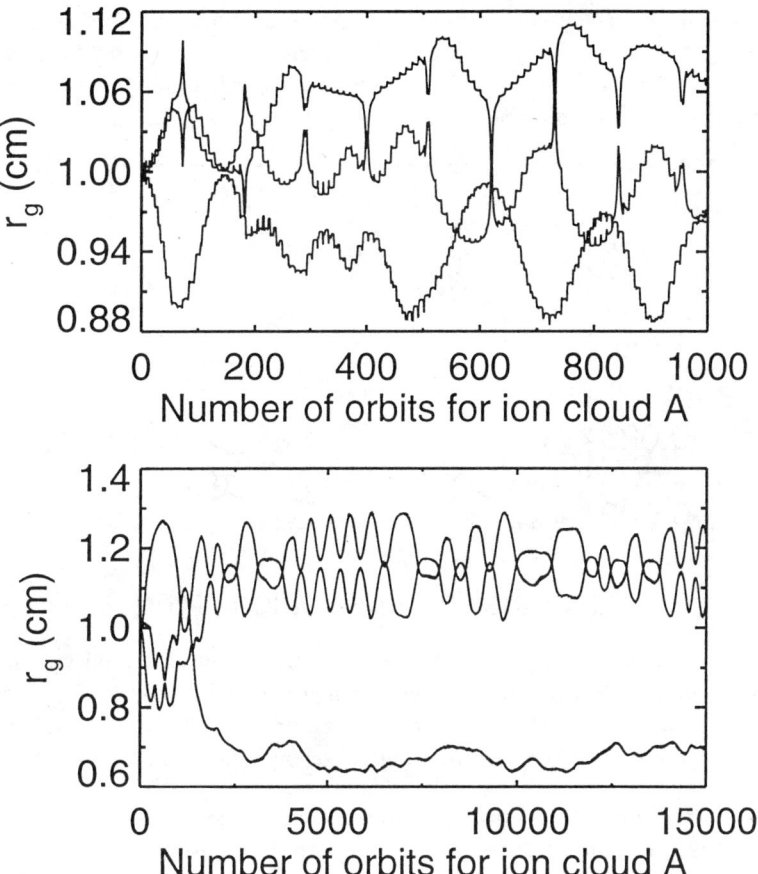

FIG. 3. Gyration radii of three commensurate ion clouds vs. number of gyration orbits for ion cloud A $(\Omega_A t/2\pi)$. Figure 3a has $N_i = 3000$ ions of mass $m = 9000, 9900, 10000$ u. Figure 3b has $m = 8183, 9000, 10000$ u and $N_i = 2000$. The ions have charge $Z = 6e$, the trap has $B = 3$ T, and the ion clouds start 120 degrees apart at $r_g = 1.0$ cm

Figure 2b shows a section of Fig. 2a expanded about the frequency ratio $\Omega_A/\Omega_B = 0.9$ for four different interaction strengths. Notice that for stronger interactions, the "resonance" at this commensurate frequency ratio is taller, broader, and has been shifted slightly.

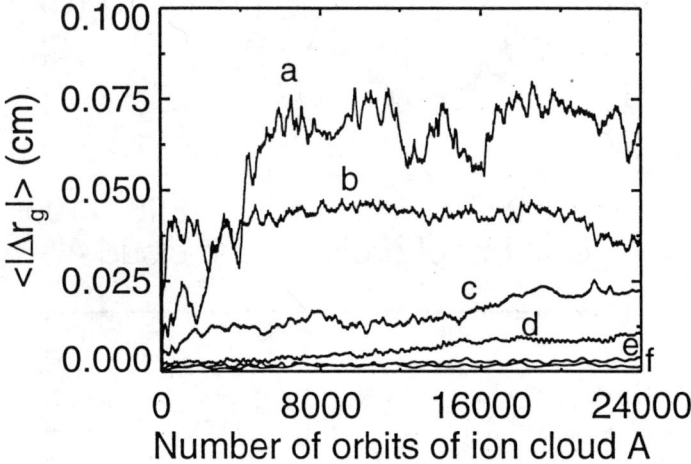

FIG. 4. The average magnitude of the change in the gyration radii vs. time for six isotopes of xenon. The total number of xenon ions is $(128, 64, 32, 16, 8, 4) \times 10^4$ for curves a) through f). The ion clouds start 60 degrees apart at $r_g = 1.0$ cm and have the mass and abundance appropriate for xenon. The trap has $B = 3$ T.

The interaction of three or more mutually commensurate gyrators typically causes chaotic motion of the individual gyrators. Figure 3 shows two examples of the types of motion that are commonly found in these systems. The frequency ratios for Figs. 3a and 3b are 90:99:100 and 90:99:110, respectively. The values of Λ are approximately 2×10^4 and 3×10^4, respectively. Note that in Fig. 3b the gyroradius of one of the ion clouds is substantially reduced for an extended period of time. Although the detailed dynamics shown in Fig. 3 are highly sensitive to the initial conditions, the amount of energy exchanged between gyrators is relatively constant as long as Λ remains unchanged. The average magnitude of the change in the gyration radii, $\langle |\Delta_{r\,g}| \rangle$, provides a useful measure of the spread in a system of ion clouds that start with the same gyration radii. Figure 4 shows the computed time-dependence of this quantity for the six most abundant isotopes of xenon. The total number of trapped xenon ions is varied so that Λ varies from 1.4×10^4 to 4.5×10^5 by factors of two for the six data sets. Since only one particular set of initial conditions is used to generate each curve, the specific details of the data are of little importance. We conclude, however, that the amount of energy (and angular momentum) exchanged between a system of gyrators depends primarily on the value of

Λ. In addition, the ion cloud spreading appears to slow as time progresses. Since this system has more than two degrees of freedom, we propose that the spreading is the result of Arnold diffusion [9].

CONCLUSION

The measured gyration frequencies of ion clouds in an actual mass spectrometer will be broadened and shifted unpredictably by the chaotic dynamics described above. Irreproducibility of mass measurements may result from the dynamical sensitivity to initial conditions. In addition, an effect similar to that shown in Fig. 3b could account for the anomalous mass measured for one of the isotopes of xenon in a recent experiment [1].

ACKNOWLEDGEMENTS

This research was supported by Associated Western Universities under Grant DE-FG06-89ER-75522 with the US. Department of Energy. Pacific Northwest Laboratory is operated by Battelle Memorial Institute for the Department of Energy under contract DE-AC06-76RL01830. We thank Steve Barlow and Greg Flynn for useful discussions on the subject of this paper.

REFERENCES

[1] J. T. Meek, W. G. Millen, G. W. Stockton, and R. T. Kouzes, Phys. Rev. C **41**, 2921 (1990).

[2] M. V. Gorshkov, S. H. Guan, and A. G. Marshall, Int. J. Mass Spectrom. Ion Processes **128**, 47 (1993).

[3] *Analytical applications of Fourier transform ion cyclotron resonance mass spectrometry*, edited by B. Asamoto (VCH publishers, Inc., New York, 1991).

[4] G. M. Alber, A. G. Marshall, N. C. Hill, L. Schweikhard, and T. L. Ricca, Rev. Sci. Ins. **64**, 1845 (1993).

[5] S. P. Chen and M. B. Comisarow, Rapid Commun. Mass Spectrom. **5**, 450 (1991).

[6] M. V. Gorshkov, A. G. Marshall, and E. N. Nikolaev, J. Am. Soc. Mass Spectrom. **4**, 855 (1993).

[7] C. L. Hendrickson, S. C. Beu, and D. A. Laude, J. Am. Soc. Mass Spectrom. **4**, 909 (1993).

[8] M. Henon and C. Heiles, Astron. J. **69**, 73 (1964).

[9] A. J. Lichtenberg and M. A. Lieberman, *Regular and Stochastic Motion* (Springer-Verlag, New York, 1983).

[10] J. Binney and S. Tremaine, *Galactic Dynamics* (Princeton University Press, Princeton, NJ, 1987), p. 149.

[11] F. Franklin, M. Lecar, and W. Wiesel, *Planetary Rings* (University of Arizona Press, Tucson, AZ, 1984), p. 562.

[12] J. M. A. Danby, *Fundamentals of Celestial Mechanics* (Willmann-Bell, Inc., Richmond, VA, 1988), p. 341.

[13] X. Xiang, P. B. Grosshans, and A. G. Marshall, Int. J. Mass Spectrom. Ion Processes **125**, 33 (1993).

[14] S. P. Auerbach, Phys. Fluids **30**, 2139 (1987).

Spin-up of an Electron Plasma - First Results

R. E. Pollock* and Francois Anderegg+

*Physics Department, Indiana University, Bloomington IN 47405 and
+Institute for Pure and Applied Physics, University of California at San Diego
La Jolla CA 92093

Abstract. The application of a rotating electric quadrupole field to a pure electron plasma confined in an elongated Penning trap of the Malmberg type operating at 1 Tesla is shown to produce an unusually long-lived state. The observations are consistent with an assumption that the rotating field suppresses the radial expansion from transport due to alignment imperfections that normally limits the lifetime of such a plasma to a few thousand seconds. With the rotating field in operation, a lifetime of 20 hours was obtained.

INTRODUCTION

A one-component plasma (OCP), confined transversely by a uniform magnetic field and axially by electrostatic fields, forms a very stable system with interesting properties (1). The plasma can persist for a very long time. For small numbers of particles in a confinement volume of small dimension, lifetimes of months to years have been reported (2). Collisions with neutral particles of background gas give rise to radial transport and a finite lifetime (3). However a dense OCP in a trap elongated along the B field direction has an anomalously-shortened confinement time (4). The elongated geometry increases difficulty in maintaining the strict cylindrical symmetry required for the angular momentum conservation which inhibits radial growth. Even with careful construction and alignment of the B field direction along the axis of symmetry of the conducting trap walls, a long plasma column is observed to expand radially and then dissipate through collisions with the side walls. The time scale is on the order of tens of minutes, reducing approximately as the square of the length of the trapped plasma

---

* Work supported in part by the U. S. National Science Foundation, grant NSF PHY 93-14783

+ Work supported by the U. S. Office of Naval Research under contract N000144-89J-1714 column. The lifetime is sufficient for studies of plasma properties, but there are applications of the trapped OCP which would benefit from further improvement.

The plasma column rotates about its own axis to generate the inward $\mathbf{v} \times \mathbf{B}$ force which counteracts radial electric repulsion. In a frame of reference rotating with the column, the conducting wall is seen to rotate in the opposite sense. Imperfections in the cylindrical symmetry of the wall can create a drag torque on the column. It is plausible to conjecture that a wall rotating in the opposite sense in this frame could produce a torque of opposite sign to counteract the imperfection torque. In the lab frame this would be a rotating field with the same sense of rotation as the column and greater angular velocity. Experiments have been carried out (5, 6) which showed a small increase in column central density when a rotating field was applied of angular velocity which exceeded the angular velocity of the column center. The long-term effect of a continued application of a rotating field was not addressed in this work. There may be a distinction between column radial compression, and the prevention of a radial expansion.

The best form of rotating field is open to debate. A rotating dipole field can interact with the dipole moment of the charge distribution, and induce undesirable growth of an m = 1 diocotron mode. For the present experiments, a rotating quadrupole field was used, in the hope of coupling more strongly to the second moment of the charge distribution. The configuration was based on a suggestion (7) that the field might induce an electric quadrupole deformation of the charge distribution, a deformation which would be static in a frame rotating at the "motor" frequency. The rotating field could then couple to the deformation to exert a torque. More recently, a mechanism based on rotational pumping via the end region deformation of the plasma column has been proposed (8).

The rotating field introduced to exert a torque may also add energy to the trapped plasma column, raising the temperature of the electron column sufficiently that some of the electrons in the tail of the kinetic energy distribution exceed the ionization threshold of residual gas in the trap volume, and liberate new electrons which can add to the total number confined. By operating at a sufficiently high B field magnitude, radiation cooling will oppose energy input from the rotating field and may then maintain the column temperature below the ionization threshold.

The paper is organized in the following way. A description of the apparatus is followed by a section on observations, first in the absence of the rotating field, second in the regime where heating is dominant and third where a long-lived state is observed which does not appear to rely on replenishment through ionization. The final section summarizes our conclusions.

APPARATUS

Experiments were conducted on the "IV" apparatus at UCSD (9). The axial magnetic field was provided by a superconducting solenoid with a uniformity better than 2% over the confinement length. An additional coil pair was adjusted to add a 0.2% transverse field to improve the alignment of the magnetic field with the axis of the trap electrodes. A baked vacuum chamber in the form of a Type 310 stainless steel cylinder with ion pump and auxiliary Ti sublimation pumping provided a base pressure of $7 \cdot 10^{-11}$ Torr in the confinement region (0.3 nTorr with filament on).

The trap electrodes take the form of an axial array of eleven hollow cylinders of gold-plated OFHC copper, with inside diameter 57.2 mm and length, including spacer, of 58.4 mm. Two of the cylinders are sectored for signal monitoring and control. One has four insulated sectors of 58° width and 90° spacing. The second has eight insulated sectors of 27° width and 45° spacing. The electron source is a filament in the form of a spiral of thoriated tungsten wire. It is operated outside the main part of the magnetic field where the B field strength is 1.1% of the interior value. A grid close to the plane of the spiral is biased positive by about 10 V to extract electrons continuously from the heated filament. With 100 W of power, the filament operates at about 1850K and emits about 0.2 mA of electron current. A set of six additional cylinders lying between the filament and the trap are biased to maximize transmission of electrons through the trap. With the filament biased to 70 V, about 40 μA can be observed in a continuous beam of diameter about 3 mm.

The electrons may be trapped in two modes. In the normal cyclic mode, a negative bias is applied to two rings forming the ends of the containment region. To fill the trap, the voltage on the "inject" ring is raised to ground by a trapezoidal pulse of duration about 5 ms. After a chosen confinement time, a rectangular pulse to ground on the "dump" ring at the other end empties the trap, allowing a measurement of total charge or, by interposing a small diameter collimator hole between the end electrode and the Faraday cup, a measurement of line-integrated density. For the long-lived plasma studies, the dump pulse was not employed, and a single inject pulse was used to refill the trap after an inadvertent loss. Between fillings the filament was not heated, for improved vacuum.

In the work reported here, nine rings lay between the inject ring and the dump ring, giving a length of 0.526 m between the equipotential planes at half the confinement potential. The plasma formed by normal injection contained about $0.5 \cdot 10^{10}$ electrons in a column of diameter about 10 mm surrounding a denser core with the 3 mm diameter of the transmitted beam. See Figure 1a.

On the four-sectored ring, one sector was connected to a high impedance amplifier for observation of coherent signal. The opposite sector received a phase-shifted and amplitude-adjusted copy of the signal for damping of m = 1 diocotron mode. The two remaining sectors were connected to pulse generators of opposite sign, which could be fired to obtain a controlled displacement of the plasma column. Negative feedback was applied most of the time. To measure the line density of the column, the feedback was interrupted, an impulse applied, followed by positive feedback for a few milliseconds, then a 0.2 s interval without feedback in which the FFT diocotron spectrum could be obtained. The ratio of second harmonic to first harmonic amplitudes gives the column displacement from the trap axis. If the period of positive feedback is increased, a sufficient displacement is reached that noticeable loss to the wall occurs. The column displacement at onset of loss thus measures the base width of the column. See Figure 2.

The eight-sector ring served as the "motor" for adding angular momentum to the column. A signal generator fed a four-way splitter through a calibrated attenuator. The splitter fed four channels of phase-shifter/amplifier to generate four drive signals equal in amplitude and phase-shifted in steps of 90°. Each signal fed opposite pairs of wall sectors. This created a rotating electric quadrupole field in the ring interior, rotating with half the frequency of the signal generator. By reversing phases of two channels, the sense of rotation could be reversed. Typical operating parameters were a frequency of 1 MHz and amplitude of 0.2 V.

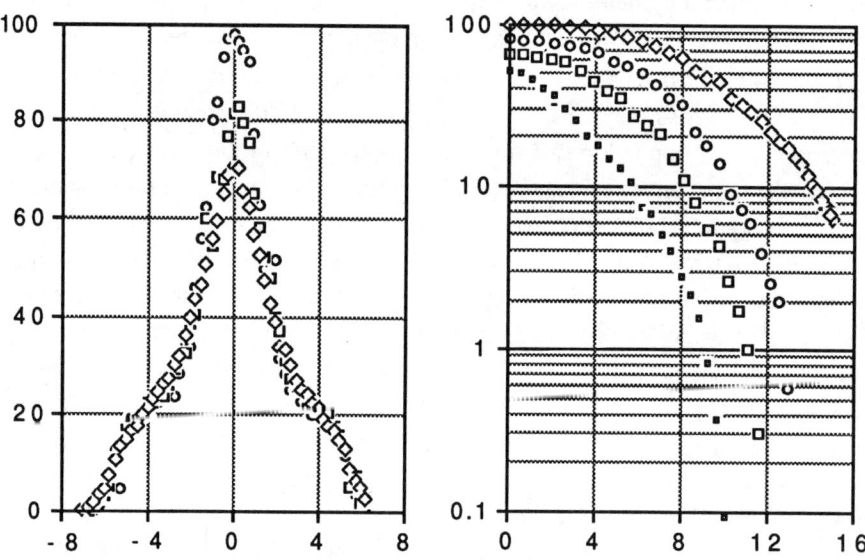

FIGURE 1.   Time evolution of radial density distributions in the absence of a rotating quadrupole field. Figure 1a- after 0.3, 1.0 and 2.0 s at 0.39 T. Figure 1b- after 1, 5, 20, and 50 s at 0.15T. Radius in mm, the wall is at 28.6 mm. Vertical scale is arbitrary.

OBSERVATIONS

Plasma Evolution with Motor Off

The evolution of the radial distribution over time was explored to establish the radial expansion rate in the absence of the rotating electric quadrupole field. Figure 1 illustrates the radial expansion over the first minute following injection. On the left, at 0.39 T, relative density scans at 0.3, 1.0 and 2.0 s show the central enhancement in density associated with the details of the injection process decaying within a few seconds. Change in the base width is small on this time scale. On the right, at 0.15 T, the radial density at 1, 5, 20 and 50 s shows a gradual expansion of the not-quite-Gaussian form. The distributions are shifted vertically for clarity. Later, after the edge of the distribution reaches the wall, both the central density and the total electron number will decay with similar mean life.

Persistent State with Ionization and Regulation

The confined electron plasma is so well isolated from its container that very small input power, for example from the noise level of an amplifier whose output is connected to one of the electrodes, is sufficient to raise the temperature and initiate ionization, signalled by a rise in the trap content with time in place of the expected decay. When the electrical environment is made sufficiently quiet that this process is not observed, then the deliberate application of a perturbing signal such as the rotating quadrupole field, can be used to generate electrons at will.

Without a regulation mechanism, the contents of a constant temperature column would grow exponentially with time until the plasma potential approached the end confinement potential. At that point electrons with highest energy will begin to spill out the ends. This is an evaporative cooling mechanism that reduces the temperature and production rate until a balance can be obtained. But note that this is a regulating mechanism only for particle number and temperature. Imperfection torques will still cause radial growth and loss to the wall. However by using the periodic kicking procedure which permits non-destructive monitoring of electron number, one may also maintain a selected column diameter. The diocotron amplitude is adjusted so that the column center reaches a desired distance from the conducting wall of the trap. Particles lying outside the radius corresponding to this separation distance are removed, and the column diameter thereby limited. In a sense this is a kind of controlled "evaporative torque" for counteracting radial growth. With losses both from the ends and the sides, all the principal plasma properties (density, column radius, temperature) can be maintained indefinitely. This is a persistent state for the plasma, but not for its constituents which are continuously replaced by scavenging electrons from the background gas.

Characteristics of this persistent state, which was observed for B fields of 0.15 and 0.4 T, included the following:

i) the state would persist with either sense of motor field rotation, showing that motor torque was not the determining feature;

ii) the number of electrons increased if the end confinement potential were increased, showing that continuous ionization was present;

iii) the number of electrons varied with the duration of positive feedback, showing that scraping of the column on the cylindrical wall was occurring.

The properties of the persistent state could be explored in a non-destructive manner. In the example shown in Figure 2 below, the motor rotation frequency was varied over the range from 0.35 MHz to 0.85 MHz. More ionization was observed at higher frequency, so the amplitude of the rotating quadrupole field was reduced at higher frequency to maintain the diocotron frequency in the range from 15 to 20 kHz. The B field was 0.39T.

The correlation between particle number and displacement after positive feedback seen Figure 2 a) is a property of the circuitry with the important consequence for the regulation mechanism that as the trap contents grow, the column is driven closer to the wall.

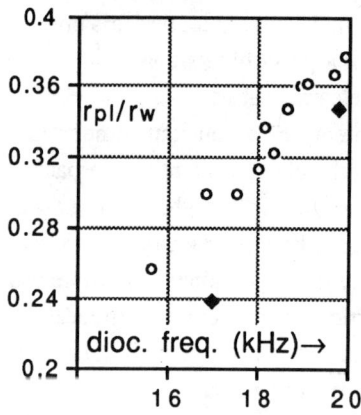 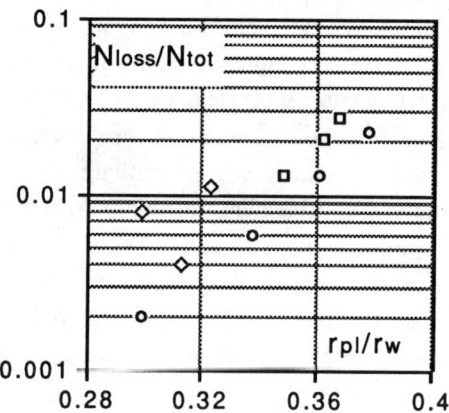

FIGURE 2. a) plasma displacement radius/wall radius vs m = 1 diocotron frequency and sense of motor rotation (circle: normal ccw, solid diamond: reversed cw). Note that the response to reversed motor sense is small but detectable.

b) fractional loss per kick vs rpl/rw. A measure of the radial distribution tail. Symbols (circle, square, diamond) denote motor power attenuator settings increasing in 5 dB steps.

By comparing the FFT spectrum measured soon after positive feedback ended, with a spectrum measured later just before negative feedback began, it was possible to obtain a measure of the fraction of the trap content removed per wall encounter. Figure 2b (right) shows that moving the column closer to the wall leads to increased loss per kick. The scatter in the points below 1% loss/kick shows the threshold sensitivity for this method. With one kick every 5 seconds, loss rates to the wall corresponding to lifetime in excess of ten minutes could be measured. The regulation of particle number and column radius in this persistent state depends on removing particles from the tail of the distribution.

The trap contents were retained overnight on two occasions in this regulating persistent state and showed the change in particle number less than 5 %. The meaning of this result is that the persistent state is readily distinguished from a true decay with mean life less than one week.

Long-lived State Without Ionization

If the rotating electric quadrupole field adds energy to the column of plasma, then a cooling mechanism other than evaporation must be present to prevent the temperature from reaching ionization threshold. For electrons and positrons, a field of 1 Tesla gives a temperature cooling time of about 4 s as the conducting cylinder is a waveguide above cutoff at the cyclotron frequency of 28 GHz (10). The input power from all sources must be low enough that ionization is halted by radiation cooling. A test for the absence of evaporative cooling is simply to raise the end confinement potential and observe that the number of particles does not increase. A test for the absence of "evaporative torque" is to reduce the amplitude of the kicked diocotron motion and show that the particle number does not increase. Finally the sense of the motor rotation may be reversed to show that the long-lived state is lost.

Figures 3 and 4 show a few hours of observation of a motor-driven plasma at 1 Tesla and 0.5 MHz. The motor amplitude could be increased to induce ionization for adjusting the number of electrons, then reduced to create a long-lived state. Left overnight on two occasions, a mean life of about 20 hours was observed. The increase of lifetime when the motor was present with the proper sense of rotation, (from 17 m to 20 h) is the magnitude of improvement expected if the presence of the rotating field eliminated the misalignment torque, leaving effects such as the background gas transport as the remaining factor in plasma lifetime determination.

This slowly and regularly decaying state was qualitatively different from the persistent state seen at lower B fields. The persistent state would remain overnight with no change in particle number, but would respond rapidly to small changes in kick amplitude or motor power. The persistent state at low field could be created

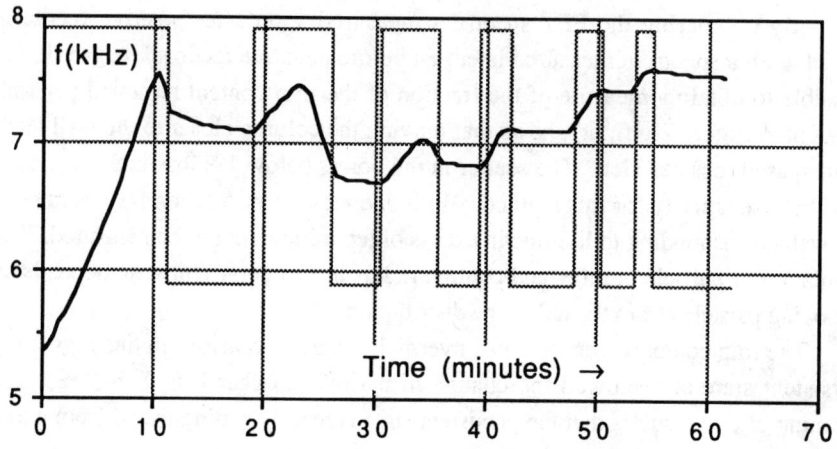

FIGURE 3.   A one-hour recording of diocotron frequency, showing the use of variable "motor" power to control build-up of trapped electron number following injection. The dotted line shows manual changes in amplitude of the rotating quadrupole field between values differing by a factor 1.7.   An upper frequency limit is set by the potential of the end rings.   Note the initial rapid growth ends with an overshoot attributed to evaporative cooling.   The rapid change of slope following each power change shows the speed of temperature equilibration at 1 Tesla.   The highest sustained number of electrons is obtained by reducing the motor power as a peak is reached.

with either sense of motor rotation, showing that energy input was more important than torque. As seen in figure 4 c), the long-lived state did not survive with the reversed sense of motor rotation.

The temperature of the long-lived plasma was measured by dropping the end potential in very small steps and recording the slope of the exponential loss onset. A longitudinal temperature of 1.8 eV was obtained under the conditions of figure 4. Measuring the charge reaching the Faraday cup at each potential step confirmed the method. This is a nearly non-destructive method as only about 1% of the charge needs to be removed to determine the temperature.

Several measurements of the depth of the plasma potential were made with 140 V and 200 V end confinement. These are logarithmically-sensitive to the column diameter and consistent with a diameter of 5 to 10 mm, perhaps increasing slightly with particle number at fixed field-rotation frequency.

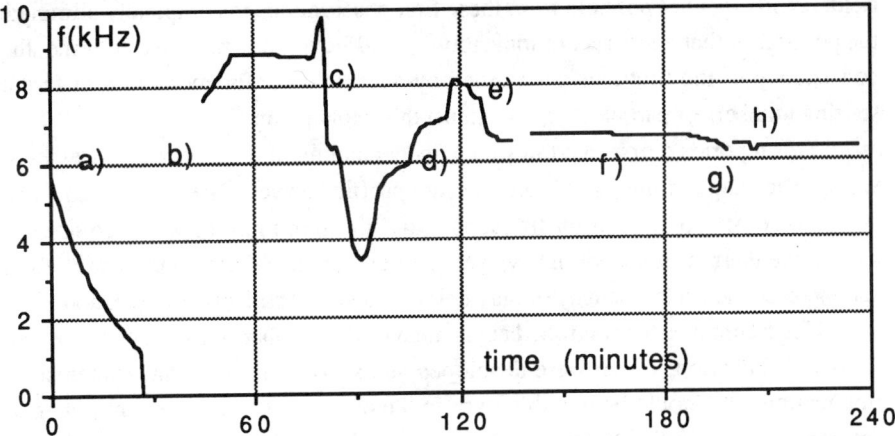

FIGURE 4. Behavior indicative of a long-lived state not accompanied by ionization. A 4 hour chart recording of diocotron frequency. A second trace (not shown here) of the ratio of second to first harmonic amplitudes showed that the column displacement during the f measurement was about 3% of the wall radius, an order of magnitude less than in the conditions where radial loss contributed to the control of radial growth.

a)with the motor off, a mean life of 17 minutes was observed;

b) after refill and controlled build-up by ionization, a very flat trace was observed. The end potential was -200 V, so the overshoot phenomena of Figure 3 were avoided by stopping before the confinement limit was reached;

c) stable state destroyed by a reversal of motor sense. Note rapid heating, a sudden loss of one-third of the electrons, then rapid decay;

d) proper motor sense restored with higher power to replenish the contents;

e) one end potential was lowered from 200 V in steps of 25 V until a noticeable loss was observed. This give a rough measure of the plasma potential of between 125 and 150 V.

f) When the higher potential was restored, the trap contents did not grow as they would if ionization were present. The plasma then was left undisturbed for one hour and a loss rate of 2±1%/hr observed.

g) second potential measurement of 85±5, with irreversible loss at 80 V

h) reversible compression as the end potential is cycled from 200 to100 to 200 V

CONCLUSIONS

The application of a rotating electric quadrupole field to an electron plasma in a confinement field of 1 Tesla gives rise to a stable state with a decay rate reduced by about two orders of magnitude over that observed in the absence of the rotating field. While it is not possible from these first observations to completely eliminate the possibility that a low rate of ionization was affecting the measured lifetime, the consistency of the lifetimes determined by overnight confinement with different starting numbers of particles argues against this explanation.

This report is a first step toward an understanding of the physical processes at work. The shape of the equilibrium density profile has yet to be determined. The best choice of motor rotation frequency, and the minimum cooling required to sustain the thermal equilibrium have yet to be explored. Whether the rotating field can increase the central density or only delay its decay is still an open question.

The techniques for controlled adjustment of the number of electrons in the trap following injection, which were developed as part of this study, have interesting implications. It should be possible to raise the electrostatic confinement potential substantially after filling the trap, and to increase the number of electrons to explore a density regime well beyond the reach of present injection devices.

ACKNOWLEDGMENT

The work described was carried out in large part while one of us (R. E. P.) was a visiting scholar at the Institute for Pure and Applied Physics at the University of California, San Diego. A sabbatical leave by Indiana University made this visit possible. Generous access to the experimental and theoretical resources of the Driscoll/O'Neil/Dubin non-neutral plasma group is gratefully acknowledged.

REFERENCES

1. O'Neil, T. M, "Plasmas with a Single Sign of Charge", in *Non-Neutral Plasma Physics*, A.I.P. Conf. Proc. **175** 1–25 (1988).
2. Dehmelt, H., *Rev. Mod. Phys.* **62** 525–530 (1990).
3. Malmberg, J. H., Driscoll, C. F., *Phys. Rev. Letters* **44** 654–657 (1980).
4. Driscoll, C. F., Fine, K. S., Malmberg, J. H., *Phys. Fluids* **29** 2015–2017 (1986).
5. Malmberg, J. H. et al, "Experiments with Pure Electron Plasmas", in *Non-Neutral Plasma Physics*, A.I.P. Conf. Proc. **175** 28–71 (1988).
6. Mitchell, T. B., PhD dissertation U.C.S.D. (1993), p 58.
7. Fine, K., private communication.
8. Crookes, S., O'Neil, T. M., these proceedings.
9. Anderegg, F. et al, *Bull. Am. Phys.Soc.* **38** 1971 (1993).
10. O'Neil, T. M., *Phys. Fluids,* **23** 725–731 (1980).

Nonlinear Dynamics and Chaoticity in an Intense Nonneutral Heavy Ion Beam Propagating Through a Periodic Focussing Field

Qian Qian and Ronald C. Davidson

Princeton Plasma Physics Laboratory
Princeton University
Princeton, NJ 08543

Abstract. Properties of an intense nonneutral heavy ion beam propagating through a periodic quadrupole focussing magnetic field are studied analytically and numerically including self-field effects. For a Kapchinskij-Vladimirskij(K-V) beam equilibrium with uniform density profile, the two nonlinear coupled envelope equations are solved for the case of a *matched* beam in the smooth-beam approximation. Space-charge effects are taken into account self-consistently. The analytical solutions for the envelope and the (space-charge-depressed) phase advance are compared with the (exact) numerical results, and the technique can be applied to a wide range of system parameters and choices of lattice functions $\kappa_q(s) = \kappa_q(s+S)$. Furthermore, a fully nonlinear study of the envelope equations for the case of a *mismatched* beam is carried out numerically. In certain parameter regimes, the beam envelope is found to exhibit bifurcation which results in chaotic behavior of the beam envelope when the focussing field strength is increased beyond a particular threshold. The (unstable) chaotic region in the parameter space $(\sigma_0, KS/\varepsilon)$ is determined numerically. Here, σ_0 is a measure of the focussing strength defined by $\sigma_0^2/S^2 \equiv \left\langle \left(\int_{s_0}^{s} ds \kappa_q(s) \right)^2 \right\rangle$, K is the self-field perveance, S is the periodicity length of the focussing lattice, and ε is the unnormalized beam emittance. The relationship between chaoticity and well-known linear instabilities is examined. In addition, for the case of a nonuniform density beam, it is found that the particle trajectory exhibits intrinsic chaoticity. The consequence of chaotic particle orbits is further investigated.

I. INTRODUCTION

Heavy ion fusion as well as industrial and medical applications of intense charged particle beams constitute important areas where space-charge effects play an important role in the evolution, confinement, and stability properties of the beam [1–9]. It is of considerable practical interest to determine the quantitative influence of self-field effects(as measured by the beam perveance K, say) on the phase advance, $\sigma \equiv \varepsilon \int_{s_0}^{s_0+S} ds/r_b^2(s)$, as the beam propagates through one period S of the quadrupole focussing lattice

$$\kappa_q(s) = \kappa_q(s + S).$$

Here, ε is the unnormalized beam emittance, and $r_b(s)$ is the radius of the beam envelope(assumed cylindrical), which varies with the axial coordinate s. Approximate methods have previously been used to estimate the phase advance σ by assuming either weak focussing field strength(e.g., the iterative method [1,2,4,10,11]) or a simple form for the lattice function $\kappa_q(s)$(e.g., the matrix method [2,3]). All of these methods have limited applicability. In this paper, a closed analytical method is described for calculating the phase advance σ, which is valid for arbitrary beam intensity and general functional form of the periodic lattice $\kappa_q(s)$.

For a uniform density beam corresponding to the Kapchinskij-Vladimirskij(K-V) equilibrium [12], the outer boundary of the elliptical beam satisfies two nonlinearly coupled *envelope* equations. The evolution of the envelope determines the confinement region of the beam. Previously, a linear stability analysis of the envelope equations was carried out indicating the presence of linear instabilities in certain parameter regimes [13], although the fully nonlinear evolution of the beam envelope had not been investigated. In the present paper, the nonlinear envelope equations including space-charge effects are studied numerically. This work not only confirms earlier linear results but also reveals new nonlinear phenomena important in the evolution of the beam envelope [14].

The dynamics of an individual test particle within the beam has been studied extensively [1-4]. For high current beams, the particle motion can be influenced significantly by space-charge effects. The Kapchinskij-Vladimirskij(K-V) distribution is the only equilibrium distribution function for a uniform density beam when space-charge effects are taken into account self-consistently. Previous studies indicate that the particle orbits are regular within a K-V beam and there are two independent invariants in the four-dimensional phase space. When the beam density profile is nonuniform, it is found that there is no longer an invariant associated with the motion of an individual particle within the beam and the particle trajectories are intrinsically chaotic [15].

This report is organized as follows. Section II introduces some important concepts related to intense beam dynamics. A new technique is described in Sec. III to estimate the phase advance for a matched beam. Section IV reports recent results concerning the nonlinear evolution of the envelope for a K-V beam. Finally, in Sec. V, the particle dynamics is discussed for the important case of a nonuniform density profile.

II. BASIC CONCEPTS IN INTENSE BEAM DYNAMICS

Here we summarize some important concepts [1-4] in preparation for the discussion of beam dynamics in Secs. III- V.

The quadrupole focussing lattices commonly used in modern accelerators were first proposed by Courant and Snyder in the 1950s [16]. The approach is to adopt a system of alternating-gradient quadrupole fields to achieve strong focussing of the particle orbits in the two transverse directions simultaneously. The quadrupole lattice function $\kappa_q(s)$ is defined by

$$\kappa_q(s) = \frac{qB_q'(s)}{\gamma_b m \beta_b c^2},\tag{1}$$

where $\kappa_q(s)$ is a periodic function with $\kappa_q(s) = \kappa_q(s + S)$, s is the axial coordinate along the beam propagation direction, and S is the axial periodicity length. The derivative of the focussing field in the transverse direction, $B_q'(s)$, is defined by $B_q'(s) \equiv (\partial B_x^q/\partial y)_0 \equiv (\partial B_y^q/\partial x)_0$. The quantities x and y measure the transverse distance from the beam axis at

$(x, y) = (0, 0)$. Moreover, q is the particle charge, m is the particle rest mass, c is the speed of light *in vacuo*, $\beta_b c$ is the average axial velocity of the beam particles, and $\gamma_b = (1 - \beta_b)^{-1/2}$ is the relativistic mass factor. A useful quantity is defined by [17]

$$\frac{\sigma_0^2}{S^2} \equiv \left\langle \left(\int_{s_0}^{s} ds \kappa_q(s) \right)^2 \right\rangle, \tag{2}$$

which is a measure of the strength of the average quadrupole focussing field-squared.

The maximum phase-space area occupied by the beam particles is called the emittance ε. The self-field perveance K is a measure of space-charge effects and is defined by

$$K \equiv \frac{2q^2}{\beta_b^2 \gamma_b^3 m c^2} N_b, \tag{3}$$

where $N_b = \int \int dx dy n_b(x, y)$ is the number of particles per unit axial length of the beam. The phase advance σ is an important physical quantity which measures the periodicity of the betatron motion of particles inside a matched beam when the beam propagates through a periodic focussing field. For the special case of a cylindrical beam with radius $r_b(s)$, the phase advance σ is defined by

$$\sigma \equiv \varepsilon \int_{s_0}^{s_0 + S} \frac{ds}{r_b^2}. \tag{4}$$

Finally, a beam is said to be *matched* if the envelope oscillation period is the same as that of the external, periodic focussing lattice. Otherwise, the beam is said to be *mismatched* if the beam envelope oscillates out of the phase with the focussing lattice.

III. PHASE ADVANCE FOR A MATCHED BEAM

For the case of a periodic quadrupole focussing lattice $\kappa_q(s) = \kappa_q(s + S)$ which is most commonly used in accelerators, the phase advance of a beam is designated by the two quantities, σ_x and σ_y, which are defined by

$$\sigma_x \equiv \varepsilon_x \int_{s_0}^{s_0 + S} \frac{ds}{a^2(s)}, \qquad \sigma_y \equiv \varepsilon_y \int_{s_0}^{s_0 + S} \frac{ds}{b^2(s)}. \tag{5}$$

Here, ε_x and ε_y are the unnormalized emittances in the x- and y-directions, respectively. Moreover, $a(s)$ and $b(s)$ are the x- and y-dimensions of the outer elliptical envelope of the charged particle beam, which for present purposes is assumed to be described by the K-V equilibrium distribution function [1,2,12], and has uniform density within the beam envelope. To determine σ_x and σ_y, the quantities $a(s)$ and $b(s)$ need to be determined self-consistently from the beam envelope equations

$$\frac{d^2}{ds^2} a(s) + \kappa_q(s) a(s) - \frac{2K}{a(s) + b(s)} - \frac{\varepsilon_x^2}{a^3(s)} = 0, \tag{6}$$

$$\frac{d^2}{ds^2} b(s) - \kappa_q(s) b(s) - \frac{2K}{a(s) + b(s)} - \frac{\varepsilon_y^2}{b^3(s)} = 0, \tag{7}$$

where the focussing coefficient $\kappa_q(s) = \kappa_q(s + S)$ is a periodic function [1]. The nonlinearly coupled differential equations (6) and (7) can be solved in the *smooth-beam* approximation, which assumes that the envelope functions, $a(s)$ and $b(s)$, undergo small-amplitude oscillations, $\delta a(s)$ and $\delta b(s)$, about the average values, $\bar{a}$ and $\bar{b}$, i.e.,

$$|\delta a| \ll \bar{a}, \quad |\delta b| \ll \bar{b}, \tag{8}$$

where $\bar{a}$ and $\bar{b}$ are assumed to be constant over one lattice period S. Here, use is made of the approximation (8) to linearize Eqs. (6) and (7) and yield the solutions [17]

$$< \kappa_q(s)\delta a(s) > -\frac{2K}{\bar{a} + \bar{b}} - \frac{\varepsilon^2}{\bar{a}^3} = 0, \tag{9}$$

$$- < \kappa_q(s)\delta b(s) > -\frac{2K}{\bar{a} + \bar{b}} - \frac{\varepsilon^2}{\bar{b}^3} = 0, \tag{10}$$

where δa_n and δb_n are defined by

$$\delta a_n = \kappa_n \frac{2K\bar{b}/(\bar{a} + \bar{b})^2 + \left[2K/(\bar{a} + \bar{b})^2 + 3\varepsilon^2/\bar{b}^4 - k_n^2\right]\bar{a}}{\left[2K/(\bar{a} + \bar{b})^2\right]^2 - \left[2K/(\bar{a} + \bar{b})^2 + 3\varepsilon^2/\bar{a}^4 - k_n^2\right]\left[2K/(\bar{a} + \bar{b})^2 + 3\varepsilon^2/\bar{b}^4 - k_n^2\right]}, \tag{11}$$

$$\delta b_n = -\kappa_n \frac{2K\bar{a}/(\bar{a} + \bar{b})^2 + \left[2K/(\bar{a} + \bar{b})^2 + 3\varepsilon^2/\bar{a}^4 - k_n^2\right]\bar{b}}{\left[2K/(\bar{a} + \bar{b})^2\right]^2 - \left[2K/(\bar{a} + \bar{b})^2 + 3\varepsilon^2/\bar{a}^4 - k_n^2\right]\left[2K/(\bar{a} + \bar{b})^2 + 3\varepsilon^2/\bar{b}^4 - k_n^2\right]}. \tag{12}$$

In Eqs. (11) and (12), δa_n, δb_n, and κ_n are the Fourier components of $\delta a(s)$, $\delta b(s)$, and $\kappa_q(s)$, respectively, defined according to

$$\delta a(s) = \sum_{n=-\infty}^{\infty} \delta a_n \exp(ik_n s), \quad \delta b(s) = \sum_{n=-\infty}^{\infty} \delta b_n \exp(ik_n s), \quad \kappa_q(s) = \sum_{n=-\infty}^{\infty} \kappa_n \exp(ik_n s), \tag{13}$$

where $k_n = 2\pi n/S$, and $< \cdots >$ is the spatial average over one lattice period S. Making use of Eqs. (11)- (13), it is straightforward to determine the average quantities $< \kappa_q(s)\delta a(s) >$ and $- < \kappa_q(s)\delta b(s) >$ [17]. For specified lattice function $\kappa_q(s)$ and normalized self-field parameter KS/ε, the solutions to Eqs. (6)- (7) are given by Eqs. (9)- (12) in the smooth-beam approximation.

Compared with other techniques [1–4,10,11,13], this approach is generally applicable for a wide range of beam ellipticity(as measured by $\bar{a}/\bar{b}$), self-field perveance(K), unnormalized beam emittance(ε), and choice of lattice function, $\kappa_q(s) = \kappa_q(-s)$ [17]. The results are found to be in very good agreement with the (exact) numerical solutions [14].

IV. NONLINEAR EVOLUTION OF THE BEAM ENVELOPE

To obtain analytical solutions to the beam envelope equations (6) and (7) is a non-trivial challenge although it is not difficult to solve the envelope equations numerically. For the case of a *matched* beam, the imposed periodic boundary conditions dramatically reduce the number of possible solutions [14]. In fact, it is found numerically that the only allowed solutions satisfy $\bar{a} \approx \bar{b}$ and $a(s) + b(s) \approx$ const. over a wide range of parameter space characterized by $(\bar{a}/\bar{b}, KS/\varepsilon, \sigma_0)$ [14,17].

For a *mismatched* beam, Eqs. (6) and (7) constitute an initial-value problem without periodic boundary conditions. The nonlinearity and choice of system parameters $(KS/\varepsilon, \sigma_0)$ renders the evolution of $a(s)$ and $b(s)$ even more complicated. Previous studies using a linearized model indicate instabilities in certain parameter regimes if the envelope is perturbed about its equilibrium [13]. Further study of the fully nonlinear equations (6) and (7) yields more interesting phenomena [14]. In particular, the beam envelope exhibits·bifurcation and chaoticity when the average focussing strength square σ_0^2 is sufficiently large. The region of parameter space $(KS/\varepsilon, \sigma_0)$ where chaotic behavior exists has been determined numerically [14]. Chaotic evolution of the beam envelope leads to poor confinement of the beam particles and limits the allowed region of parameter space $(KS/\varepsilon, \sigma_0)$ for stable beam propagation.

V. NONUNIFORMITY AND NONLINEAR PARTICLE DYNAMICS

In a space-charge-dominated beam(where K is large), the shape of the beam density profile plays a critical role in determining the transverse particle dynamics. For a uniform density beam which corresponds to the K-V equilibrium [12], the self-field forces in the equations of motion for a test particle within the beam are linear, and the particle motion is regular and decoupled between the two transverse directions, i.e., the $x-$ and y-directions. When the density profile is nonuniform, which is more realistic and of practical interest, not only are the particles subject to nonlinear self-field forces but also their motions in the x- and y-directions are coupled. Based upon the Hamiltonian for a test particle inside the beam, it is found that a nonuniform density profile results in chaotic particle orbits. This is evident both from Poincaré surface-of-section plots and an examination of the Liapunov exponents of a test particle trajectory [15,14]. The Lie-Bäcklund symmetry approach has been used to investigate analytically the nonintegrability of the Hamiltonian [15].

Further implications of the chaotic particle orbits due to a nonuniform density profile are being investigated [14]. One current area of research is to relate the chaotic particle mixing to the homogenization of an initially-nonuniform density profile and to determine the time scale for this homogenization process. The other related problem is to study the mechanism of halo formation due to chaotic particle orbits inside and outside the beam. It is found at sufficiently high current, that a nonuniform beam can eject a cluster of particles in the form of a halo surrounding the beam core. Halo formation can result in a degradation of beam current and quality, and considerable effort is being devoted to exploring the detailed mechanisms by which halos are produced [18–22]. Our preliminary studies indicate that a nonuniform density profile can cause the breakdown of invariant curves in phase space and result in the escape of halo particles from the beam interior.

VI. CONCLUSIONS

Some results of theoretical investigations of a space-charge dominated beam have been summarized in this report. A new technique for determining the phase advance σ is found to be applicable over a wide range of system parameters and choice of focussing lattices. The nonlinear evolution of the envelope for a K-V beam equilibrium has been studied numerically and chaotic behavior of the beam envelope is found in some regions of parameter space. Based upon the Hamiltonian for an individual test particle, it concluded that the particle orbits are intrinsically chaotic within a beam with nonuniform density.

ACKNOWLEDGMENTS

The authors would like to express their appreciation to Dr. Chiping Chen for stimulating discussions. This work was supported in part by U.S. Department of Energy Contract No. DE–AC02–76–CHO–3073 and in part by the Office of Naval Research.

REFERENCES

[1] R. C. Davidson, *Physics of Nonneutral Plasmas* (Addison-Wesley Publishing Co., Reading, MA, 1990), Chapter 10, and references therein.

[2] J. Lawson, *The Physics of Charged-Particle Beams* (Oxford Science Publications, New York, 1988), and references therein.

[3] D. Edwards and M. Syphers, *An Introduction to the Physics of High Energy Accelerators* (John Wiley & Sons, Inc., New, York, 1993).

[4] J. S. Humphries, *Principles of Charged Particle Acceleration* (Wiley, New York, 1986), and references therein.

[5] P. Lapostolle, IEEE Transactions on Nuclear Science **NS-18**, 1101 (1971).

[6] R. Jameson, IEEE Transactions on Nuclear Science **NS-28**, 2408 (1981).

[7] J. Struckmeier, J. Klabunde, and M. Reiser, Particle Accelerators **15**, 47 (1984).

[8] T. Wangler, K. Crandal, R. Mills, and M. Reiser, IEEE Transactions on Nuclear Science **NS-32**, 2196 (1985).

[9] I. Hofmann and J. Struckmeier, Particle Accelerators **21**, 69 (1987).

[10] M. Reiser, IEEE Transactions on Nuclear Science **NS-24**, 1009 (1977).

[11] M. Reiser, Particle Accelerators **8**, 167 (1978).

[12] I. Kapchinskij and V. Vladimirskij, in *Proceedings of the International Conference on High Energy Accelerators and Instrumentation* (CERN Scientific Information Service, Geneva, 1959), p. 274.

[13] J. Struckmeier and M. Reiser, Particle Accelerators **14**, 227 (1984).

[14] Q. Qian and R. C. Davidson, "Nonlinearity and Chaos in a Nonuniform Density Beam", to be published.

[15] Q. Qian, R. C. Davidson, and C. Chen, Physics of Plasmas **1**, 1328 (1994).

[16] E. Courant and H. Snyder, Annals of Physics **3**, 1 (1958).

[17] R. C. Davidson and Q. Qian, Physics of Plasmas **1**, 3104 (1994).

[18] M. Reiser, in *IEEE Proceedings of the 1991 Particle Accelerator Conference*, Vol. 3(Inst. Electrical and Electronic Engineers, New York, NY, 1001), p.2497.

[19] A. Cucchetti, M. Reiser, and T. Wangler, in *IEEE Proceedings of the 1991 Particle Accelerator Conference*, Vol. 1(Inst. Electrical and Electronic Engineers, New York, NY, 1991), p.251.

[20] J. O'Connell, T. Wangler, R. Mills, and K. Crandall, in *IEEE Proceedings of the 1993 Particle Accelerator Conference*, Vol. 4(IEEE Service Center, Piscataway, NJ, 1993), p.3657.

[21] J. Lagniel, Nucl. Instrum. and Methods Phys. Res. A **345**, 405 (1994).

[22] R. L. Gluckstern, Phys. Rev. Lett. **73**, 1247 (1994).

BOHM-CONDITION FOR NONNEUTRAL PLASMA STREAMS

H. Ramachandran, G.J. Morales, and V.K. Decyk
Dept. of Physics, University of California, Los Angeles, CA 90024-1547

ABSTRACT

It is demonstrated through a time-dependent particle simulation that an extended sheath forms spontaneously when a nonneutral plasma reflects from a potential barrier. An analytical description shows that extended sheaths develop when the flow velocity is smaller than the Gould-Trivelpiece speed, hence a generalization of the Bohm-condition to nonneutral systems is obtained.

INTRODUCTION

One of the characteristic properties of neutral plasmas is the formation of localized sheaths at the interface with external boundaries. These self-consistent charged layers cause the reflection of the bulk of the electron population and result in the outward flow of ions. However, the localization of the sheath (typically a few Debye lengths) requires that the ion flow velocity be sufficiently large, i.e., it must satisfy the Bohm-condition[1]. If the ion flow is subsonic, then a class of spatially extended solutions is mathematically feasible, but in neutral plasmas they are not commonly observed because the extended fields regulate the flow to meet the required speed. By contrast, the present analytical and simulation study illustrates that single species plasmas (i.e., nonneutral plasmas) spontaneously develop extended sheaths when the flow velocity to the confining potential barrier falls below the speed of the Gould-Trivelpiece modes[2]. Since this is the effective sound speed in single species column plasmas, a generalization of the Bohm-condition is obtained.

SIMULATION

We proceed to describe the result of a specific particle simulation to illustrate the nature of this nonlinear effect. We have surveyed a variety of systems and have found that the features described here are ubiquitous to the reflection of a nonneutral stream from an external potential.

The particle simulation code[3] consists of an electrostatic, bounded, magnetized, 2-1/2 dimensional system. The magnetic field points along the z-direction and confines particles in the x-direction, with the y-coordinate ignorable, as illustrated in Fig. 1. The plasma is bounded in the x-direction by walls whose potential is analytically prescribed to generate localized potential barriers along the z-direction, as in the traps developed[4] at the University of California, San Diego. Initially the plasma is confined within a small region, between the barriers labelled a and b in Fig. 1, and is allowed to equilibrate. At a prescribed time ($t \equiv 0$) the barrier b is removed and the plasma is allowed to expand until it encounters the barrier c.

We do not describe here the early expansion of the column, which has been experimentally studied and modelled by Moody and Malmberg[5]. Instead, we focus on the dynamics that results when the expanding plasma encounters the trapping barrier c. The important new features are illustrated in Fig. 2, at a time $t = 300 \, \omega_{po}^{-1}$, where $\omega_{po} = \sqrt{4\pi e^2 n_0/m}$ with n_0 the initial plasma density. The axial phase-space (v_z, z) is shown in Fig. 2a, while the self-consistent axial potential $\phi(x = 0, z)$ is found in Fig. 2b. It is evident that on the right-hand side a spatially extended potential barrier (or sheath) has developed spontaneously. The high-velocity, multiple bands in Fig. 2a correspond to those particles that reflected from the barrier c before the external sheath was formed (i.e., $t < 150 \, \omega_{po}^{-1}$). The characteristic feature of the extended sheath is a flat region of high potential that connects on one side to the vacuum potential barrier while on the other it drops abruptly, on a scale of the half-width of the column, a, to the nearly uniform potential of the bulk plasma. The radial dependence within the flat region is in Fig. 3b.

ANALYTICAL MODEL

To understand the simulation results we analyze a model problem that contains the essential physics. The expanding plasma is taken to be a warm Lorentzian beam along z, with the drift velocity independent of x, and cold in the transverse direction (conditions well satisfied in the simulations). The potential structure $\phi(x, z)$ is assumed to be in quasistatic equilibrium described by the scaled Poisson's equation:

$$\frac{\partial^2 \psi}{\partial \xi^2} + \frac{\partial^2 \psi}{\partial \eta^2} = -\text{ph}(\psi; \Delta)\theta(d - |\xi|), \tag{1}$$

where $\xi = x/a$, $\eta = z/a$, $\psi = \left(q\phi - q\phi(-\infty)\right)/E_0$, and E_0 the streaming energy. Δ is the ratio of the spread in beam energy to its mean value, and $p = m\omega_p^2 a^2/E_0$ is the key parameter of the problem. It is proportional to the square of the ratio of the Gould-Trivelpiece speed to the axial speed of the stream. h is the scaled charge density.

The boundary conditions are: $\psi \to 0$ at $|\xi| = 1$ and also as $\eta \to -\infty$;

$\partial\psi/\partial\xi = 0$ at $\xi = 0$; ψ and $\varepsilon_\perp \partial\psi/\partial\xi$ continuous at the plasma-vacuum interface ($|\xi| = d$), with $\varepsilon_\perp$ the dielectric coefficient.

An approximate solution to Eq. (1) is extracted by assuming a nearly separable dependence $\psi(\xi, \eta) = g(\eta)\psi_\perp\left(\xi;\delta g(\eta)\right)$, where $\psi_\perp = 1$ at $\xi = 0$, and is weakly dependent ($\delta \ll 1$) on g, so Eq. (1) becomes

$$\left(\frac{g''}{g}\right)\psi_\perp + \frac{\partial^2\psi_\perp}{\partial\xi^2} = -p\,\frac{h(g\psi_\perp; \Delta)}{g}\,\theta(d - |\xi|) \ . \tag{2}$$

The associated perpendicular eigenvalue problem

$$\frac{d^2\psi_\perp}{d\xi^2} = -p\,\frac{h(g\psi_\perp; \Delta)}{g}\,\theta(d - |\xi|) - \lambda\psi_\perp \ , \tag{3}$$

is solved subject to $\psi_\perp(1) = 0$, $\psi_\perp(0) = 1$, $(d\psi_\perp/d\xi)(0) = 0$, and $\psi_\perp$ and $\varepsilon_\perp\, d\psi_\perp/d\xi$ continuous at $|\xi| = d$. This yields the eigenvalues $\lambda(g; p, \Delta, d)$ and a family of solutions $\psi_\perp(\xi; g, p, \Delta, d)$ which is substituted back into Eq. (2) to obtain $g'' = \lambda g$. This can be reduced to conservation form

$$\frac{1}{2}\left(\frac{dg}{d\eta}\right)^2 = \int_0^g g\lambda(g; p, \Delta, d)dg \equiv -V_g(g; p, \Delta, d) \ , \tag{4}$$

from which the key physical properties can be deduced. To obtain V_g values are chosen for p, Δ, and d, Eq. (3) is numerically solved for a continuous range of g to obtain $\lambda(g)$, and finally the integral in Eq. (4) is evaluated numerically.

The effective potential V_g is displayed in Fig. 4a for three different values of p; the corresponding solutions $g(\eta)$ are shown in Fig. 3a. For small p (curves A), V_g is monotonic and corresponds physically to spatially localized sheaths. As p increases, V_g develops a local maximum, as seen in curve B in Fig. 4a, and the corresponding slope (axial electric field, g') becomes smaller. As p is increased further, the value of V_g at the local maximum becomes nearly zero and results in the extended potential structure C in Fig. 3a.

The value of p which results in $\partial V_g/\partial g = V_g = 0$ is defined as the critical value, p_c, for development of an extended sheath. This quantity increases as the effective temperature, Δ, increases.

Figure 4b shows the time evolution of the parameter $p(t)$ (solid curve) during the expansion phase of the simulation leading to the results shown in Fig. 2. The various horizontal dashed-lines labelled A, B,

C correspond to the effective potentials shown in Fig. 4a. It is clearly seen that the expanding stream reaches a saturation at a critical value p=2.2, as predicted by the analysis. Simultaneously with this saturation, one observes the formation of an extended sheath, of the type shown in Fig. 2b. The saturation of the expanding plasma to a value lower than the critical p is a self-consistent response to the presence of the extended sheath.

The radial dependence of the potential within the axially flat portion of the extended sheath is shown in Fig. 3b. The crosses are the result of the particle simulation and the solid line is the prediction of the analytical model. Near the axis (ξ = 0) the profile is flat because few particles are able to penetrate the extended barrier. Near the edge the profile is as expected from the linearized response, while the curvature in the transition region is determined by the energy spread Δ.

We have found that the analytical results reported here for a slab also hold for cylindrical geometry, but with suitable cylindrical scaling.

CONCLUSION

In summary, it has been demonstrated through a time-dependent particle simulation that when an expanding nonneutral plasma reflects from a potential barrier it spontaneously develops an extended sheath. An analytical description of this process shows that the extended sheaths develop when the critical parameter $m\omega_p^2 a^2/E_0 > 2$ for slab (6.3 for cylinder), thus yielding the nonneutral plasma generalization of the Bohm-sheath condition for neutral plasmas.

ACKNOWLEDGMENT

This work is sponsored by the Office of Naval Research.

REFERENCES

1.   D. Bohm, "The Characteristics of Electrical Discharges in Magnetic Fields," A. Guthrie and R.K. Wakerling, Eds. McGraw-Hill (1949). Chapter 3, p 27.
2.   A.W. Trivelpiece and R.W. Gould, J. Appl. Phys. 14, 1784 (1959).
3.   H. Ramachandran, G.J. Morales, and V.K. Decyk, Phys. Fluids B 5, 2733 (1993).
4.   C.F. Driscoll, J.H. Malmberg, and K.S. Fine, Phys. Rev. Lett 60, 1290 (1988).
5.   J.D. Moody and J.H. Malmberg, Phys. Rev. Lett. 69, 3639 (1992).

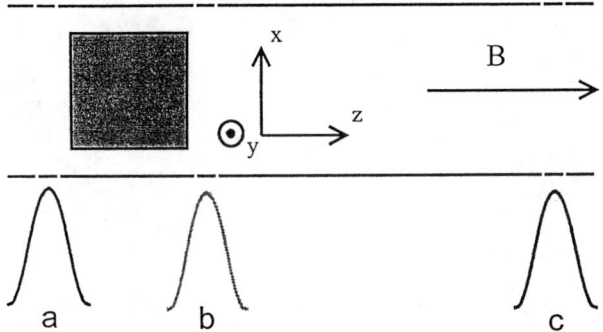

Fig. 1.   Schematic of simulation geometry. A nonneutral plasma is initially confined between barriers a, b. At t = 0 barrier b is removed; plasma expands and reflects from barrier c where sheath is formed. At t = 0 the aspect ratio ($\Delta z/\Delta x$) is 11 and after expansion it is 56.

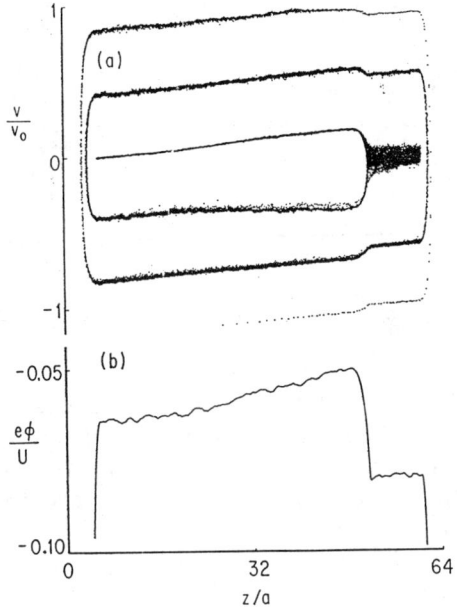

Fig. 2 (a)   Axial phase-space at $\omega_{po}t = 300$. Velocities are scaled to $v_0 = \omega_{po}a$. (b) Axial dependence of self-consistent potential $\phi(x = 0, z)$ scaled to $U = mv_0^2/2$. Extended sheath is flat region on right-hand side.

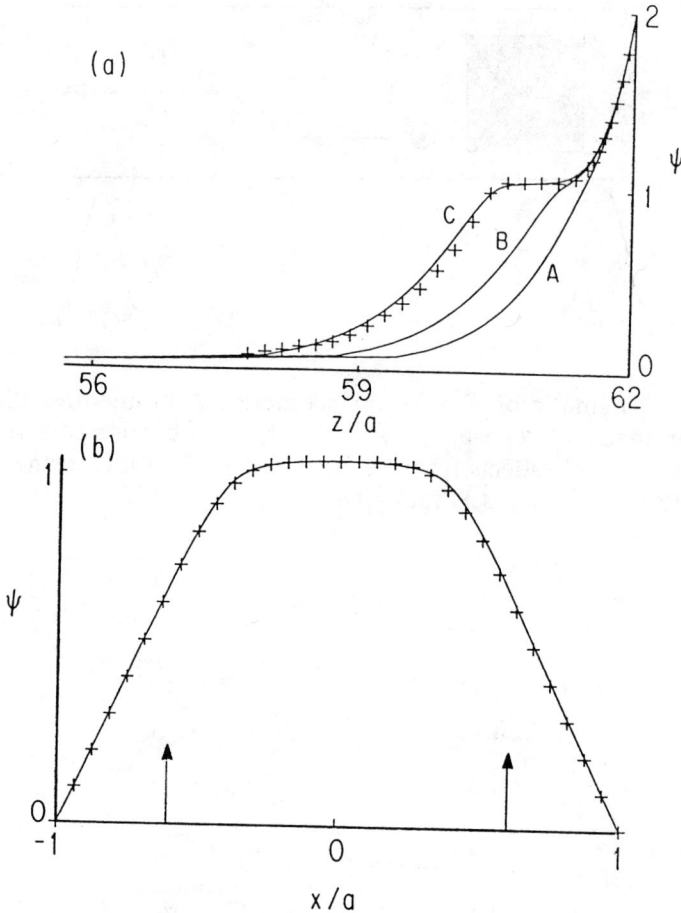

Fig. 3 (a)   Axial dependence of sheath. Crosses indicate simulation result. Curves A, B, C correspond to analytical solutions for p=1.3, 2.1, 2.2, Δ = 0.025. b) radial dependence of potential in flat region of curve C. Solid curve is analytical prediction and crosses are the simulation result. Vertical lines mark the plasma edge (d = 0.6).

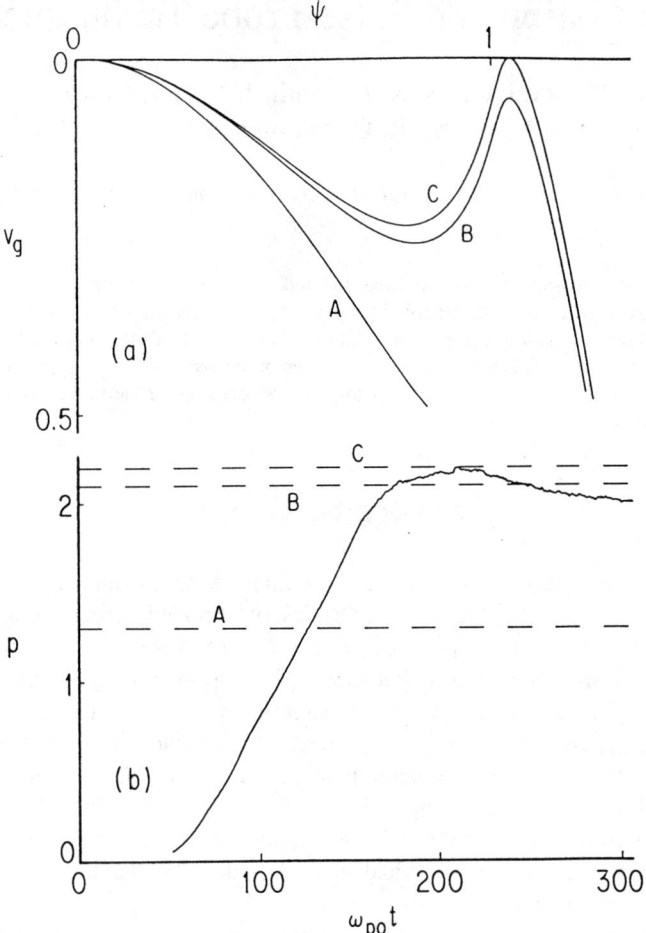

Fig. 4 (a) Effective potential of Eq. (4) for p = 1.6, 2.1, 2.2, Δ = 0.025. A, B, C correspond to the profiles of Fig. 3a. (b) Time evolution of the parameter p in simulation. Dashed lines A, B, C correspond to labels in Figs. 3a. (b) Time evolution of the parameter p in simulation. Dashed lines A, B, C correspond to labels in Figs. 3a and 4a.

A Study of Trapped Ion Dynamics by Photon-Correlation and Pulse-Probe Techniques

J. Rink, K. Dholakia, G. Zs. K. Horvath, J. L. Hernandez-Pozos, W. Power, D. M. Segal, R. C. Thompson and T. Walker

Blackett Laboratory, Imperial College, Prince Consort Road, London SW7 2BZ, UK

Abstract. We demonstrate non-invasive methods for observing ion and ion cloud oscillation frequencies in a quadrupole ion trap. These trap resonances are measured for small clouds using a photon correlation technique. For large clouds the rotation frequency can be detected with the help of an additional pulsed probe laser. We show applications of the photon correlation method such as estimating the dynamic properties of a combined trap and detecting ion crystals.

INTRODUCTION

Since the introduction of laser cooling in the late seventies, ion traps have had a formidable impact on many branches of physics, most notably spectroscopy (1) and quantum optics (2). It became possible to optically detect small ion clouds down to a single ion. For many trap applications, for instance high resolution laser spectroscopy (3), frequency standard experiments (4) and comparison of theoretical descriptions of traps (5), it is necessary to study ion trap dynamics in detail. One of the most common methods to determine oscillation frequencies is to apply a small additional oscillating voltage to the trap electrodes. However, this technique measures only centre-of-mass modes of the ion cloud and not characteristic frequencies of individual ions. Additionally, the ions are disturbed during the measurement (6).

Recently, we have developed two non-invasive methods for measuring ion trap oscillation frequencies which are based on a statistical analysis of the ion's fluorescence (7)(8). It is then possible to determine quickly and accurately all the ion's oscillation frequencies at the same time. The ion motion in a trap can take the ion in and out of the light beam. Also, due to the Doppler effect, it moves the ion in and out of resonance with the laser frequency. Both effects modulate the resonance fluorescence of the ion. Subsequent photons from the same ion show characteristic time delays related to the oscillation frequencies in the trap. The probability that consecutively detected photons will have been emitted from the same ion becomes smaller with larger ion clouds. Therefore this technique is applicable to relatively small clouds only, but measures all trap resonances.

The second method enables us to measure the ion-cloud rotation frequency in the Penning trap. In this sense it is complementary to the one described above, since there is no limitation on cloud size. In addition to the normal cooling laser beam, a pulsed laser beam tags a section of the ion cloud. This is done by optically pumping them into a metastable state and therefore removing them from the cooling cycle. As a result, the fluorescence intensity as a function of time decreases when the tagged portion of the ion cloud moves through the cooling laser beam. The frequency of the fluorescence modulation then gives the ion cloud rotation frequency or a simple multiple of it.

EXPERIMENT

The apparatus has been described in detail elsewhere (7). A cloud of Magnesium ions is stored in a hyperbolic trap with r_0=5 mm, which is situated in a vacuum system at a pressure of $2 \cdot 10^{-10}$ mbar. If run as a Penning trap, the dc voltage between the endcaps and the ring electrode is 10 V and the magnetic field is about 1 T. During operation as a Paul trap, an ac voltage (V_0=235 V, $\Omega/2\pi$=2.57 MHz) is applied. The 280 nm light required for laser cooling is now generated in a BBO crystal inside a dye laser cavity. Typically, a 100 µW beam is focused to a waist of 100 µm at the position of the ion cloud. Its polarisation is perpendicular to the magnetic field (In previous experiments, an AD*P crystal outside the laser cavity was used). In order to cool the ions in the Penning mode, the laser beam is shifted away from the trap centre towards the side where the magnetron motion causes the ions to move away from the laser. The fluorescence is imaged by a focusing system onto a cathode of a photomultiplier. For the photon-photon correlation data, photomultiplier pulses were converted to TTL pulses. The arrival of the first pulse starts a time-to-amplitude converter and the next pulse stops it. An output pulse whose amplitude is proportional to the delay between the two pulses is sent to a multichannel analyser card in a computer. The ion oscillation frequencies are extracted from the multichannel analyser data by performing a fast Fourier transform.

For the pulse probe experiment radiation near 280 nm is generated by frequency doubling of a multimode pulsed dye laser pumped by the second harmonic of a Q-switched Nd:YAG laser. Typically, 10 ns light pulses of 1-10 µJ energy and a repetition rate of 10 Hz are used. The pulsed beam counterpropagates with the cooling beam.

PHOTON CORRELATION MEASUREMENT OF ION OSCILLATION FREQUENCIES

Technique

The method is described in detail in (7). Fluorescence from a small ion cloud is collected for several minutes with the multichannel analyser, as described above. For randomly arriving pulses the analyser output, which displays the time intervals between consecutively collected photons, would show an exponential decay. The periodic modulation of the fluorescence, caused by the ions changing velocity and moving in and out of the beam, can be seen in the modulation of the trace. Its Fourier transform gives the ion oscillation frequencies. Figure 1 shows a trace measured on ions in a combined trap. Frequencies (i), (ii), (iv) and (v) correspond to the three trap resonances and the frequency of the applied ac voltage. The other two peaks are sum frequencies which are artefacts of the Fourier transform process. If only one ion is present in the trap, each photon of the fluorescence is correlated with the next. The photon-correlation trace of the time intervals between successive photons then has a maximum modulation. This shows that we indeed measure the individual ion oscillation frequencies and not the ion cloud oscillations. With increasing number of ions the modulation gets weaker since only successive photons from the same ion contribute to the modulation on the correlation trace. For clouds of more than 100 ions the modulation is washed out. However, it is worth noting that there is one exception. In a pure Paul trap there is a constant phase of the micromotion between the ions. The micromotion frequency can therefore be detected from any cloud size and it is convenient to use this for frequency calibration of the signal.

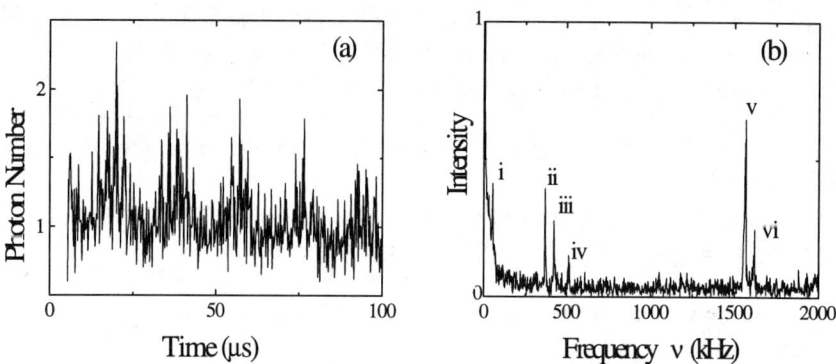

FIGURE 1. Photon correlation trace of ions in a combined trap (a). Its Fourier transform (b) gives all the ion oscillation frequencies.

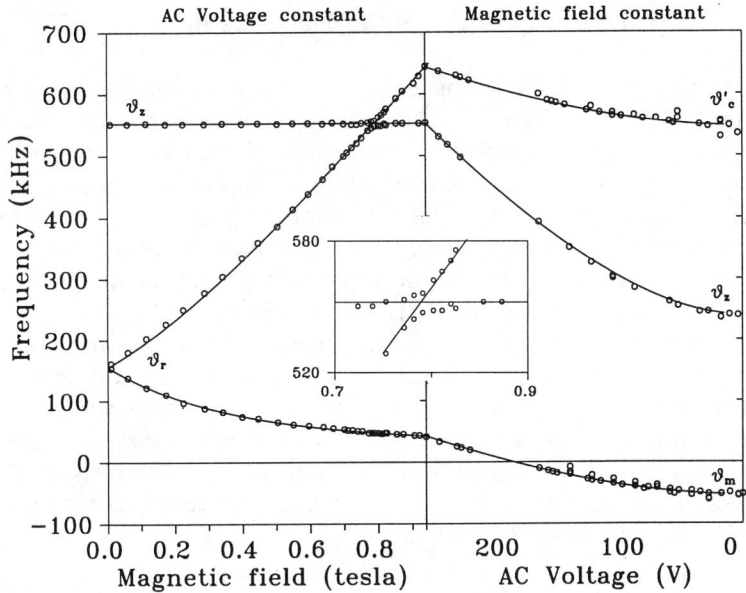

FIGURE 2. Oscillation frequencies in a combined trap. The Penning mode is shown on the left and the Paul mode on the right of the diagram.

The combined trap

The vast majority of work done with trapped ions has been performed using either the Penning trap or the Paul trap (9). However, dependent on the particular application, there are some advantages in confining charged particles in a hybrid trap where a magnetic field and dc and ac potentials are applied. The stability regions are larger than those of a Paul trap and it is possible to trap ions of opposite charge and different charge to mass ratio. It has for instance been suggested that one could create and study anti-hydrogen in such a trap (10).

Figure 2 shows our experimental data obtained with the photon correlation technique (11). If one starts from a pure Paul trap on the left, the radial frequency is shifted and splits into two branches. The axial frequency remains unshifted. This is due to the fact that the magnetic field is parallel to the z-direction. Since the ac potential provides a trapping force in any direction, all three oscillation frequencies decrease with reduced amplitude of the ac voltage and eventually become the magnetron, axial and modified cyclotron motion of a pure Penning trap. The solid lines represent the theoretical treatment of the combined trap and show that the basic principles of ion motion in a combined trap are well understood.

Precise alignment of the magnetic field

The inset of figure 2 shows an interesting feature of the combined trap at a magnetic field of about 0.8 T. The axial frequency and the upper radial frequency are expected to be degenerate according to theoretical predictions. Experimentally, we observe an 'avoided crossing' between the two oscillation frequencies. This coupling is due to the fact that the magnetic field is slightly misaligned with respect to the axial direction. The resultant angle between the magnetic field direction and the trap axis is proportional to the frequency difference between these two frequencies near the crossing point (12). This can be seen in figure 3, where avoided crossings are observed for two different angles. The graph obtained for an angle of two degree shows the closest gap of the avoided crossing with respect to the direction where the angle was varied. This indicates that the trap is still misaligned in the direction perpendicular to the one which was changed. In experiments where a precise alignment of the magnetic field with respect to the trap is necessary, the crossing of the two oscillation frequencies in a combined trap can provide an adjustment better than one degree.

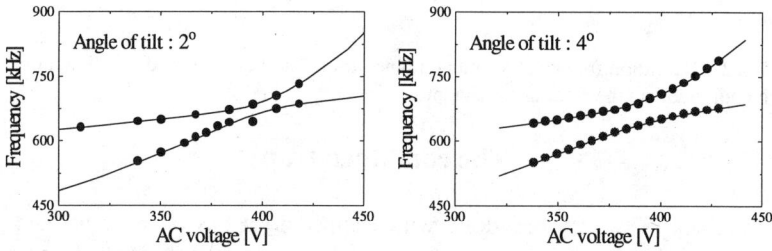

FIGURE 3. Avoided crossings of the axial and the upper radial oscillation frequency in a combined trap at two different angles between the trap axis and the magnetic field.

Ion crystals in a Penning trap

The most common way to observe a crystalline behaviour of ions in a Paul trap is to take an image of the ion crystal. Field imperfections prevent the ion crystal from rotating and an image can be taken (13). This technique is not applicable in a Penning trap because the ions rotate with the magnetron oscillation frequency. It is possible with the photon-correlation method to detect and to study crystallisation in a Penning trap. When the ions are very cold they form a regular pattern where their relative positions is fixed. Because of this correlation in position we expect their emitted photons to be correlated in time as well. One of the simplest cases is shown in figure 4, where three ions of a crystal are held in the radial plane. Because of their correlation we expect a modulation in the photon-correlation trace at three times the magnetron frequency. Since the number of fluorescence photons per ion per magnetron cycle is larger than one in the case of strong cooling most successive photons come from two ions next to each other. The photon-correlation trace, which in principle measures the time intervals between two photons, therefore would show in the above case the presence of higher multiples of the magnetron frequency and the absence of the fundamental. This is indeed what we have seen in one of the multichannel traces taken from ions in a pure Penning trap. After the dc potential applied to the trap was increased on the same ion crystal the measured oscillation frequency changed from a three-fold to a four-fold symmetry and the oscillation frequency corresponds to four times the (new) magnetron frequency (Fig. 5). That indicates a change in the crystal structure.

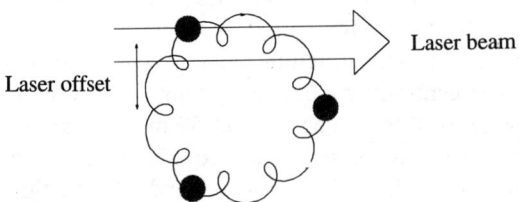

FIGURE 4. Detection of an ion crystal in a Penning trap with the photon correlation technique.

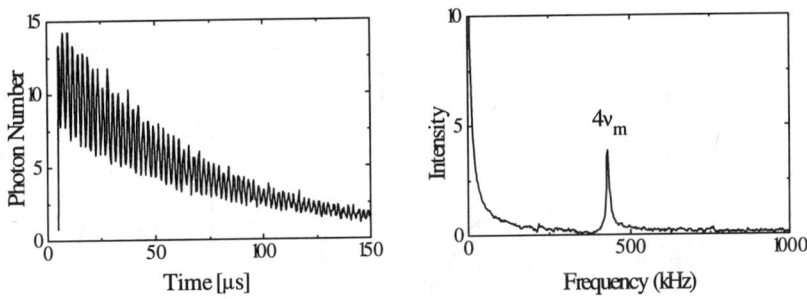

FIGURE 5. Photon correlation trace and its Fourier transform of a crystal in a Penning trap.

These preliminary data show that the photon correlation method is suited for observing ion crystals. Nevertheless, the interpretation is not straightforward. One needs additional information about the number of ions present in the trap. This can be done by observing the fluorescence intensity as a function of the laser frequency. Another problem arises because different crystal formations could give rise to a similar correlation trace, because this technique can only give the symmetry properties of a crystal.

PULSE-PROBE DETECTION OF THE ION CLOUD ROTATION FREQUENCY

In order to study the motion of ions within a large cloud, the ions are pumped into a metastable state with a light pulse short compared with the magnetron frequency. The cooling laser is tuned as usual to the closed cycle. The pulsed beam excites a fraction of the ions into the metastable $M_J=+1/2$ level, consequently removing them from the cooling cycle. However, there is a probability for the ion to undergo a Raman process back to the ground state. The pulsed laser beam counterpropagates with the cooling beam. As a result, the fluorescence as a function of time shows dips when the tagged portion of ions interact with the cooling beam. The rate at which this occurs gives the rotation frequency of the cloud. Apart from this magnetron period two other timescales are important. One is the time it takes for the ions to return to the ground state. This was measured by averaging the fluorescence over a time long compared with the magnetron period. This repumping time is of the order of 10 to 50 ms and varies with cloud size and temperature. The minimum repumping time of about 10 ms would occur for very small, cold clouds in which the ions would spend all their time in the laser beam and are resonant with the laser frequency.

The other timescale describes the diffusion of the tagged ions within the cloud and in our case is shorter than the repumping time. It can be determined by looking at the decreasing amplitude of the dips.

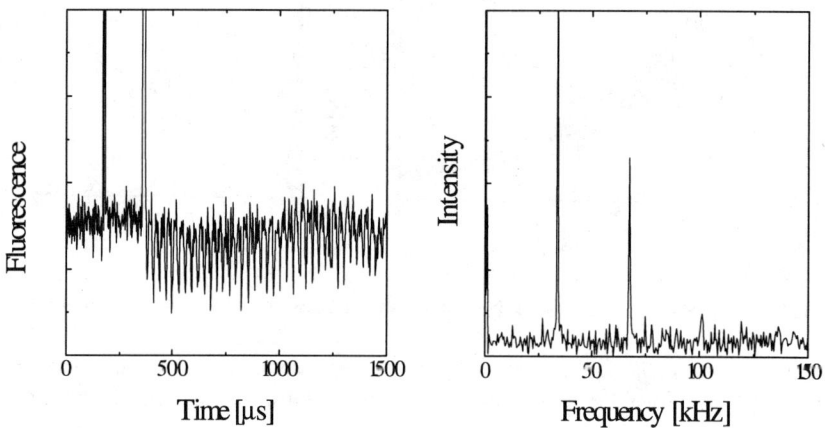

FIGURE 6. Fluorescence trace modulated by the pulsed laser. The two spikes at the beginning of the trace result from pickup noise by the detection system associated with the pulsed laser trigger pulse. The Fourier transform gives the cloud rotation frequency. The second harmonic visible is an artefact of the Fourier transform process.

A typical fluorescence trace modulated by the pulsed "tagging" laser is shown in figure 6. Its Fourier transform gives the cloud rotation frequency. The second harmonic visible in the trace results from the fact that the dips are non-sinusoidal and has little to do with ion dynamics. Note, that this method works better with large clouds while the photon-correlation technique is applicable only to small clouds. In this sense, the two methods are complementary. For a more detailed discussion and applications of the pulsed-probe technique such as measuring the space-charge shift of the cloud rotation frequency see (8).

CONCLUSION

We have developed two complementary optical techniques for measuring ion oscillation frequencies in traps. One is based on photon-correlation and the other is based on tagging a fraction of the ions using an additional pulsed laser. The photon correlation technique was used to measure all the ion oscillation frequencies in a combined trap for a wide range of parameters. The observed avoided crossing, caused by a slight misalignment of the trap axis with respect to the magnetic field direction, has been studied and can be used for precision trap alignment. We showed that ion crystals can be detected with the photon-correlation method and we will study ion crystals in more detail in future.

REFERENCES

1. Nagourney, W., Yu, N., Dehmelt, H., Opt. Comm. **79**, 176-180 (1990)
2. Bollinger, J. J., Heinzen, D. J., Itano, W. M., Gilbert, S. L., Wineland, D. J., Phys. Rev. Lett. **63**, 1031-1034 (1989)
3. Kälber, W., Meisel, G., Rink, J., Thompson, R. C., J. Mod. Opt. **39**, 335-348 (1992)
4. Bollinger, J. J. , Prestage, J. D., Itano, W. M., Wineland, D. J., Phys. Rev. Lett. **54**, 1000-1003 (1985)
5. Vedel, F., Int. J. Mass. Spectrom. Ion Processes **106**, 33-61 (1991)
6. Brown, L. S., Gabrielse, G., Rev. Mod. Phys. **58**, 233-311 (1986)
7. Dholakia, K., Horvath, G. Zs. K., Segal, D. M., Thompson, R. C., Warrington, D. M., Wilson, D. C., Phys. Rev. A **47**, 441-448 (1993)
8. Dholakia, K., Horvath, G. Zs. K., Power, W., Segal, D. M., Thompson, R. C., Appl. Phys. B, in press, (1994)
9. Blatt, R., Gill, P., Thompson, R. C., (ed.), *Special Issue: Physics of Trapped Ions*, J. Mod. Opt. **39**, 193-443 (1992)
10. Li, G.-Z., Commun. Theor. Phys. (China) **12**, 355 (1989)
11. Dholakia, K., Horvath, G. Zs. K., Segal, D. M., Thompson, R. C., J. Mod. Opt.

39, 2179-2185 (1992)

12. Horvath, G. Zs. K., Doctoral thesis, Imperial College London, (1995) (in preparation)

13. Wineland, D. J., Bergquist, J. C., Itano, W. M., Bollinger, J. J., Manney, C. H., Phys. Rev. Lett. **59** , 2935-2938 (1987)

Trapping of Dust Particles in a Kingdon Trap

Scott Robertson

Department of Astrophysical, Planetary and Atmospheric Sciences
University of Colorado, Boulder CO 80309-0391

Abstract. Experimental techniques for charging, trapping, and observing macroscopic particles (dust) are described. Particles with a sufficiently large charge-to-mass ratio (4×10^{-3} C/kg) are created by using hollow glass microspheres about 50 μm in diameter charged to a potential of -30 volts by electrons from a filament. The charged spheres drop by gravity into a Kingdon trap and are trapped in near-circular orbits by a rising potential. A spherical version of the trap with a Keplerian potential is also used. Illumination from a halogen lamp is sufficient for videotapes to be made of particle dynamics. Phenomena observed in the Kingdon trap include libration due to a resonant perturbation and the resonant coupling of two adjacent orbiters. Precession due to gravity is observed in the spherical trap. Confinement times in the Kingdon trap approach an hour. Briefly discussed are the use of charged liquid droplets, a planar trap based upon the ponderomotive force and the collective phenomena of a non-neutral plasma in a Kingdon trap.

I. INTRODUCTION

The first observations of crystal-like structure due to strong electrostatic coupling in a non-neutral plasma were made in a Paul trap with small particles of aluminum.[1] The coupling is described by the parameter $\Gamma = Q^2/4\pi\varepsilon_0 Td$, where Q is the charge on the particles, T is the temperature in energy units, and d is the mean particle spacing. The charge which can be put on macroscopic particles, typically about 10^{-13} C, results in coupling which is about 10 orders of magnitude larger than for singly-charged ions with the same temperature and spacing. For example, at room temperature with a charge of 5×10^5 electrons and a spacing of 1 cm, one finds $\Gamma \approx 10^6$. This value suggests that macroscopic particles ("dust") offer a means of easily obtaining a strongly-coupled, one-component plasma. The value of experiments with dust, however, has been limited by the spread in the charge-to-mass ratio. The experiments described here are aimed at developing new experimental techniques for preparing, trapping, and observing macroscopic particles which are more uniform in charge and mass than suggested by the label "dust".

An advantage of dust is that photography is simplified by the relatively large cross section for light scattering. An incandescent source rather than a laser gives sufficient illumination for the particles to be imaged by CCD cameras. The large cross section and the relatively low velocity of dust particles compared with that of ions allows the trajectories of individual particles to be recorded. This feature has allowed the study of dynamics in a Paul trap[2] and suggests that experiments on dynamics with other types of potentials should be possible. The electrical detection of particles is also made easier. Individual dust particles create an image charge in a nearby conductor which is easily detected. The laser cooling applied to ions can be replaced with collisional cooling of dust by a background gas.

In Sec. II, techniques are described for preparing charged macroscopic particles with a known charging potential and a narrow range of size. Sec. III and IV describe experiments in which these particles are trapped in orbits in electrostatic potentials which are cylindrically or spherically symmetric. Sec. V describes a ponderomotive trap which confines particles to a planar region. A summary and conclusion is presented in Sec. VI.

II. DUST PREPARATION, CHARGING AND DETECTION

Several new techniques have been developed for launching, charging and detecting macroscopic particles which will be reviewed here without the details given elsewhere.[3,4] A difference between dust particles and ions is the much larger force of gravity. The charge-to-mass ratio of the particles should be made as large as possible to maximize the ratio of the electrostatic force to the gravitational force. Hollow glass microspheres with an effective density of 0.3 g/cm^3 are available in large quantities as a light-weight additive for plastic. These spheres have diameters from 20 to 200 μm and a wall thickness of about 1.6 μm. The size distribution is reduced to a narrow range by using screen mesh sieves. Trapping experiments are usually performed with a fraction having passed through sieves with 53 μm openings and not having passed through 40 μm openings (Fig. 1). The samples are sieved several times to remove smaller spheres and pieces of spheres which adhere to the desired fraction. The size is determined by balancing the needs for a large scattering cross section and for a reasonable trapping potential. A hollow sphere with a diameter of 50 μm has a calculated mass of 2×10^{-11} kg.

A particle stream is created by vibrating a reservoir of spheres causing them to drop through a hole in the bottom (Fig. 2). The reservoir is vibrated by an electromagnet which is activated by a rectangular current pulse or a train of pulses. The amplitude and width of the pulses determines the approximate number of spheres dropped per pulse. The falling spheres are charged as they pass a filament emitting electrons. The bias potential on the filament determines the energy of the

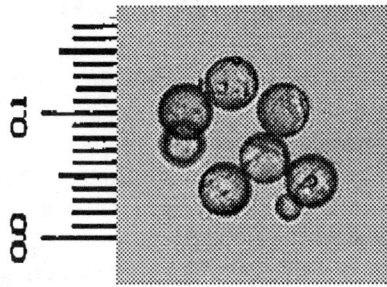

FIGURE 1. Hollow glass microspheres used in the trapping experiments. The spheres have been sieved to select a fraction with diameters in the range 40-53 microns. The scale measures 0.1 mm.

incident electrons. The angular deflection of the electrons by the charge on the spheres results in a charging current which decreases to zero asymptotically as the filament potential is approached. The charging current as a function of potential may be derived from the theory of Langmuir probes.[5] The current from the filament must be made sufficiently large (0.2 mA) for the charging time to be shorter than the time the particles are exposed to the electrons. The spheres are observed to be deflected from a downward trajectory by the potential on the filament unless shielded by a grounded mesh. The velocity of the spheres is determined by the acceleration of gravity.

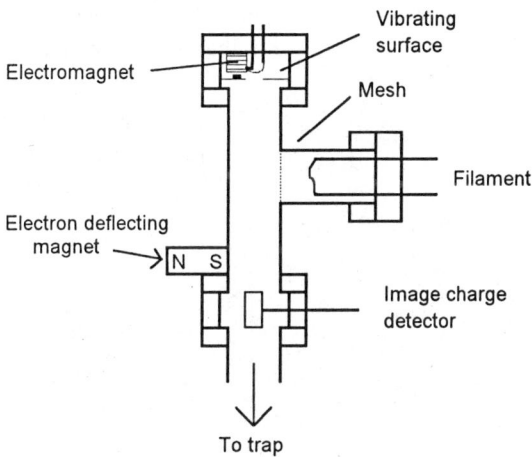

FIGURE 2. Schematic diagram of the particle launcher.

The charge on the spheres is determined by $Q = CV$, where Q is the charge, C is the capacitance of an isolated sphere and V is the filament potential. A typical capacitance is 3 fF and a typical charge for a -30 volt charging potential is 5×10^5 electrons. The charge on the spheres is significantly reduced from the expected value if the incident electrons are sufficiently energetic to cause the emission of secondary electrons. The transition to this behavior occurs near -40 volts for glass which sets a limit on the filament potential. The charge is measured by having the particles fall into a Faraday cup connected to a sensitive amplifier. The passage of dust may be detected without interrupting the stream by replacing the Faraday cup with an open cylindrical tube in which the particles induce an image charge. A typical charge-to-mass ratio is 4×10^{-3} C/kg and the scaling of charge and mass with radius imply that this ratio varies inversely with size for hollow spheres. The spread in charge-to-mass ratio results from the known variation in size and the unknown variation in wall thickness.

The trajectories of particles are recorded by video photography. A CCD still camera is used (Electrim EDC-1000) as well as a small video camcorder (Sony FX-620). The cameras are placed 30 cm from the center of the Kingdon trap and have their focal length adjusted for viewing an area approximately 5 cm in width. The still camera is usually used with an exposure time of 0.2 sec which records about six orbits. Still images with 1/30 second exposure or less may be obtained from the videotapes. Images may be enhanced by background subtraction and other manipulations. The orbits are viewed either perpendicular or parallel to the orbital plane.

A source of charged liquid droplets is under development as a possible replacement for the hollow microspheres. An image charge can be placed on a liquid stream with a biased cylindrical conductor (Fig. 3). The image charge on the stream is determined from the formula for the charge per unit length on biased

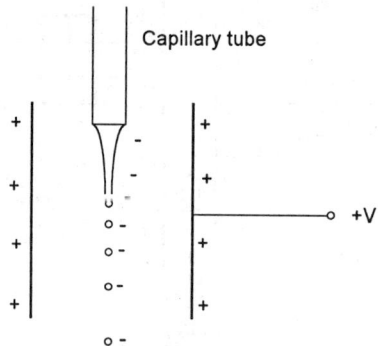

FIGURE 3. Illustration of the technique for creating charged liquid droplets. A biased cylinder is placed around the flow from a capillary tube which spontaneously breaks into droplets.

coaxial conductors. Droplets carry the charge and mass of the length of the fluid stream which coalesces to form the droplet. This suggests that the charge-to-mass ratio is insensitive to variations in the size of the resulting droplets. The stream may be allowed to break into droplets spontaneously or may be aided by launched acoustic waves. The liquid must have a minimum electrical conductivity; however, nonconductors such as diffusion pump oil may be used if seeded with an appropriate additive.[6]   Coulomb fission of the droplets[7,8] or the onset of electrospraying at the nozzle[9] sets an upper bound on the charge. Charges as large as 10^{-12} C have been reported on droplets from a 55 μm diameter capillary.[10] Droplets created in this way have been suspended in a Paul trap for fluorescence spectroscopy.[11]   We have generated streams of charged water droplets from a hypodermic needle with an inner diameter of 120 μm. The size of droplets was not measured but is typically 1.5 times the stream diameter. The largest charge obtained is estimated from the measured droplet current to be 2 x 10^{-12} C. The velocity and mass of these droplets is too large for confinement in either the Kingdon trap or the ponderomotive trap described in Sec. V. These droplet experiments are continuing with smaller capillary tubing in order to obtain an increased charge-to-mass ratio.

III. KINGDON TRAP

The confinement of macroscopic particles in the Kingdon trap[12,13,14] has been discussed previously and the experimental apparatus has been described in detail.[3] The design of the apparatus (Fig. 4) will be briefly reviewed and the results of recent experiments will be presented. A radially confining electrostatic field is created by a biased central rod electrode and a grounded concentric outer conductor. Axial confinement is provided by conducting end caps on the outer cylinder which generate curved equipotentials concave toward the axis. The radial variation of the potential is approximately logarithmic and orbits are rosettes. The period of the axial motion is much longer than the period of the orbital motion. The nature of orbits in axisymmetric potentials has been discussed in detail in the astrophysical literature.[15,16]

The cylindrical outer conductor is 7.5 cm in diameter and 6.5 cm long. There are four holes in the conductor to allow the entry and exit of dust particles and of the light beam. There is a beam dump and a viewing dump to reduce scattered light and to increase the contrast of the images. The light source is a collimated beam from a 100 W low-voltage halogen lamp. The trapping potential of 8 to 12 kV is applied by a mechanical vacuum relay after a delay of 225 ms which is the time for the dust to fall from the dropper to the level of the center rod. A mesh sector electrode is located at the bottom of the outer conductor to allow the application of electrostatic perturbations. The dust dropper is located above the trap and offset laterally 1 cm from the trap axis so that the particles do not

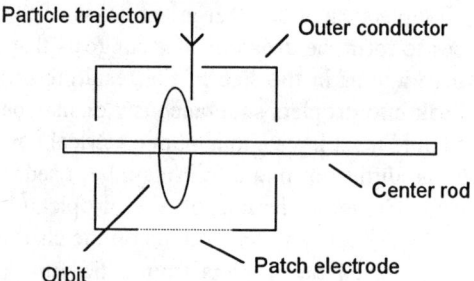

Particle trajectory

Outer conductor

Center rod

Patch electrode

Orbit

FIGURE 4. Schematic diagram of the Kingdon trap. The particles enter from the top and are trapped by a rising potential on the center electrode.

strike the center rod. A typical orbital radius of 1 cm and the particle velocity within the trap of 2 m/s determine an orbital frequency of 30 Hz. The variation in the time that the particles leave the dropper results in considerable variation in the size and eccentricity of orbits.

Orbits usually begin as precessing ovals and circularize after a few minutes (Fig 5). The orbits slowly decay in radius and eventually strike the center conductor. The orbital decay time is determined by molecular drag. The ionization gauge reading is 1×10^{-6} Torr which gives a calculated decay time of few hours if a residual gas of water vapor is assumed. The observed decay time is from several minutes to one hour ($\sim 10^5$ orbits) and is shorter than the calculated time perhaps due to the uncertainty in the vacuum conditions. Numerical integration of the equations of motion with drag show that orbits decay with constant eccentricity; thus the circularization is not explained by molecular drag. Eccentric orbits induce varying image charges on the center and outer electrodes

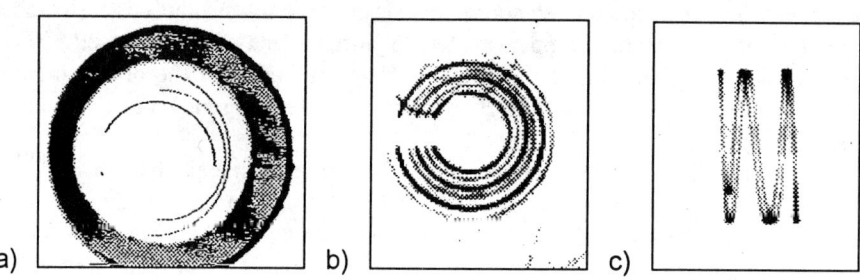

a) b) c)

FIGURE 5. Negative images of particles in the Kingdon trap. (a) Orbit segments of three orbiters recorded in one video frame. (b) Exposure of 0.2 seconds recording six orbiters. The shadow of the central rod is cast to the left. (c) Exposure of 0.2 seconds recording a single orbiter viewed perpendicular to the axis. The orbital motion and axial bounce motion produce the Lissajous pattern. The orbital radii are 1 to 2 cm.

which may account for the circularization. However, estimates made using the model developed for Paul traps[17] indicate a decay constant of the order of a day. The determination of the mechanisms for circularization and decay will require additional experiments with variable pressures and circuit parameters.

Experiments have been performed to observe the effects of sinusoidal perturbations. A sine wave generator is used to apply an alternating voltage to the mesh patch electrode at the bottom of the trap. The applied potential is typically 200 volts with a frequency of 30 to 80 Hz. Similar perturbations occur in celestial mechanics; for example, the perturbation of the orbits of asteroids by Jupiter may lead to dynamical chaos.[18] A perturbation at the orbital frequency (1:1 resonance) will accelerate particles (as in a cyclotron) if phased properly. The resulting larger orbit will have a smaller orbital frequency and the phase relationship will change so that the particle is alternately decelerated and accelerated. This libration (Fig 6a) is a slow oscillation in radius about the resonant orbit defined by the applied frequency.

Experiments have also been performed with several particles orbiting simultaneously. A half-dozen orbiters can be trapped if a much larger number of particles is launched by the dropper. The orbits of the particles usually appear to develop independently of one another; however, on several occasions motion has been observed indicating that particles with small radial separations are electrostatically coupled. For example, radial oscillations of adjacent orbiters have been observed apparently due to their perturbations upon one another (Fig 6b). Coupling of adjacent orbiters has been analyzed to explain the motion of co-orbital satellites observed by the Voyager spacecrafts.[19,20] The gravitational and electrostatic forces have different signs, of course, which changes the phase of the perturbation by 180 degrees. Unequal charge-to-mass ratios can result in particles

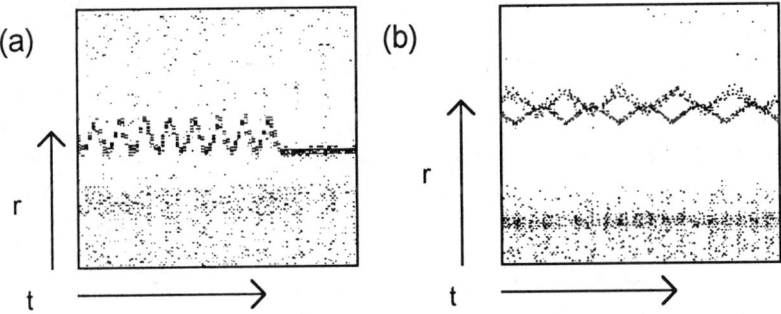

FIGURE 6. Negative streak photographs constructed from vertical lines from successive video images. (a) Oscillations in radius due to a 1:1 perturbation. The radial oscillation is at 4 Hz and the orbital frequency is approximately 30 Hz. The image scale is approximately 2 cm vertically and 2.5 s horizontally. (b) Oscillations due to coupled motion of two orbiters. The image scale is approximately 2 cm vertically and 5 s horizontally. The orbits do not cross.

at slightly different radii having the same orbital frequency and consequently a 1:1 resonance. This is apparently the case in Fig. 6b where the mean orbital radii of the particles are slightly different. In the present apparatus, axial and orbital motion are not simultaneously observed thus the coupled motion cannot be interpreted unambiguously.

A non-neutral plasma of orbiting macroscopic particles is possible if the number of orbiters can be further increased. In the Kingdon trap, this plasma is a disk with sheared rotation and collective effects will be analogous to those in planetary rings and accretions disks.[21] Expected wave phenomena include spiral density waves which propagate radially outward and spiral bending waves which result in motion perpendicular to the orbital plane. Coulomb collisions cause dissipation of orbital energy and result in the spreading of ring particles to larger and smaller radii. The number of particles necessary to be a plasma is approximately 10^3 if the transition is defined to occur where the plasma frequency is comparable to the orbital frequency.

IV. SPHERICAL TRAP

The literature on celestial mechanics primarily concerns Keplerian potentials varying inversely with radius rather than logarithmic potentials. A Keplerian electrostatic potential can be created by the use of concentric spheres. An obvious experimental difficulty is the distortion in the spherical potential caused by the supporting stalk for the central sphere. The distortion is reduced to a small level by placing around the stalk biased surfaces which have potentials which nearly match the desired potential (Fig. 7). Three electrically biased brass tubes are placed over the supporting stalk and the potentials are arranged in descending order from the center outward. The cylinders extend into the trap to a radius corresponding to the distance at which the cylinder potential matches a $1/r$ potential. The potentials are 0.6, 0.4, and 0.2 times the potential on the central sphere. A cylinder with 0.8 of the central potential is not used because of the small spacing between the central sphere and the cylinder with 0.6 of the central potential. The computed potential in cylindrical geometry shows that the outer conductor can be a cylinder with end caps (as in the Kingdon trap) without a significant perturbation to the $1/r$ dependence in an orbital plane perpendicular to the stalk.

A static perturbation to an orbit will result in precession which may cause the orbiter to strike the supporting stalk. In initial experiments with the spherical trap, the orbiters made only a few orbits before being lost. Computer simulations with both the electrostatic and gravitational forces showed that the orbits rapidly precessed so that the orbital plane intersected the stalk. This problem could be solved by using a central sphere supported from below and particles injected into a horizontal plane by the use of electrostatic deflection. In this case precession

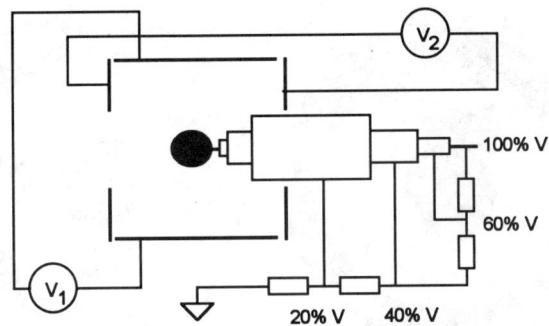

FIGURE 7. Schematic drawing of the Kingdon trap with a spherical center conductor. The potential on the supporting stalk is matched to a 1/r potential by biased concentric cylinders. A potential difference V_1 (typically ±60 V) is applied to two sides of the outer cubical conductor to cancel the force of gravity. The potential V_2 (typically ±260 V) is applied to ensure that the mean orbital plane is approximately perpendicular to the stalk. The two remaining sides of the box are grounded.

causes the orbital plane to make small excursions about the horizontal plane with the maximum tilt determined by the initial tilt. Gravity causes the plane of the orbit to be below the plane of central sphere. The geometry of the vacuum chamber (vacuum pump mounted below) made this solution difficult. The solution adopted is the cancellation of gravity by an additional upward electrostatic field. A third electrostatic field parallel to the stalk is applied to define a "preferred" axis so that the precession consists of small excursions about the desired orbital plane perpendicular to the stalk.

The application of vertical and horizontal electric fields requires that the outer conductor be split into six pieces. The outer conductor is a cube with electrically isolated sides (Fig 7). The top and bottom are biased to approximately cancel gravity, the sides perpendicular to the stalk are biased so that the residual perturbing force vector is nearly parallel to the stalk and the remaining sides are grounded. The sides of the cube are 5 cm in length and the center sphere is 1.2 cm in diameter. The light scattered from the central sphere is reduced by imaging the lamp filament onto a small occulting disk before it is imaged within the trap. The reduced efficiency of the optical system requires the use of a 500 W projection lamp. A second occulting disk is placed between the central sphere and the camera if long exposures are made.

Typical orbits are shown in Fig 8. Elliptical orbits initially precess slowly perhaps due to the lack of spherical symmetry of the outer conductor. As in the Kingdon trap, the eccentricity is damped and the orbits become stationary circles after a few minutes. In these initial experiments, the confinement time is shorter than in the Kingdon trap for reasons which have not yet been determined.

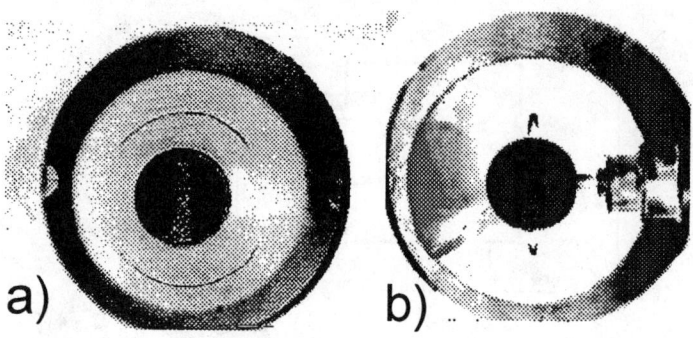

FIGURE 8. (a) Negative photograph of an orbit in the spherical trap viewed perpendicular to the orbital plane. The central electrode is occulted by the disk in the center. The outer dark annulus is the solid rear surface of the trap. The orbit, a thin black circle, is seen against a blackened viewing dump which appears white. The missing sections of the orbit at the left and right are due to the annular light beam. (b) A different orbiter viewed nearly parallel to the orbital plane.

V. PONDEROMOTIVE PICKET FENCE

The ponderomotive picket fence, Fig. 9, is a planar confinement system under development for studying a two-dimensional, strongly-coupled, one-component plasma. (Planar arrays of particles with hexagonal symmetry have recently been observed in a multicomponent dusty plasma.[22]) The trap has its name derived from the picket fence plasma confinement system[23] in which parallel wires carrying oppositely directed currents create a confining magnetic field at the surface of a plasma. In the electrostatic system, confinement is from the ponderomotive force due to the spatial variation of an oscillatory electrostatic field. The static field of an alternating wire grid in the far-field approximation is

$$\vec{E}(x,y) = [\, D\, e^{-ky} \sin(kx),\ D\, e^{-ky} \cos(kx),\ 0\,], \tag{1}$$

where the wires are parallel to the x axis, D is a constant, $k = \pi/\delta$ and δ is the spacing between adjacent wires. The error in this approximation is of order e^{-2ky}. Near the wires, the potential is logarithmic. Dust particles trapped in the near field settle preferentially into the region between wires. The ponderomotive acceleration for potential sinusoidal in time is[24]

$$\ddot{y} = -\frac{Q^2}{2m^2\omega^2} \frac{\partial}{\partial y}\left\langle E^2 \right\rangle = \frac{D^2 k Q^2}{m^2\omega^2} e^{-2ky}, \tag{2}$$

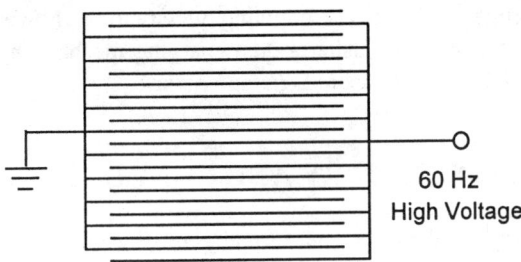

FIGURE 9. Schematic diagram of the picket fence trap.

where m is the mass and ω is the frequency of the applied voltage. There is no dependence on x or z in the far field. Dust particles are suspended in the plane where the ponderomotive acceleration cancels the gravitational acceleration.

The device is modeled with a two-dimensional computer program which integrates the equations of motion. A collisional drag is included to remove momentum from the vertical and horizontal motion. Particles are given some initial downward velocity from gravity. Particles are observed to settle into a stable equilibrium position with micromotion, to pass through the wires without being trapped, or to bounce in a seemingly chaotic manner until lost through the wires. There is a lower bound on the voltage arising from the requirement of there being a net potential well from the combined gravitational and ponderomotive potentials. The model also shows that there is a lower bound on the frequency arising from the requirement of having many oscillations in the electrostatic field before the particle falls through the mesh.

The concept has been tested at atmospheric pressure with the dust charging and injection techniques developed for the atmospheric Paul trap.[25] The 10 cm x 10 cm grid is constructed on a standard circuit board with a spacing of 5.1 mm between adjacent wires. Each grid element is of stiff piano wire and is soldered at the ends to supporting pins. Odd-numbered elements are grounded and even-numbered elements are connected to a high voltage 60 Hz, AC transformer. The output can be adjusted from 0-1500 rms volts by means of a variable transformer. Dust is injected from a plastic hypodermic syringe. The charging occurs by friction and is not controlled or measured. The dust is glass microballons which have been sieved to remove particles larger than 30 microns. The dust is injected into the air about 15 cm above the picket fence and slowly settles onto the grid. The device is extremely sensitive to air currents and a clear cover must be placed around the experiment. A small hole in the top of the cover is used for injecting the dust. The illuminating beam is from a 35 mm slide projector.

Video and still photography show that the dust is suspended in the near field and is concentrated in the regions between wires. This indicates that the ponderomotive force is insufficient for confinement in the far field. Particles might be moved to the far field by the addition of a static field to partially cancel gravity.

Meaningful experiments on strong coupling will require, however, moving the experiment into a vacuum and using a dust charging mechanism which generates more uniform particles with measured characteristics.

VI. SUMMARY AND CONCLUSION

Techniques for charging and trapping macroscopic particles have been developed which have applications in dynamics of many-body systems and in the static structure of strongly-coupled plasmas.   A new charging technique has generated streams of hollow particles charged to a fixed potential.   The use of hollow glass microspheres maximizes both the charge-to-mass ratio and the light scattering cross section and minimizes the perturbation of gravity and the required trapping potentials.   The Kingdon trap allows the study of orbital dynamics, resonant perturbations and the electrostatic coupling of adjacent orbits.   The long confinement time ($>10^5$ orbits) permits the study of small perturbations applied for long periods which may lead to dynamical chaos.   A spherical version of the trap has created a Keplerian confining potential with more direct applications in celestial mechanics.   A planar trap, the ponderomotive picket fence, offers an opportunity to study a two-dimensional, strongly-coupled, one component plasma. These confinement systems may exhibit waves and other collective phenomena when filled with more particles.   The trapping of a larger number of particles will require further developments in experimental techniques.

ACKNOWLEDGMENTS

The author is indebted to numerous colleagues.   Mihály Horányi and Bob Walch suggested and carried out, respectively, the initial experiments on dust charging with partial support from the National Aeronautics and Space Administration.   Ted Biewer and Greg Newton, who were supported by the National Science Foundation (Research Experiences for Undergraduates), developed the three traps and optical diagnostics.   David Alexander and Bob Walch also assisted in trap development.

REFERENCES

[1] R. F. Wuerker, H. Shelton and R. V. Langmuir, J. Appl. Phys. **30**, 342 (1959).
[2] J. Hoffnagle and R. G. Brewer, Phys. Rev. Lett. **71**, 1828 (1993).
[3] T. Biewer, D. Alexander, S. Robertson and B. Walch, Am. J. Phys. **62**, 821 (1994).
[4] B. Walch, M. Horányi and S. Robertson, IEEE Trans. Plasma Sci. **22**, 97 (1994).
[5] N. Hershkowitz, in *Plasma Diagnostics*, edited by O. Auciello and D. Flamm, (Academic Press, San Diego, 1989), vol. 1, p. 113.

[6] A. R. Jones and K. C. Thong, J. Phys. D **4**, 1159 (1971).

[7] Lord Rayleigh, Philos. Mag. **14**, 184 (1882).

[8] A. Gomez and K. Tang, Phys. Fluids **6**, 404 (1994).

[9] M. Cloupeau and B. Prunet-Foch, J. Electrostatics **22**, 135 (1989).

[10] J. M. Schneider, N. R. Lindblad, C. D. Hendricks, Jr. and J. M. Crowley, J. Appl. Phys. **38**, 2599 (1967).

[11] S. Arnold and L. M. Folan, Rev. Sci. Instrum. **57**, 2250 (1986).

[12] K. H. Kingdon, Phys. Rev. **21**, 408 (1923).

[13] R. D. Knight, Appl. Phys. Lett. **38**, 221 (1981).

[14] The Malmberg-Penning trap is not useful for macroscopic particles because of the large field required for confinement. The drift induced by the gravitational force and the angular $E \times B$ drift results in an offset from the geometric axis analogous to the Shafranov shift of a tokamak. The requirement of a shift much smaller than the radius sets the lower bound on the field.

[15] J. Binney and S. Tremaine, *Galactic Dynamics* (Princeton University Press, Princeton, 1987), chaps. 5 and 6.

[16] F. H. Shu, in *Planetary Rings*, R. Greenberg and A. Brahic, eds. (University of Arizona Press, Tucson, 1984), p. 513.

[17] H. G. Dehmelt and F. L. Walls, Phys. Rev. Lett. **21**, 127 (1968).

[18] G. J. Sussman and J. Wisdom, Science **257**, 56 (1992).

[19] J. A. Burns, in *Satellites*, edited by J. A. Burns and M. S. Matthews, (University of Arizona Press, Tucson, 1986), p. 117.

[20] S. J. Peale, in ref. 19, p. 159.

[21] S. Robertson and D. Alexander, "Collective behavior of non-neutral plasma in a Kingdon trap," to appear in *Physics of Plasmas*.

[22] H. Thomas, G. E. Morfill, V. Demmel, J. Goree, B. Feuerbacher and D. Mohlmann, Phys. Rev. Lett. **73**, 652 (1994).

[23] G. Schmidt, *Physics of High Temperature Plasmas*, 2nd ed., (Academic Press, New York., 1979), p. 95.

[24] G. Schmidt, ibid., p. 49.

[25] H. Winter and H. W. Ortjohann, Am J. Phys. **59**, 807 (1991).

Cyclotron Modes of a Multi-Species Ion Plasma

E. Sarid,* F. Anderegg and C. F. Driscoll

Institute for Pure and Applied Physical Sciences
University of California at San Diego, La Jolla, CA 92093

Abstract. Cyclotron modes varying as $\exp(il\theta)$, with $l = 1$, 2 and 3, have been observed in an unneutralized Mg ion plasma. The $l = 1$ mode is observed to be down-shifted from the corresponding cyclotron frequency, while the $l \geq 2$ modes are found to be up-shifted. Good agreement is found between the observed down-shifts of the $l = 1$ modes of Mg^+ and Mg^{++} and the predictions of a multi-species cold plasma theory. The down-shifts depend on the composition and size of the the plasma, and the relative abundance of each ion can thus be determined.

INTRODUCTION

In this paper we describe measurements of the cyclotron modes of a non-neutral multi-species Mg ion plasma in a cylindrical Penning-Malmberg trap. Several high frequency modes with angular dependence $\exp(il\theta)$, with $l \geq 1$, are observed near the cyclotron frequencies of the various Mg ions. The $l = 1$ modes are downshifted from the cyclotron frequencies of the various ions, while the higher l-modes are up-shifted. These frequency shifts are due to $\mathbf{E} \times \mathbf{B}$ rotation of the plasma. This phenomenon is known as the magnetron shift in short trapping cells [1,2], arising there from the confinement electric fields. In our longer trap and denser plasma, the electric fields that cause the frequency shifts arise mainly from the space charge of the plasma and the image charges induced on the walls.

While the observed cyclotron modes are qualitatively similar to those studied by Gould and Lapointe [3] in a pure electron plasma, the frequency shifts are different due to the multiple ion species. Only one species supports the mode near a particular cyclotron resonance, but the space charge from the other species contributes to the frequency shift of the mode.

The main emphasis in this work is on the downshifts of the $l = 1$ modes of the Mg^+ and Mg^{++} ions. A multi-species, cold plasma theory is found to explain the observed downshifts for the majority species $^{24}Mg^+$, but frequency shifts of the minority species ($^{25}Mg^+$, $^{26}Mg^+$, $^{24}Mg^{++}$, etc.) remain

---

*Present address: NRCN, P.O. Box 9001, Beer Sheva, Israel 84190

unexplained. The downshifts of the $l = 1$ modes are larger than the diocotron frequency, and they depend on the plasma size and composition. Measurements of the plasma size thus test the theory. The dependence of the shifts on the plasma composition can be an important diagnostic tool, since the measured shifts determine the relative abundance of the various ions in the plasma.

THE ION PLASMA EXPERIMENT

Figure 1 gives a schematic of the cylindrical trap. Mg ions from a metal vapor vacuum arc [4] are stripped of their electrons and trapped in a Malmberg-Penning trap with wall radius $R_w = 2.9$ cm and full length $L_{max} = 45$ cm. The uniform axial magnetic field $B = 20$ kG gives radial confinement, and positively-biased inject and dump rings contain the ion column axially. The initial ion plasma diameter is $2R_p \approx 1.5$ cm, with density of $n \lesssim 1 \times 10^8 \text{cm}^{-3}$. The ion density is diagnosed by dumping the column axially onto charge collectors.

The cyclotron modes are observed with a single-frequency transmission technique. An oscillating voltage (typically 1.5 V) is applied to the sectored ring with four 90° sectors, and the response of the plasma is detected on the sectored ring with eight 45° sectors, 30 cm distant. If the frequency of the applied voltage matches a coherent oscillating mode in the plasma, the mode is excited. The received image charge signals are processed by a spectrum analyzer tuned to the transmission frequency. In addition, the frequency

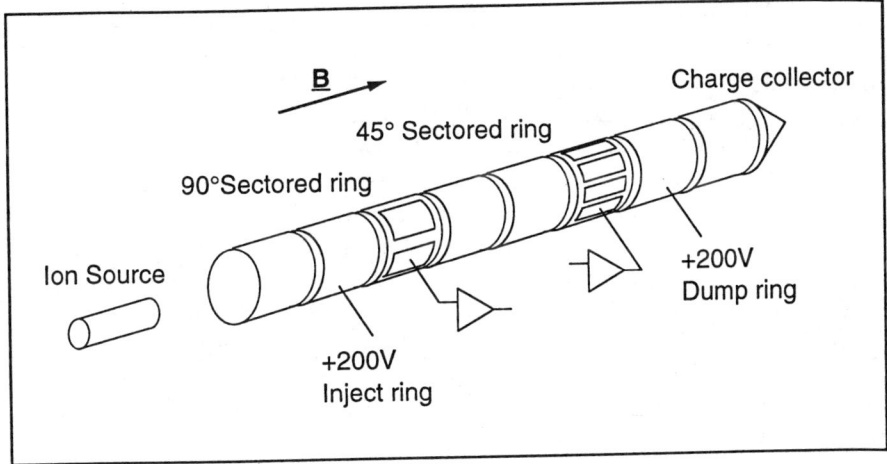

Figure 1: Schematic of the cylindrical confinement electrodes. The sectored rings are used for excitation and detection of waves.

$f_d \lesssim 10$ kHz of the low-frequency "diocotron" mode is simultaneously monitored by a separate spectrum analyzer.

THE CYCLOTRON MODES

Figure 2 shows repeated scans on the same plasma, covering a frequency range of 30 kHz around the cyclotron frequency of Mg_{24}^+. Each scan took about 3 seconds, and they were repeated every $6-30$ seconds. Three different modes can be observed in most of these scans. Analysis of the phases received on different sectors confirms that these have azimuthal mode number $l = 1$, 2 and 3, with no axial dependence, $i.e.$ $k_z = 0$. While the $l = 1$ mode is down-shifted relative to the cyclotron frequency, the $l = 2$ and $l = 3$ modes are up-shifted. The frequency shifts of all three modes are decreasing with time, because the plasma is evolving with time. Similar behavior was also observed in electron plasmas [3], and was related to the evolution of the $l = 1$ diocotron mode frequency.

Figure 3(a) shows the frequencies of the $l = 1$ and $l = 2$ cyclotron modes of Mg_{24}^+ versus the simultanously-measured $l = 1$ diocotron frequency; the $l = 1$ cyclotron mode of Mg_{24}^{++} is similarly shown in Fig. 3(b). These $l = 1$ ion cyclotron modes are downshifted from the single particle cyclotron frequency by about 1 to 3 times f_d. In contrast, in the electron plasma experiments the $l = 1$ electron cyclotron mode was down-shifted by exactly f_d. This difference

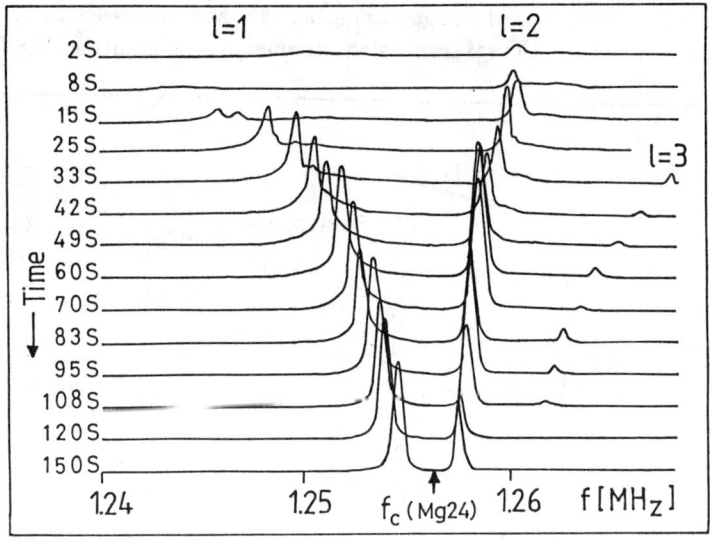

Figure 2: Repeated frequency scans near the cyclotron frequency of Mg_{24}^+, showing the $l = 1$, $l = 2$, and $l = 3$ azimuthal modes varying with time.

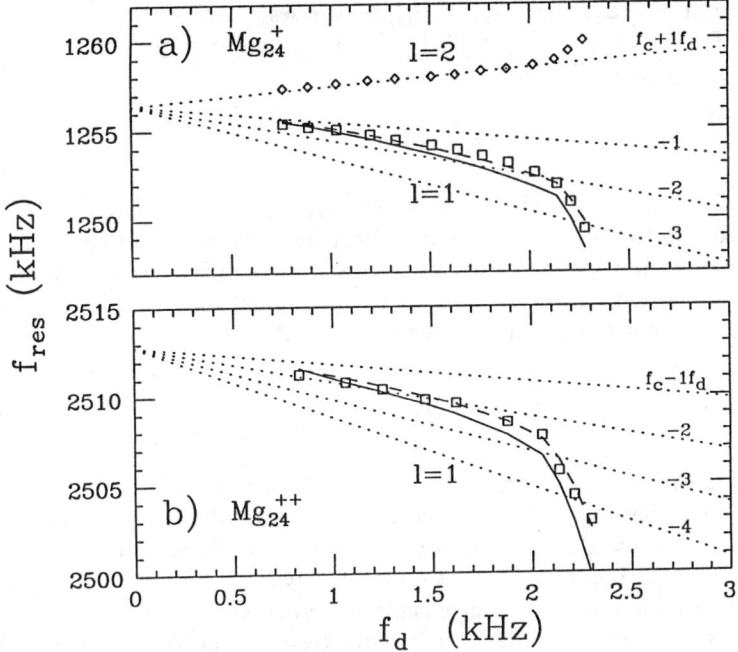

Figure 3: The frequencies of the $l = 1$ and $l = 2$ cyclotron modes of Mg_{24}^+ and the $l = 1$ cyclotron mode of Mg_{24}^{++}, versus the simultanously-measured diocotron frequency f_d. The solid and long-dashed lines are the calculated shifts, based on a multi-species cold-plasma theory.

results from the multi-species nature of the ion plasma. The time-dependence of the ratio between the shifts and f_d results from the expansion of the ion plasma, as discussed in the next section.

THEORY OF THE CYCLOTRON MODES

The electrostatic dispersion relation for $k_z = 0$ modes in a constant density, multi-species plasma column is given by [5]

$$0 = 1 - \sum_j \frac{\omega_{pj}^2 \left[1 - (R_p/R_w)^{2l} \right]}{2(\omega - l\omega_{rj})\left[(\omega - l\omega_{rj}) + (-\Omega_{cj} + 2\omega_{rj}) \right]} \ . \tag{1}$$

The sum in the dispersion relation is over the species j of the plasma, and $\omega_{pj}^2 = 4\pi n_j e_j^2/m_j$ is the square of the plasma frequency corresponding to component j. We have taken a sign convention such that the cyclotron frequencies $\Omega_{cj} \equiv |q_j|B/m_jc$, the rotation frequency ω_{rj}, and the mode frequency ω, are all positive.

For rotation frequencies $\omega_{rj} \ll \Omega_{cj}$, all species rotate with nearly the same rotation frequency $\omega_R \equiv \sum_j \omega_{pj}^2/2\Omega_{cj}$. In this case the dispersion relation has low frequency ($\omega \ll \Omega_{cj}$) solutions

$$\omega = \omega_R \left[l - 1 + (R_p/R_w)^{2l} \right] \ . \tag{2}$$

The $l = 1$ solution is the diocotron frequency, $\omega_d \equiv \omega_R \cdot (R_p/R_w)^2$.

In addition to these low-frequency diocotron solutions, there exist solutions of Eq. (1) near the cyclotron frequency of each of the plasma components. For the solution near Ω_{ci}, $i.e.$ $\omega_i \equiv \Omega_{ci} + \Delta_i$, with $\Delta_i \ll \Omega_{ci}$, we retain in Eq. (1) only the term corresponding to species i and obtain:

$$\Delta_i \equiv \omega_i - \Omega_{ci} = \omega_R \left\{ (l-2) + \delta_i \left[1 - \left(\frac{R_p}{R_w} \right)^{2l} \right] \right\} \ . \tag{3}$$

Here, δ_i is the relative charge fraction of species i in the plasma, $i.e.$ $\delta_i \equiv q_i n_i / \sum_j q_j n_j$. Equation (3) predicts, in agreement with the experimental results, that the $l = 1$ resonances are downshifted while the $l \geq 2$ resonances are upshifted relative to the cyclotron frequency Ω_{ci}. Furthermore, it predicts single-species ($\delta_i = 1$) $l = 1$ downshifts equal to ω_d, as appropriate for the electron plasma experiments; and it correctly predicts larger downshifts for multi-species plasmas ($\delta_i < 1$). This analysis has also been extended to the case of parabolic density profiles [6,7]. The measured shifts can also be used to calculate the composition of the plasma. The results are consistent with a fixed composition in time, and with Mg^+ ions density which is $2.5 - 4.5$ times larger than the Mg^{++} density.

Equation (3) accurately models the observed shifts of the $l = 1$ cyclotron resonances of Mg_{24}^+ and Mg_{24}^{++}. Here, we used an independent measure of R_p, presumed a parabolic profile, and took a fixed ratio of 3.0 between the densities of Mg^+ and Mg^{++}, and included isotopic abundances. The result of this calculation is shown as solid lines in Figs. 3(a) and 3(b). Since the plasma profile is not exactly parabolic, our definition of the plasma diameter might be biased. The frequency shifts were recalculated with 2.5 mm added to the measured plasma diameters, and are shown as dashed lines in Figs. 3(a) and 3(b). As can be seen, there is good agreement between the calculated and measured shifts, especially when the uncertainty in the plasma diameter is taken into account.

Alternately, the measured shifts can be used to obtain the plasma size and composition from Eq. (3). Based on the down-shifts measured for Mg_{24}^+ and Mg_{24}^{++}, Figure 4 shows the predicted diameters for a flat (F's) and parabolic (P's) density profile. The agreement with the measured diameters (squares) can be seen to be good.

The $l = 2$ and $l = 3$ modes are up-shifted from the cyclotron frequency, as predicted by Eq. (4). However, the size of shifts can not be quantitatively

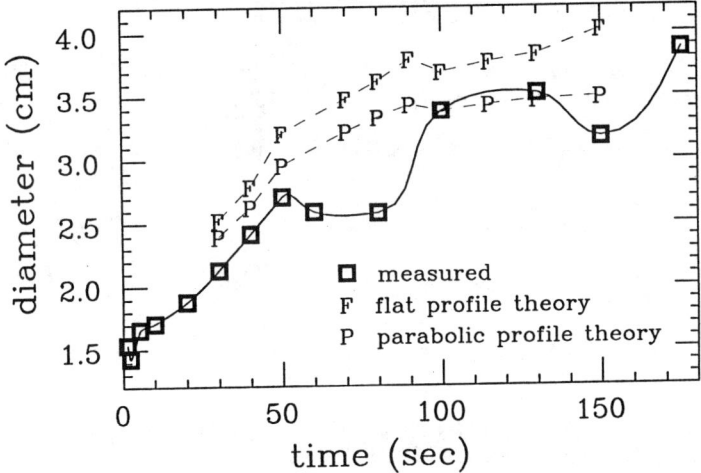

Figure 4: Measured plasma diameter versus time (squares); and diameter predicted from theory using the measured cyclotron mode shifts.

explained by the present theory. The theory also fails to describe the observed $l = 1$ down-shifts for minority species like Mg_{25}^+ and Mg_{26}^+. While the theory predicted large down-shifts (close to the rotation frequency of the plasma), the observed down-shifts for these minority species are systematically about twice the diocotron frequency. These topics are further discussed in Ref. [7].

Although most of our experiments were performed in a parameter regime which is very different from the low-density short-cells FTMS measurements [1], we believe that our results are significant for understanding of the role of space-charge and image-charge corrections in these experiments.

ACKNOWLEDGMENTS

We wish to thank R.W. Gould and T.M. O'Neil for their contributions to the theory, I.G. Brown for his help in designing the ion source, and R. Bongard for technical assistance. The research is supported by ONR N00014-89-J-1714.

REFERENCES

1. *Fourier Transform Mass Spectrometry* (M.V. Buchanan, editor), ACS Symposium Series 359, Washington, D.C.: ACS, 1987.

2. Gabrielse, G., Fei, X., Orozco, L.A., Tjoelker, R.L., Haas, J., Kalinowsky, H., Trainor, T.A., and Kells, W., *Phys. Rev. Lett.* **65**, 1317 (1990).

3. Gould, R.W., and LaPointe, M.A., *Phys. Fluids B* **4**, 2038 (1992).

4. Brown, I.G., Galvin, J.E., MacGill, R.A., and Wright, R.T., *Appl. Phys. Lett.* **49**, 1019 (1986).

5. Davidson, R.C., *Physics of Nonneutral Plasmas*, Redwood City, CA: Addison-Wesley, 1990, p. 258.

6. Gould, R.W., "Theory of Cyclotron Resonance in a Cylindrical Non-Neutral Plasma," *Phys. Plasmas*, submitted (1994).

7. Sarid, E., Anderegg, A., and Driscoll, C.F., "Cyclotron Resonance Phenomena in a Non-Neutral Multi-Species Ion Plasma," *Phys. Plasmas*, submitted (1994).

Ordered One-Component Plasmas: Phase Transitions, Normal Modes, Large Systems, and Experiments in a Storage Ring

J. P. Schiffer

Physics Division, Argonne National Laboratory, Argonne, IL 60439
and The University of Chicago, Chicago, IL 60637

The property of cold one-component plasmas, confined by external forces, to form an ordered array has been known for some time both from simulations (1) and from experiment (2). The purpose of this talk is to summarize some recent work on simulations and some new experimental results.

As a reminder of the qualitative features of this ordering, I would like to show the result of, what I believe, is the largest number of ions followed in a simulation. This is a system of 20,000 ions in isotropic confinement with a constant harmonic force, that was allowed to find its lowest potential energy in a Molecular Dynamics simulation -- whose essential features have been described before. This calculation was carried out over a period of almost a year by shepherding it through the NERSC Crays at low priority. The simulation requires computing the forces between all particles: $1/2 \times (20,000)^2$ calculations at each time step for many thousands of time steps. The total computer time invested represents on the order of 1000 actual Cray cpu hours. The motivation behind this calculation was, in part, to try and understand how far the layered structures seen in the simulation of smaller systems (1) continued in larger ones. After all, one knows that in an infinite system a bcc configuration has the lowest potential energy. Dubin (3), had estimated that possibly bcc order may win out in the interior after about 20 layers -- but this estimate did not include the potential energy in forming an interface between the two forms of ordering.

The type of order seen is shown in figure 1. The upper part of the figure shows the ordering of ions on the outer shell of the cold 20,000 ion system, with the triangular pattern that is characteristic of two-dimensional Coulombic systems. The lower part of the figure shows the radial density and indicates that the shell structure continues all the way to the interior, forming 18 distinct shells. The ordering of particles in each shell is similar to that in the outermost one -- though the thickness of the shells is somewhat greater. Table 1 indicates the data on all the shells seen in this simulation -- and several features are evident:

a) that the shells are equally spaced,
b) that the number of particles in each shell is proportional to the shell's radius (in other words that the surface density in each shell is constant, except that the outermost shell has about 1% higher surface density than all the others,
c) that the thickness of the shells increases gradually inward.

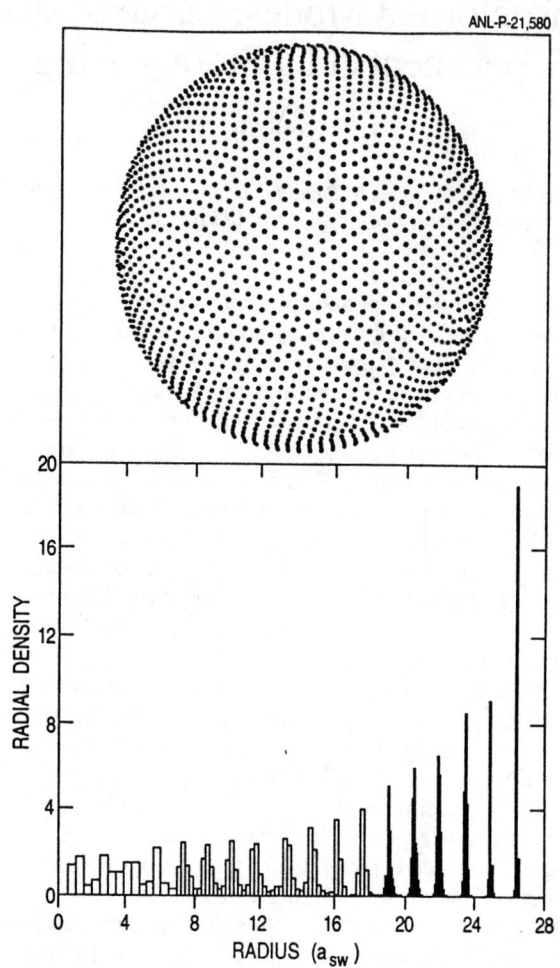

FIGURE 1. Simulation of 20,000 ions in an isotropic trap. On top, the outer shell of ions is shown, on the bottom the radial density. The distance scale is in units of the Seitz-Wigner radius ($4/3 \, \pi a_{SW}^3 = 1/\rho$).

The simulation reached a minimum in the potential energy which corresponds to the configuration presented -- but it is not clear whether this minimum is the true minimum or whether it is a local one. Thus one cannot really speak of a value of the coupling parameter Γ to characterize the system -- in a sense Γ is infinite within the accuracy of the simulation -- the classical temperature is zero. It would seem that this structure therefore is still stable for the 18-shell structure seen here. In the future one might try such a simulation with a bcc seed cell in the interior to see whether a lower potential energy might be reached.

TABLE 1. Shell Structure in the 20,000 Ion System[a]

Shell Number	Shell Radius	Shell Spacing	Shell Width	Number in Shell	Excess[b]
1	26.399	1.4753	.0195	3116	38
2	24.924	1.4701	.0547	2743	0
3	23.454	1.4707	.0848	2423	-6
4	21.983	1.4695	.107	2138	4
5	20.514	1.4652	.125	1847	-11
6	19.049	1.4705	.148	1599	-3
7	17.578	1.4683	.167	1366	2
8	16.110	1.4602	.184	1138	-8
9	14.650	1.4694	.206	940	-8
10	13.180	1.4732	.243	773	6
11	11.707	1.4693	.244	602	-3
12	10.238	1.4683	.270	464	1
13	8.770	1.4686	.274	338	-2
14	7.301	1.443	.285	235	0
15	5.858	1.4178	.253	144	-8
16	4.440	1.4787	.289	86	-1
17	2.962	1.4754	.294	39	0
18	1.486		.209	9	-1

[a]All distances are in units of the Seitz-Wigner radius. [b]The "excess" is the number in a shell over (or under) what one would expect if the surface density in each shell were a constant.

The correlations between particles in these systems may be studied. For instance, the correlation between particles in one shell shows very pronounced structure that is characteristic of this two-dimensional order, but the correlation between particles in different shells does not show pronounced peaks. Figure 2a and 2b show the correlation functions g(r) for a single shell, and for the whole system. The correlation as a function of the coordination, the total number of particles within a given distance from an initial particle, is also shown in figure 3.

While there is no obvious longer-range correlation between particles in different shells, other than that between nearest neighbors and successive shell layers, careful inspection of the adjacent layers indicates that the *orientation* of the two-dimensional lattices does tend to be similar from shell to shell. Figure 4 shows the correlation between the angular orientation of the lattices as a function of shell separation; note that the structure is symmetric for 60 degree rotations and so an uncorrelated system should yield 15 degrees for the value of this correlation (since 30 degrees is equivalent to - 30 degrees, and an uncorrelated system would imply the mean between 0 and 30). The figure indicates that the value of correlation function between adjacent shells is 4 degrees, and increases to 15 degrees in 8 shells. So clearly there is a correlation in the orientation of these systems that persists for many layers, even when the radial correlation function g(r) does not reflect it.

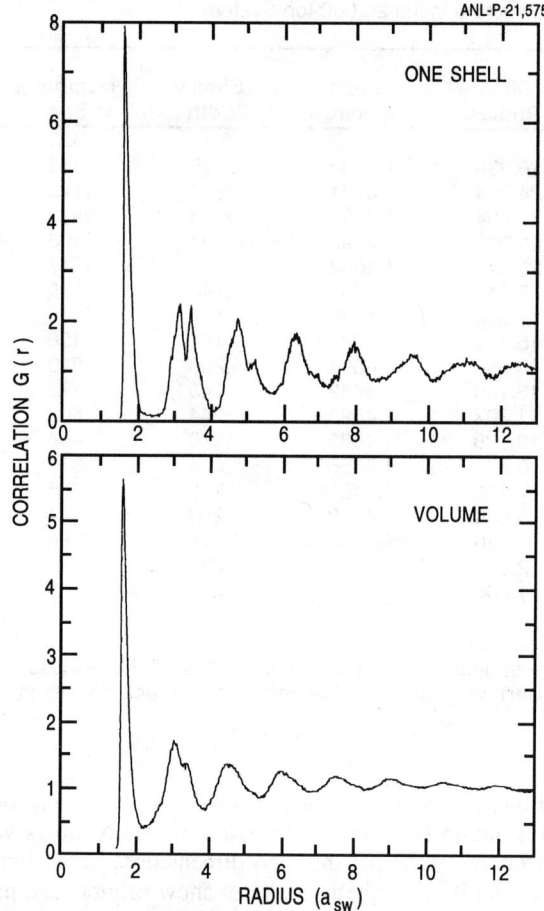

FIGURE 2. Correlation function in the spacing of ions. The top plot is the correlation function restricted to one shell -- on the bottom that for the whole volume -- but excluding the ions in the outer shells, in order to avoid surface effects.

This set of coordinates for the 20,000 particles under isotropic confinement could be the test bed for other calculations and the numbers are available on request.

Next I would like to show some simulations of the same number of particles under anisotropic confinement -- changing the confining force in the z-direction by a factor of 5. These simulations were not followed quite to a local minimum though Γ is at least on the order of 10^4 -- an average kinetic energy is difficult to estimate when the system is being cooled, because generally one does not wait for the system to become completely equilibrated before lowering the temperature (it may take several thousand time steps to establish some sort of equilibrium value

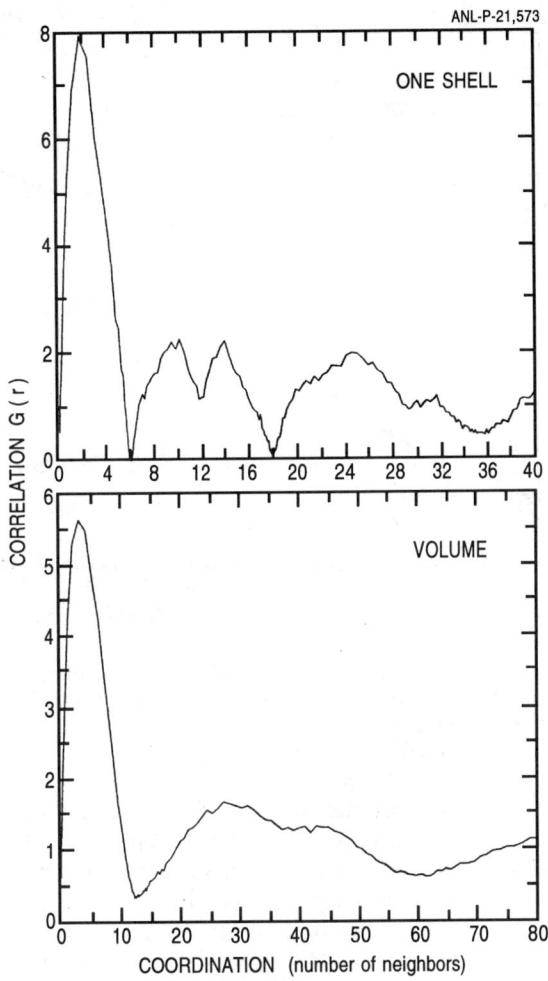

ANL-P-21,573

FIGURE 3. The same correlation function as in figure 2, but plotted against the integral of nearest neighbors. For the single shell the first peak corresponds to a total of 6 nearest neighbors. For the volume correlation it is about 14.

for the kinetic energy). The results for these anisotropically confined systems are shown in figure 5. What one might expect naively is that the ion cloud would take on a spheroidal shape along equipotentials and that the interior layers would form additional spheroids. However, this is <u>not</u> the case, as may be seen by a closer examination of the data in figure 5. The outer shell of the system is indeed spheroidal. But the interior shells are not spheroids. One might have expected that the interior shells would be more deformed spheroids than the outer one, since if the shells are equally spaced (as in the spherical case), an equal step in the major and minor axes would lead to increasing eccentricities in these shells.

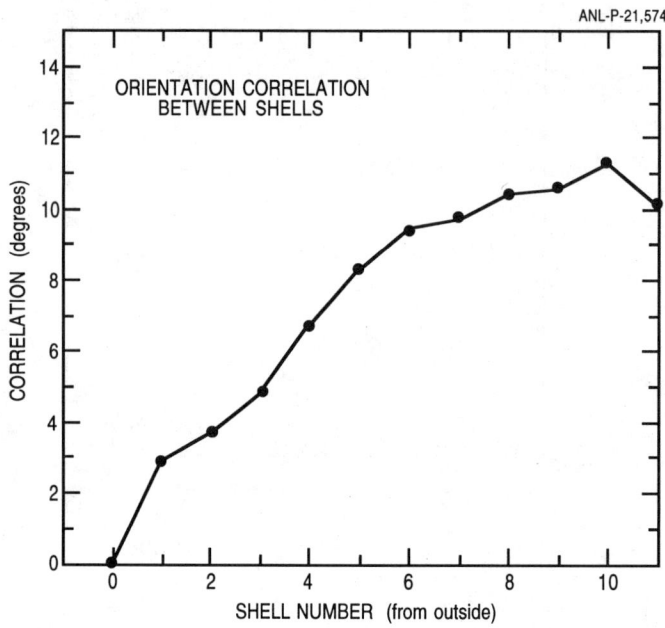

ANL-P-21,574

FIGURE 4. Correlation in the directional orientation of the two-dimensional structures in successive shells. Random orientations would yield 15 degrees -- so some correlation persists for 10 or more shells.

Instead of spheroids the particles seem to form shells that are at all points equidistant from the outermost shells by a constant perpendicular distance. This indeed leads to the formation of interior shells of increasing eccentricity but they are not spheroidal, but develop sharp points at the narrowest end. This feature of the calculations was certainly not expected a priori.

Next, I would like to review another aspect of asymmetric confinement that was found in simulations, that of dimensional phase transitions (4). Consider a system of few ions confined in a trap -- say 4 as is illustrated in figure 6. If the trap parameters are similar to the "linear trap" configuration (5), with the confining force in the z-direction much weaker than in the radial direction, then the 4 ions will form a one-dimensional array and all sit on the z-axis. As the force in the z-direction is increased the ions will first pop out to form a rhomboid in 2 dimensions with all four simultaneously leaving the z axis at $F_z > 0.747$, then form a rhombus with two ions on the z axis, and two ions at $z = 0$ away from the axis at $F_z = 0.872$, then the ions form a three-dimensional tetrahedron with one edge parallel to the z axis and one perpendicular starting at $F_z = 0.912$, the tetrahedron becomes a regular tetrahedron with no preferred orientation but only for $F_z = 1$, then the ions will form a tetrahedron with two edges parallel to the z axis, and finally collapse into a two-dimensional square for $F_z > 1.47$. Thus even this simple 4-ion system will have undergone 6 shape transitions as the trap parameters (the anisotropy in the confining force) are changed continuously, three of these are changes in the dimensionality of the system. These dimensional

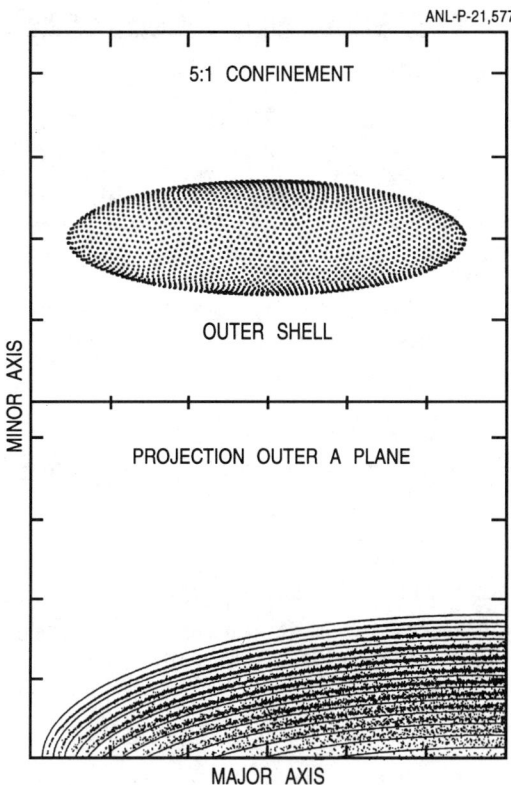

ANL-P-21,577

5:1 CONFINEMENT

OUTER SHELL

PROJECTION OUTER A PLANE

MINOR AXIS

MAJOR AXIS

FIGURE 5. Simulation of 20,000 ions in anisotropic confinement with the restoring force in the z-direction 5 times less than in x and y. The ions in the outer spheroidal shell are shown on top, while on the bottom a projection of particles onto the z-ρ plane is plotted. The lines are drawn to emphasize the fact that in the interior the ions do not settle onto spheroidal surfaces, but on ones that are a fixed perpendicular distance from the outer (spheroidal) shell.

changes are a general feature of any number of ions; the aspect ratio of a trap where the configuration changes from 1 to 2, from 2 to 3, and from 3 back to 2 dimensions, changes smoothly with the number of ions. This behavior was found in simulations and very shortly thereafter explained in terms of simple hydrodynamic considerations by Dubin (6).

Next I would like to show some results on normal modes of spherical clouds. This is an ongoing collaboration with Dan Dubin where we are trying to understand the effect of the correlations (of the crystalline structure) on the hydrodynamic normal modes that are apparent in these systems. Some results are shown in figure 7. It is evident, for instance, that the simplest (monopole) mode, which is a shape-conserving volume oscillation of a spherical cloud, shows no damping. On the other hand, a quadruple mode, which is a shape oscillation, shows appreciable damping -- after several hundred oscillations. We are trying to

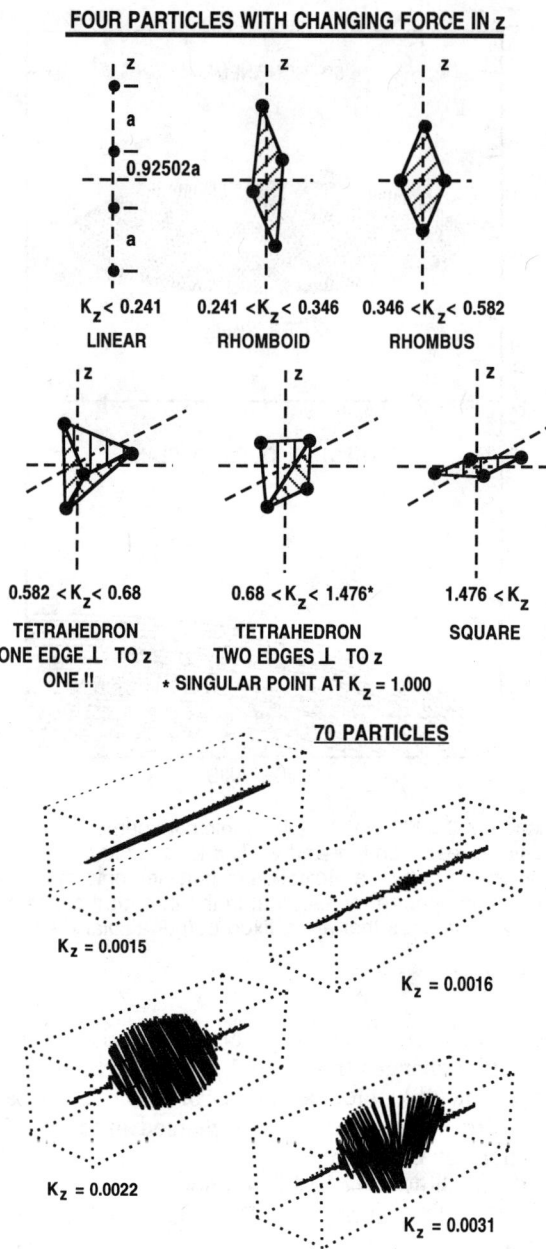

FIGURE 6. Illustration of dimensional phase transitions in the simulation of ions under anisotropic confinement. On top six distinct classes of shapes are shown as the focusing in the z-direction is changed (relative to the focusing in x and y). On the bottom, some of the changes in the configurations for 70 ions are shown. Lines are drawn to connect the ions to emphasize the shapes of the configurations, and on the bottom curves the transverse scale is enlarged by a factor of 10.

ANL-P-21,582

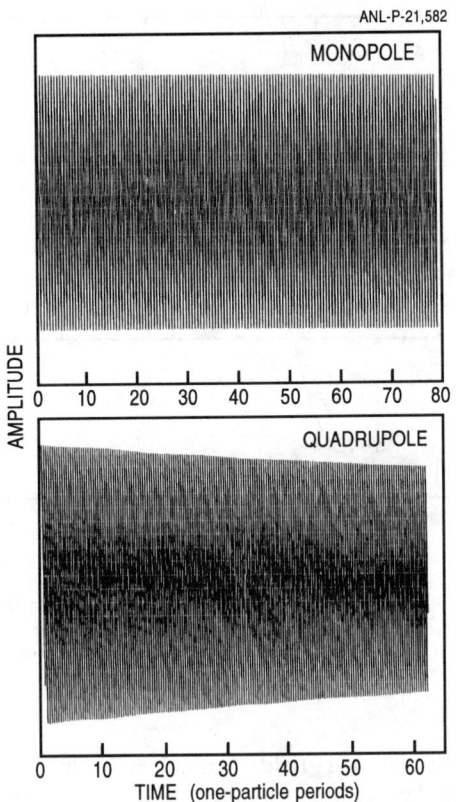

FIGURE 7. Oscillations of a cold ion cloud of 1000 ions in isotropic confinement, when subjected to a perturbation the excites a pure radial mode (top) and when a quadrupole mode is excited. Time is in units of the period of one particle oscillating in the same confining field.

understand this damping, and small deviations in the frequencies of these modes, more quantitatively. Then there are modes that are only possible in a crystalline system -- torsional modes may exist only from the interparticle interactions in an ordered system -- since a liquid has no resistance to shear. An example of this is given in figure 8.

I would now like to turn to some experimental work on real storage rings, magnetic storage devices in which particles circulate with large kinetic energies and for which laser cooling is used on partially ionized ions to attain temperatures ten or more orders of magnitude lower than their uniform kinetic energies. I cannot review the subject here and only want to mention two recent experiments at the ASTRID storage ring in Aarhus. In one of these, a single laser was used to cool a beam that was bunched by an rf system in the ring. This geometry is essentially that of a linear trap discussed earlier, with an aspect ratio on the order of 1000 or more. Without laser cooling, particles that are bunched undergo

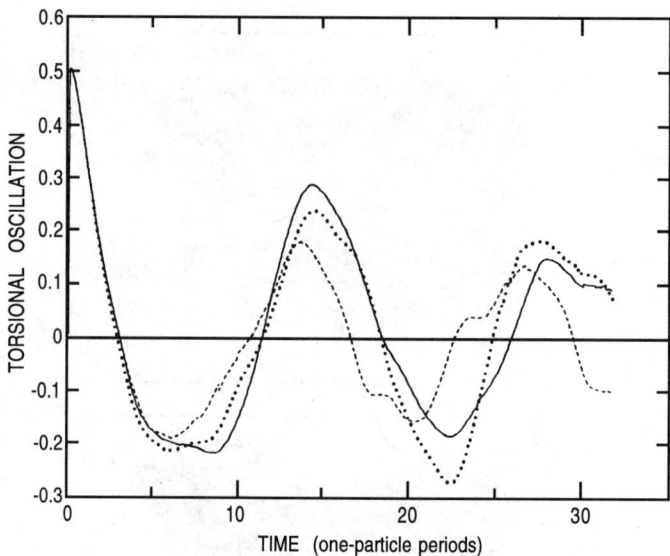

FIGURE 8. Torsional oscillations of a cold 1000 ion cloud. The three curves correspond to the response to the same perturbation applied about the x, y, and z axes, they are slightly different because the finite cloud is not exactly symmetric, yet the oscillations are similar.

longitudinal oscillation in the rf fields, with a period that is large compared to the circulation period in the ring (the so-called synchrotron oscillations). When laser cooling is applied from one direction, these oscillations are damped out and the particles should be confined to a much tighter bunch in the ring. This has now been observed experimentally at ASTRID as shown in figure 9.

Another line of investigation is to use two lasers and cool a continuous ion beam between the two cooling forces. This we have also attempted recently and found to work. In figure 10 are shown two sets of signals -- one is the velocity spread of the ions as the laser cooling progresses, and the other is the 'Schottky signal', which is the Fourier transform of the noise signal from irregularities in the particle distribution. The splitting in the Schottky signal is from longitudinal wave propagation along the beam. Note that as the beam gets colder the logarithmic Schottky signal decreases, as it should. These results are both very new -- and indicate substantial progress in cooling these moving plasmas -- however we have still quite a way to go before talking of crystallization in these systems.

I hope this report of various aspects of the field gives you a flavor of the many fascinating aspects of such condensed systems.

This research was supported by the U.S. Department of Energy, Nuclear Physics Division, under Contract W-31-109-Eng-38. The calculations reported were done on the NERSC CRAYs.

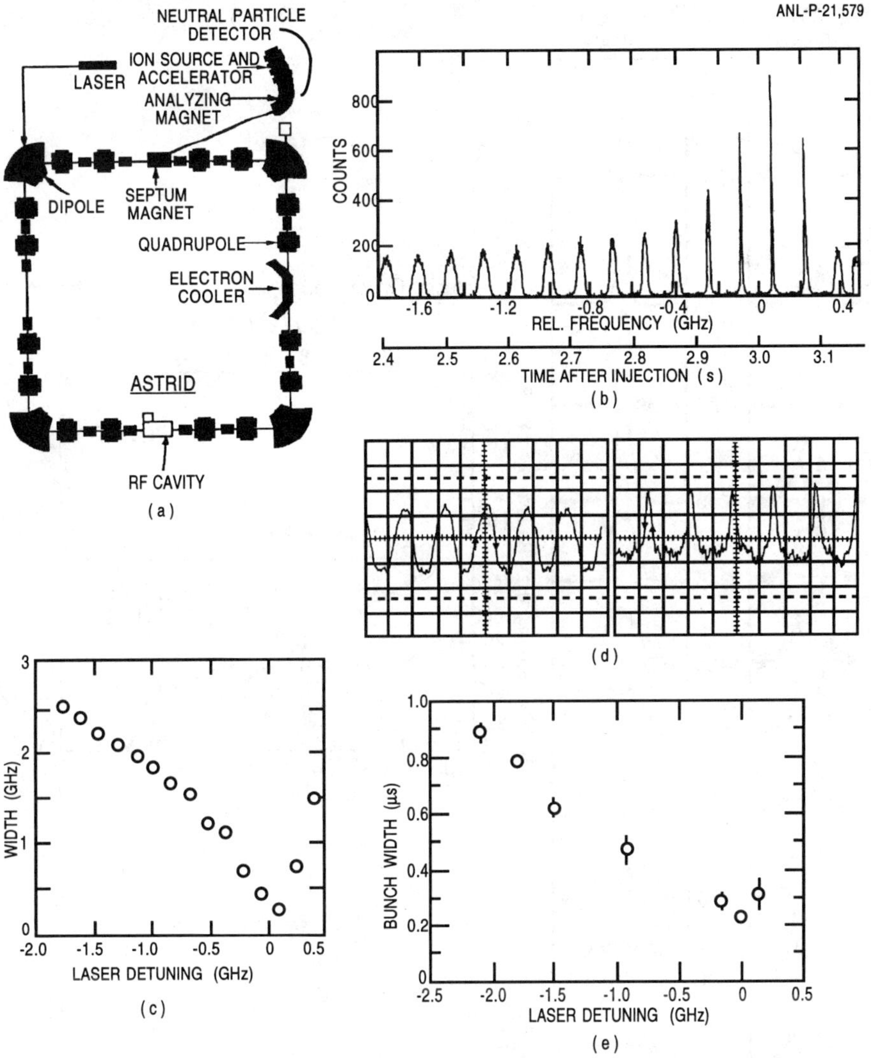

FIGURE 9. Results from laser cooling Mg$^+$ ions in a storage ring bunched into 23 longitudinal bunches while traveling around the ring with kinetic energies of 100 keV, obtained with J. S. Hangst, J. S. Nielsen, O. Poulsen, P. Shi, and B. Wanner. The ASTRID storage ring in Aarhus is shown schematically in (a). The laser scans with a probe laser of the beam are shown in (b), showing the sharpening velocity distribution of the ions as the laser cooling progresses; (c) shows the widths extracted from these scans. In (d) the physical length of the bunched beams is shown at two stages in laser cooling from the signal on pickup electrodes, and in (e) the bunch widths (lengths) extracted from these data are shown.

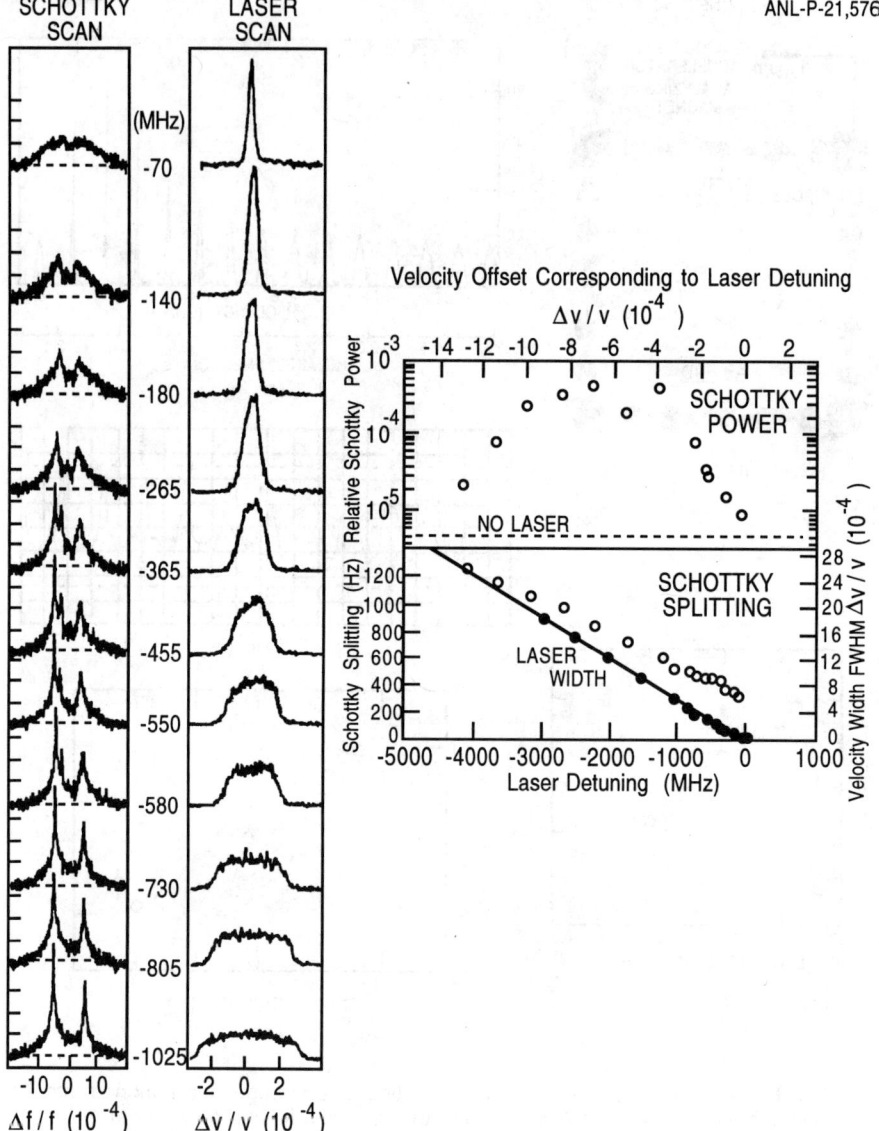

FIGURE 10. Results of laser cooling a continuous beam of Mg ions in the ASTRID storage ring (with J. S. Hangst, A. Labrador, V. Lebedev, N. Madsen, J. S. Nielsen, O. Poulsen, and P. Shi). On the left the velocity scans from a probe laser are shown as a function of laser cooling, along with the Schottky signal from pickup electrodes. On the right the total power in the Schottky signal, and the widths extracted from the data on the left are plotted.

REFERENCES

1. Rahman, A. and Schiffer, J. P., *Phys. Rev. Lett.* **57**, 1133–1136 (1986); Dubin, D. H. E. and O'Neal, T. M., *Phys. Rev. Lett.* **56**, 728–731 (1986).
2. Gilbert, S. L., Bollinger, J. J., and Wineland, D. J., *Phys. Rev. Lett.* **60**, 2022–2025 (1988); Birkl, G., Kassner, S., and Walther, H., *Nature* **357**, 310–313 (1992).
3. Dubin, D. H. E., Phys. Rev. A **40**, 1140–1143 (1989).
4. Schiffer, J. P., *Phy. Rev. Lett.* **70**, 818–821 (1993).
5. Raizen, M. G., Gilligan, J. M., Bergquist, J. C., Itano, W. M., and Wineland, D. J., *Phys. Rev. A* **45**, 6493–6501 (1992).
6. Dubin, D. H. E., *Phys. Rev. Lett.* **71**, 2753–2756 (1993).

Numerical Modeling of Non-Neutral Plasmas

Ross L. Spencer

Department of Physics and Astronomy
Brigham Young University
Provo, Utah 84602

Several numerical tools for modeling non-neutral plasmas have been developed at Brigham Young University, including codes for computing equilibria, for simulating plasmas, and for computing mode frequencies with numerical eigenvalue methods. Our hope is that these programs will allow us to make careful comparisons between theory and experiment and allow us also to investigate the differences between various plasma models. This talk will summarize this work and give examples of physical applications.

I. INTRODUCTION

Non-neutral plasma experiments are performed in a wide variety of confinement geometries and on plasmas with widely varying parameters. As in the rest of plasma physics, it is unlikely that analytic theory alone will be capable of handling all of the interesting physics in these systems. Furthermore, plasma theory routinely produces equations that are practically impossible to solve, making it difficult to test the utility of our theoretical models. So it seems that the numerical work performed by various members of our community will continue to be important to the future of non-neutral plasma physics.

Our plasma theory group has developed, and is continuing to develop, codes that can be used to study non-neutral plasmas. Among the codes that have been developed are the following: (a) a general purpose axisymmetric equilibrium code, (b) an axisymmetric (rz) particle-in-cell simulation, (c) an $r\theta$ simulation for studying $\mathbf{E} \times \mathbf{B}$ drift dynamics, and (d) an eigenvalue code that finds linear mode frequencies for two-dimensional (rz) equilibria with perturbations proportional to $\exp{(im\theta)}$. In the remainder of this talk these codes will be briefly described and examples of their application to physical problems will be given.

II. EQUILIBRIUM

The equilibrium code that we have developed uses Successive-Over-Relaxation to solve non-linear Poisson equations of the form

$$\nabla^2 \phi = -\frac{q}{\epsilon_o} n(r, \phi) \tag{1}$$

in cylindrical geometry.[1] The boundary conditions can be specified in a variety of ways, including confining rings of arbitrary length and location as well as internal conducting structures of any shape (so long as they are axisymmetric). It also computes equilibria in three different ways. (1) It computes global thermal equilibria, in which both the radial and axial density profiles are self-consistently determined.[2] (2) It computes equilibria with a specified midplane radial profile. And (3), it computes equilibria whose line-density profile $h(r) = \int n(r,z)dz$ is specified, as would be appropriate for analyzing experiments where the plasma is diagnosed by dumping it onto charge collectors at the end of the confinement region.

As an example, Fig. 1 shows contours of constant density and electrostatic potential for a plasma equilibrium that matches a measured line-density profile in the positron trap experiments of Surko, Greaves, and Tinkle at the University of California at San Diego.[3]

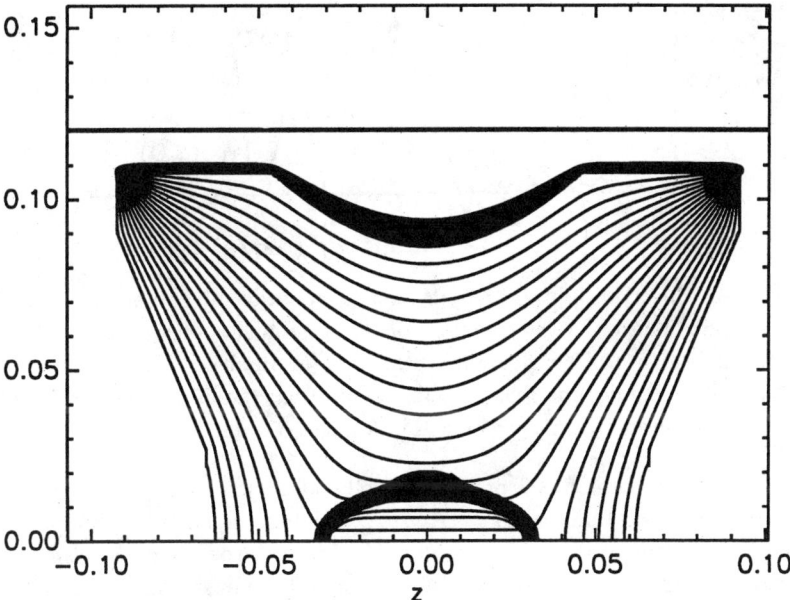

FIG. 1. Contours of constant density (clustered in the center) and constant electrostatic potential for a computed equilibrium whose line-density profile matches an experimentally measured profile.

III. SIMULATIONS

Two types of two-dimensional simulations have been developed for use with non-neutral plasmas: $r\theta$ simulations for infinitely-long plasmas, and an rz simulation for finite-length axisymmetric plasmas. Only the latter will be discussed here.

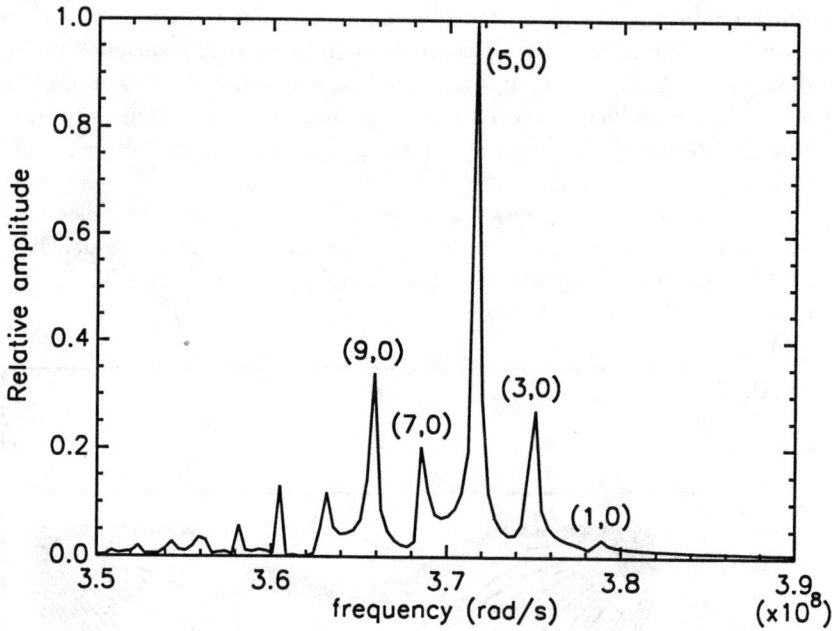

FIG. 2. Spectrum obtained from an rz simulation of "drumhead" modes in a pancake-like plasma in the geometry of the experiments of Weimer, et. al., at the National Institute of Science and Technology at Boulder. The peaks are labeled with their (ℓ, m) designations in Dubin's notation; the initial perturbation for the run was based on Dubin's eigenfunction for the (5,0) mode.

The rz simulation is a particle-in-cell code, with the particles moving in a three-dimensional phase space (r, z, v_z), although since the plasma is axisymmetric and drift motion is assumed in the plane perpendicular to the confining magnetic field, the r-positions of the particles never change once they are loaded. This code takes as input the geometry and plasma computed by the equilibrium code, after which perturbations are added to study various physical processes. This program simulated the axisymmetric vibrational motions of the plasmas produced in the positron trap experiments of Surko, Greaves, and Tinkle[3] where it was used to study the possibility of using ratios between mode frequencies as a non-destructive diagnostic. The simulation was able to give information about the variation of mode frequencies with temperature, functioning as a very sophisticated, but very slow, digital thermometer. This simulation has also been used to study the modes of vibration of the very thin pancake-like plasmas studied by Weimer, et. al., at the National Institute of Science and Technology at Boulder.[4] By seeding the plasma with offsets in the z-direction appropriate to the modes calculated by Dubin for these plasmas[5], the simulation can detect several of these modes in the real geometry of the experiment (to the extent that the experiment was axisymmetric, of course). Figure 2 shows the spectrum obtained from the simulation when a mode with $\ell = 5$ and $m = 0$ (Dubin's notation) was seeded. (The modes we can study are like the axisymmetric vibration modes of a drumhead.) The detection diagnostic picked up not only the mode that was seeded, but several neighboring modes as well, which gives a fortunate saving of computer time. This work is part of an ongoing study by Grant Mason on the suitability of mode frequency ratios in these pancake-like plasmas to measure their aspect ratios.

IV. EIGENVALUE CALCULATION

Simulations are a nice tool, but, like experiments, they are sometimes hard to interpret. For instance, it is difficult to find mode eigenfunctions, and it can be tricky to properly identify the peaks that come from a spectrum like that given in Fig. 2. The best thing is to be able to compare simulations with other kinds of theory, like perturbation analysis, to more fully explore the models used to describe plasmas.

To try to get a better theoretical description of the normal modes of non-neutral plasmas, and to explore the predictions of various models, we have begun to develop an eigenvalue code. Our hope is that we will be able to explore many different plasma models with this tool, but we have begun with a very simple one: the finite-temperature drift-fluid model. It assumes $\mathbf{E} \times \mathbf{B}$ drift motion perpendicular to the confining magnetic field and an adiabatic fluid response parallel to the confining field. The fluid equations corresponding to this description are

$$\frac{\partial n}{\partial t} + \mathbf{v_d} \cdot \nabla n + \frac{\partial}{\partial z}(n v_z) = 0 \quad ; \quad \frac{\partial p}{\partial t} + \mathbf{v_d} \cdot \nabla p + v_z \frac{\partial p}{\partial z} = -\gamma p \frac{\partial v_z}{\partial z} \qquad (2)$$

$$\mathbf{v_d} = \frac{-\nabla \phi \times \hat{z}}{B} \quad ; \quad M n \left[\frac{\partial}{\partial t} v_z + \mathbf{v} \cdot \nabla v_z \right] = -q n \frac{\partial \phi}{\partial z} - \frac{\partial p}{\partial z} \quad , \qquad (3)$$

where n is the particle density, $\mathbf{v_d}$ is the drift velocity perpendicular to the magnetic field $\mathbf{B}$, v_z is the fluid velocity parallel to $\mathbf{B}$, ϕ is the electrostatic potential, M is the particle mass, q is the particle charge, p is the fluid pressure give information about the variation of mode frequencies with temperature, functioning as a very sophisticated, but very slow, digital thermometer. This simulation has also been used to study the modes of vibration of the very thin pancake-like plasmas studied by Weimer, et. al., at the National Institute of Science and Technology at Boulder.[4] By seeding the plasma with offsets in the z-direction appropriate to the modes calculated by Dubin for these plasmas[5], the simulation can detect several of these modes in the real geometry of the experiment (to the extent that the experiment was axisymmetric, of course). Figure 2 shows the spectrum obtained from the simulation when a mode with $\ell = 5$ and $m = 0$ (Dubin's notation) was seeded. (The modes we can study are like the axisymmetric vibration modes of a drumhead.) The detection diagnostic picked up not only the mode that was seeded, but several neighboring modes as well, which gives a fortunate saving of computer time. This work is part of an ongoing study by Grant Mason on the suitability of mode frequency ratios in these pancake-like plasmas to measure their aspect ratios.

IV. EIGENVALUE CALCULATION

Simulations are a nice tool, but they are fairly crude. For instance, it is difficult to find mode eigenfunctions from simulations, and it can be tricky to properly identify the peaks that come from a spectrum like that given in Fig. 2. In fact, perhaps the best way to think about simulations is as well-diagnosed experiments in a world with simplified and controllable physical laws. Simulations can guide theory, but they are not theory.

To try to get a better theoretical description of the normal modes of non-neutral plasmas, and to explore the predictions of various models, we have begun to develop an eigenvalue code. Our hope is that we will be able to explore many different plasma models with this tool, but we have begun with a very simple one: the finite-temperature drift-fluid model. It assumes $\mathbf{E} \times \mathbf{B}$ drift motion perpendicular to the confining magnetic field and an adiabatic fluid response parallel to the confining field. The fluid equations corresponding to this description are

$$\frac{\partial n}{\partial t} + \mathbf{v_d} \cdot \nabla n + \frac{\partial}{\partial z}(nv_s) = 0 \quad ; \quad \frac{\partial p}{\partial t} + \mathbf{v_d} \cdot \nabla p + v_z \frac{\partial p}{\partial z} - -\gamma p \frac{\partial v_z}{\partial z} \quad (2)$$

$$\mathbf{v_d} = \frac{-\nabla \phi \times \hat{z}}{B} \quad ; \quad Mn \left[\frac{\partial}{\partial t} v_z + \mathbf{v} \cdot \nabla v_z \right] = -qn \frac{\partial \phi}{\partial z} - \frac{\partial p}{\partial z} \quad , \quad (3)$$

where n is the particle density, $\mathbf{v_d}$ is the drift velocity perpendicular to the magnetic field $\mathbf{B}$, v_z is the fluid velocity parallel to $\mathbf{B}$, ϕ is the electrostatic potential, M is the particle mass, q is the particle charge, p is the fluid pressure parallel to **B**, and where γ is the adiabatic exponent, which we take to be 3 because the strong magnetic field limits the kinetic response of the plasma to just one dimension.

When these equations are linearized to describe small perturbations about a non-neutral plasma equilibrium described by unperturbed density $n_o = n_o(r, z)$ and unperturbed electrostatic potential $\phi_o = \phi_o(r, z)$ the following mode equation results (assuming perturbations proportional to $\exp\left(im\theta - i\omega t\right)$:

$$\gamma\lambda_D^2 \frac{\partial}{\partial z}\left[V\frac{\partial}{\partial z}\nabla^2\phi\right] + \gamma\lambda_D^2 \frac{\partial}{\partial z}\left[V\frac{\partial F_o}{\partial z}\nabla^2\phi\right] - $$

$$\frac{(\gamma-1)mv_{th}^2}{r\omega_c}\frac{\partial}{\partial z}\left[V\frac{\partial}{\partial z}\left(\frac{\partial n_o/\partial r}{n_{oo}(\omega-m\omega_o)}\phi\right)\right] - \frac{(\gamma-1)mv_{th}^2}{r\omega_c}\frac{\partial}{\partial z}\left[V\frac{\partial F_o}{\partial z}\frac{\phi}{(\omega-m\omega_o)}\right] + $$

$$\frac{(\omega-m\omega_0)}{\omega_p^2}\nabla^2\phi - \frac{\partial}{\partial z}\left[V\frac{n_o}{n_{oo}}\frac{\partial\phi}{\partial z}\right] - \frac{m\partial n_o/\partial r}{r\omega_c n_{oo}}\phi \qquad = 0 \quad , \qquad (4)$$

where λ_D is the Debye length obtained from the central density n_{oo}, kT is the temperature (assumed independent of r and z) in energy units, $F_o = q\phi_o/kT$, $v_{th} = \sqrt{kT/M}$ is the thermal speed, ω_c is the cyclotron frequency, and where $\omega_o = \omega_o(r, z) = v_d/r$ is the equilibrium-drift rotation frequency. The quantity V which appears throughout the equation is given by

$$V(r, z) = \left[\omega - m\omega_o + (\gamma - 1)v_{th}^2\frac{\partial}{\partial z}\left(\frac{\partial F_o/\partial z}{(\omega - m\omega_o)}\right)\right]^{-1} \qquad (5)$$

It is perhaps worth noting that this mode equation in infinitely-long geometry works very well, giving essentially the same dispersion relation as the corresponding kinetic-theory calculation until Landau damping becomes important. Note also that as the temperature approaches zero only the last three terms in Eq. (4) survive and V approaches unity, recovering the cold mode equation given by Prasad and O'Neil[6].

We have so far only solved the mode equation for axisymmetric modes $(m = 0)$. The solution method we use has been developed by Johnny Jennings and K. C. Hansen, and consists of replacing the homogeneous problem given in Eq. (4) with an inhomogeneous problem, and then looking for the values of ω for which the problem becomes singular, as described below. Equation (4), which is of the form

$$L(\omega)\phi = 0 \quad , \qquad (6)$$

is made inhomogeneous by replacing the zero on the right hand side by some convenient function $\rho(r, z)$:

$$L(\omega)\phi = \rho(r, z) \qquad (7)$$

The function ρ is chosen to be similar to the structure of the mode being sought, i.e., ρ has the same number of radial and axial nodes as the mode.

This removes the difficulty that the equation is homogeneous, tempting the computer always to return $\phi = 0$ as the solution, and also shades the solutions found by the computer toward the desired mode. The mode frequencies ω are simply determined by varying ω, solving Eq. (7), and finding the values of ω which make its solution infinite (or at least very large). This works because the eigenvalues of a matrix are the values that make it singular.

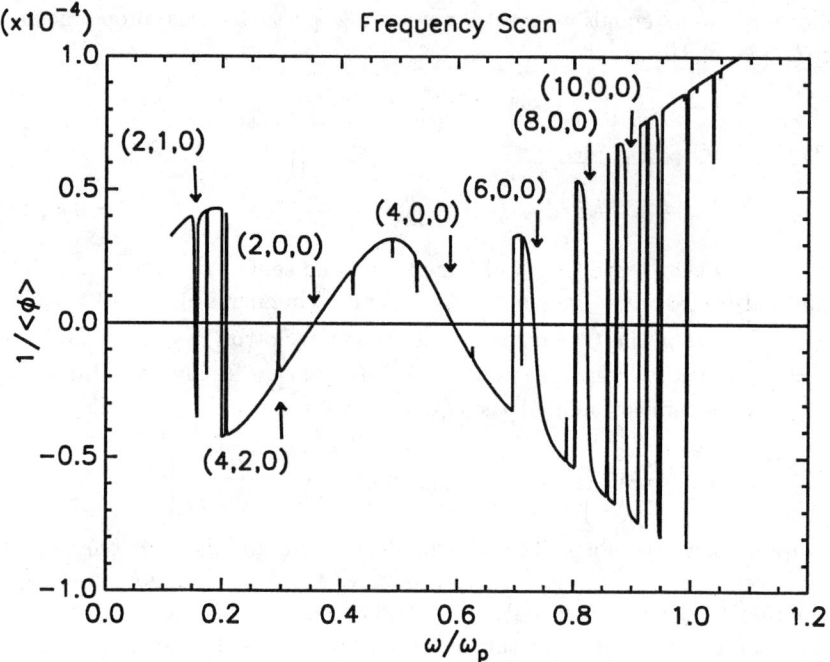

FIG. 3. The result of solving the finite-temperature mode equation for 1200 different values of ω and watching for singularities (zeroes of $1/<\phi>$) where $<\phi>$ is the volume average of the mode potential over the computing region. The labelling of the various modes follows the pattern (ℓ_z, ℓ_r, m), where ℓ_z is the number of axial nodes along the cylinder axis and where ℓ_r is the number of radial nodes in the plasma midplane.

We use a finite-difference approximation to turn the operator $L(\omega)$ into a large square matrix with side-dimension equal to the number of points in the two-dimensional rz grid used to describe the plasma equilibrium. Fortunately the matrix is banded, but it is still huge, and because the system is nearly singular, the usual iterative solution methods that work so well for elliptic problems can't be used here. The best we have been able to do so far is just to use a banded system solver and use lots of memory. The memory required by the calculation on a grid with n_r radial points and n_z axial points is $n_r \times n_z \times (4n_r + 1) \times 8$ bytes (double precision). The calculations we have done so far are limited by the 128 Mbytes of memory on our computer to grids of order 100×200.

Figure 3 gives an example of this procedure of scanning the frequency and watching for singularities. The plasma equilibrium used is in a long cylinder with the radius of the plasma half as large as the cylinder radius and with the plasma half-length 5 times the plasma radius (the aspect ratio is 5). It is a global thermal equilibrium with a Debye length that is 6% of the plasma radius. The function ρ was chosen to have 2 axial nodes ($\ell_z = 2$) and no radial nodes ($\ell_r = 0$), corresponding to Dubin's ℓ-value having the value ($\ell = \ell_z + 2\ell_r = 2$). The mode identifications given in the figure are based on the appearance of the computed eigenfunction as a singular point is approached. As can be seen, the reciprocal of the volume-averaged mode potential $1/ < \phi >$ passes through zero several times, and one of these prominent zeroes corresponds to the $(2,0)$ mode. Choosing the function ρ appropriately can make certain modes stand out, but the system will be singular near other modes as well, as can be seen in the figure. The modes sought don't always stand out as cleanly as the main ones do in Fig. 3, but they often do, making it possible sometimes to use zero-finding methods to find ω instead of scanning.

We have tested the eigenvalue code against other methods and typically find agreement to within 1-2%. For example, we have numerically computed the modes found analytically by Dubin (for the case of cold plasmas[5]) by analyzing several warm equilibria and extrapolating down to $T = 0$, and find agreement to about 1%. We have also compared with the experimental results and the simulations reported by Tinkle, et. al..[3] Using the same equilibria employed in the simulations the eigenvalue code gives mode frequencies that are within 2% of the frequencies observed in the experiments.

A. Cold Cylindrical Plasmas

The eigenvalue calculation has been used to study the cold axisymmetric modes of non-neutral plasmas surrounded by cylindrical conducting walls. A similar calculation was done by Prasad and O'Neil, but without taking into account the correct equilibrium shape of the plasmas.[6] We employed the technique of computing modes for several different temperatures, then extrapolating to $T = 0$. In all we performed about 400 singularity searches to find about 100 cold mode frequencies for various plasma radii, various plasma lengths, and for modes with $\ell_r = 0$ and $\ell_r = 1$. We find that for plasmas with lengths much shorter than the wall radius, Dubin's analysis[5] gives accurate mode frequencies, i.e., image charge effects are unimportant for such plasmas. For plasmas whose lengths are about the same as the conducting wall radius, things are complicated and numerical calculations are probably required. But for plasmas whose lengths are longer than the wall radius, we find that a simple adjustment to the Trivelpiece-Gould dispersion relation for infinitely-long plasmas[7] gives the same frequencies as the complicated numerical calculations to within 1%, or better. The Trivelpiece-Gould dispersion relation is obtained by assuming an infinitely long cold plasma with constant density so that the modes are proportional to $\exp{(ikz)}$, resulting in the relation

$$\frac{I_o'(kr_p)K_o(kr_w) - K_o'(kr_p)I_o(kr_w)}{I_o(kr_p)K_o(kr_w) - K_o(kr_p)I_o(kr_w)} = \beta \frac{J_o'(\beta kr_p)}{J_o(\beta kr_p)} \quad , \tag{8}$$

where the I's and K's are modified Bessel functions, where r_p is the plasma radius, where r_w is the conducting wall radius, and where $\beta = \sqrt{\omega_p^2/\omega^2 - 1}$. Given a choice for k, this equation can be solved for the mode frequency ω. The adjustment uses the idea of an effective length for the plasma, similar to the use of an effective length for an open tube to accurately compute its resonant acoustic frequencies. To lowest order, there is a potential antinode at the end of the plasma,[6] so the k of the mode is approximately

$$k = \frac{\pi \ell_z}{2z_p} \quad , \tag{9}$$

where z_p is the half-length of the plasma. An analyis of the numerical results for modes with either $\ell_r = 0$ or $\ell_r = 1$ shows that the Trivelpiece-Gould dispersion relation can reproduce all of the numerical results for plasmas with $r_p/r_w = 0.25, 0.5, 0.75$ and with aspect ratios z_p/r_p greater than 3 if the formula for the wavenumber to use in Eq. (8) is modified by adding a simple correction to the plasma length, empirically determined from the numerical data:

$$k = \frac{\pi \ell_z}{2z_p + c_1 r_w + c_2 r_p} \quad , \tag{10}$$

with

$$c_1 = 0.3 \quad ; \quad c_2 = 0.7 \quad \text{for} \quad \ell_r = 0 \quad \text{and} \quad c_1 = -0.2 \quad ; \quad c_2 = 0.9 \quad \text{for} \quad \ell_r = 1 \tag{11}$$

This makes it much easier accurately to compute mode frequencies for such plasmas, and future work will concentrate on trying to find similar simple connections to infinitely-long cylinder results when finite temperature and radial profile effects are included.

B. Acoustic Resonances

A natural calculation to perform with the eigenvalue code is the extension of Dubin's spheroidal study to finite temperature. To this end the equilibrium and eigenvalue codes were modified to handle spheroids in infinite space and ω was scanned to look for modes. Figure 4 shows the result for a spherical plasma with the right-hand side of the mode equation chosen to look for the mode with $(\ell_r, \ell_z) = (2, 0)$. It looks like there should be a prominent mode near 0.9×10^8 s^{-1}, but a bunch of extra modes seems to have gotten in the way.

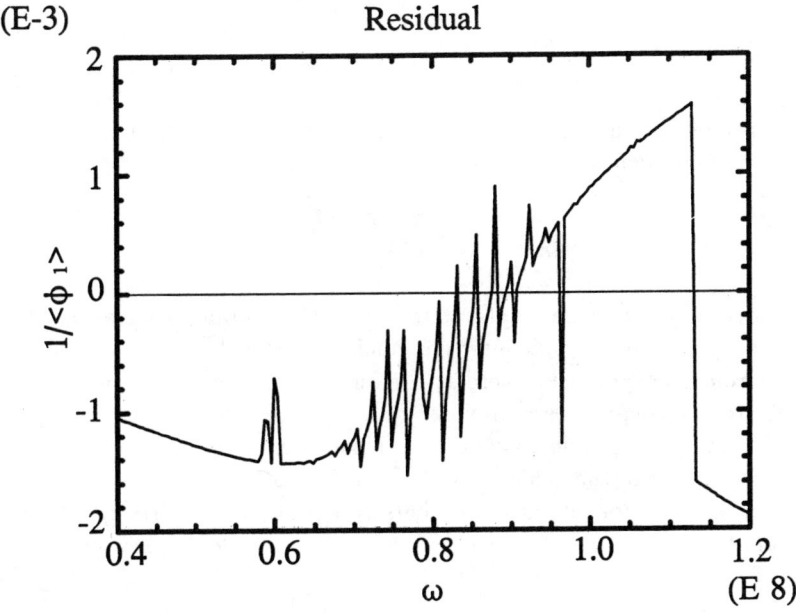

FIG. 4. The reciprocal of the volume-averaged potential vs. frequency is shown for a spheroidal plasma with aspect ratio 1 and a ratio of the Debye length to the plasma radius of 0.15. The mode being sought is the (2,0,0) mode, and its frequency at zero temperature is $0.775 \times 10^8 \text{s}^{-1}$. The zig-zags in the middle are associated with continuum modes, and each would have appeared as a zero-crossing if a finer scan had been used.

If the grid resolution is doubled, the number of closely-spaced modes doubles, suggesting perhaps that the model has encountered a continuous spectrum which the grid is attempting to resolve by associating a mode with each grid point. This is, in fact, the case.

The continuous spectrum is connected with radii at which the mode frequency ω has a value for which the equation

$$\gamma \lambda_D^2 \frac{\partial}{\partial z} \left[V \frac{\partial}{\partial z} \nabla^2 \phi \right] + \gamma \lambda_D^2 \frac{\partial}{\partial z} \left[V \frac{\partial F_o}{\partial z} \nabla^2 \phi \right] - \frac{\omega}{\omega_p^2} \nabla^2 \phi = 0 \qquad (12)$$

is satisfied for a non-trivial $\nabla^2\phi$. The terms in this equation are the terms from Eq. (4) containing the highest radial derivatives of the perturbed potential, and when this equation has a non-trivial solution at some radius, the mode equation is singular, similar to what happens in one-dimensional continuous spectrum problems when the coefficient of the highest derivative vanishes. The physical meaning of the singularity can be seen by taking the low-temperature limit, for which F_o is constant inside the plasma, transforming Eq. (12) into

$$\gamma \frac{kT}{M} \frac{\partial^2}{\partial z^2} \nabla^2\phi + \nabla^2\phi = 0 \quad , \tag{13}$$

the equation for one-dimensional sound waves. The continuum resonances occur roughly whenever

$$\omega = \sqrt{\frac{\gamma kT}{M}} \frac{\pi \ell_z}{L(r)} \quad , \tag{14}$$

where $L(r)$ is the plasma length at radius r. For rectangular plasmas $L(r)$ is nearly constant and doesn't interfere much with the electrostatic modes. But for spheroids, L approaches zero at the outer radius of the plasma, sweeping the acoustic resonances up through the modes of interest, as shown in Fig. 4, and ruining the mode calculation. This is an unphysical effect from the fluid model used in the calculation, for purely acoustic standing waves do not occur in one-component plasmas. This is a particularly striking case where the fluid theory simply gets it wrong, and to fix the calculation a better model (like kinetic theory or a non-local fluid theory) must be used. Future work will focus on using the solution techniques developed for the fluid problem to handle more complicated models.

---

[1] R. L. Spencer, S. N. Rasband, and Richard R. Vanfleet, Phys. Fluids B **5**, 4267 (1993).

[2] S. A. Prasad and T. M. O'Neil, Phys. Fluids **22**, 278 (1979)

[3] M. D. Tinkle, R. D. Greaves, C. M. Surko, R. L. Spencer, and G. W. Mason, Phys. Rev. Lett. **72**, 352 (1994).

[4] C. S. Weimer, J. J. Bollinger, F. L. Moore, and D. J. Wineland, Phys. Rev. A, **49**, 3842 (1994).

[5] Daniel H. E. Dubin, Phys. Rev. Lett. **66**, 2076 (1991).

[6] S. A. Prasad and T. M. O'Neil, Phys. Fluids **26**, 665 (1983).

[7] A. W. Trivelpiece and R. W. Gould, J. Plasma Physics **30**, 1784 (1959).

Laser-Cooled Trapped-Ion Experiments at NIST[1]

Joseph N. Tan, J. J. Bollinger, A. S. Barton, and D. J. Wineland

Time and Frequency Div., National Institute of Standards and Technology, 325 Broadway, Boulder, CO 80303

Abstract. We describe some recent experiments that use a new Penning trap to trap and laser-cool more than 10^5 Be^+ ions. In this work the Coulomb coupling parameter was estimated to be typically greater than 100 and in some cases greater than 300. Shell structure was difficult to observe in the ion fluorescence from these large clouds. An initial Bragg scattering experiment to obtain information on the ion correlations is described. We also briefly review our work on the electrostatic modes of a Penning trap plasma, including the use of the modes to diagnose the shape and density of a cryogenic electron plasma. Some applications of laser-cooled plasmas in Penning traps, such as atomic clocks, are discussed.

I. INTRODUCTION

At the National Institute of Standards and Technology (NIST) in Boulder, laser-cooled ions stored in Penning (1) and rf or Paul (1) traps are used for plasma studies, high precision spectroscopy, and atomic clocks. In this manuscript we discuss aspects of our trap work which involve the nonneutral plasma physics of large systems of confined ions. We concentrate on experiments in Penning traps because these studies involve plasmas of a few hundred thousand trapped ions while in our rf trap work, only a few or perhaps a few tens of ions are trapped. Penning traps have been used at NIST to trap electrons (2), Mg^+ (3,4), Be^+ (5-9), and Hg^+ (10) ions. Reference 11 reviews some of this work. Here we discuss some of our recent work on $^9Be^+$ ions (12) and electrons.

Ion confinement in a Penning trap can be extremely long (> 1 day). Therefore the trapped ions have sufficient time to evolve to a state of global thermal equilibrium (13,14,15). (Measurements indicate that our trapped ion plasmas evolve to a spatial distribution consistent with global thermal equilibrium (6). However, in some cases

---

[1]Work of the U. S. Government, not subject to copyright.

motion parallel and perpendicular to the magnetic field may have different temperatures (5,6).) In thermal equilibrium, the rotation (at frequency ω_r) of the ions about the magnetic field axis is uniform or "rigid" (6,14). An important advantage of using ions in plasma studies is that the atomic structure of the ions can be used to manipulate the equilibrium state of the plasma. Doppler laser-cooling (16), for instance, is now routinely used to reach temperatures less than 10 mK for small Be^+ plasmas. The Doppler-cooling limit for Be^+ is 0.5 mK. In addition to cooling the ions, laser beams are used to apply a torque on the ion plasmas and change or control the ion density (8).

Laser-cooling yields ion plasmas in which the Debye length λ_D is much less than the plasma dimensions. In this case the plasma has a uniform density $n_0 = 2\epsilon_0 m \omega_r (\Omega - \omega_r)/q^2$ with a sharp boundary. Here q and m are the charge and mass of the ion, $\Omega = qB/m$ is the ion cyclotron frequency, ϵ_0 is the permittivity of the vacuum, and B is the magnetic field of the Penning trap. In our experimental work, the plasma dimensions are very much smaller than the trap dimensions, typically by more than a factor of 10. In this case, the trap potential is quadratic in the region of the ion plasma. We let ω_z denote the harmonic oscillation (or axial bounce) frequency of a single ion in the trap. When λ_D is much less than the plasma dimensions and the plasma dimensions are much less than the trap dimensions, a Penning trap plasma is a uniform density spheroid (an ellipse rotated about one of its symmetry axes) (6,17,18). Let $2z_o$ denote the axial extent of the plasma spheroid at r = 0 and $2r_o$ denote the plasma diameter at $z = 0$. The aspect ratio $\alpha = z_o/r_o$ of the spheroid is determined (6,17) by ω_r, Ω, and ω_z.

The aspect ratio of an ion plasma as a function of ω_r can be understood by examining the radial confining force acting on ions as they rotate through the magnetic field in a Penning trap. Due to the plasma rotation, the ions experience a $\vec{v} \times \vec{B}$ Lorentz force which pushes radially inward. For low rotation frequencies (slightly greater than the single ion magnetron frequency ω_m), this radial confining force is weak resulting in a pancake-shaped or oblate plasmas. As ω_r increases, the radial confining force grows, and the plasma density and aspect ratio increase. The highest density and largest plasma aspect ratio are obtained at the Brillouin limit where $\omega_r = \Omega/2$. For higher rotation frequencies, the centrifugal acceleration, which increases as ω_r^2, overcomes the radial binding force and the plasma once again becomes an oblate spheroid.

II. EXPERIMENTAL APPARATUS

Figure 1 shows a diagram of the electrode structure of the Penning trap we are currently using to trap Be^+ ions. The apparatus, which consists of two separate traps, is enclosed in a room-temperature vacuum jacket (10^{-8} Pa) and centered in the vertical bore of a homogeneous, 4.5 T superconductive magnet. With the use of plasma modes (8,9), the symmetry axis of the electrode assembly is aligned to within

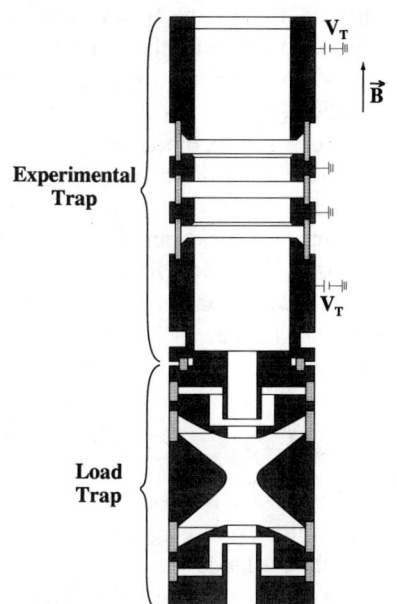

Figure 1. Cross section of the Penning trap used to trap plasmas of Be⁺ ions. The dark shaded parts are copper electrodes. The light shaded parts are ceramic insulators. Ions are created in the load trap and then transferred to the experimental trap. The experimental trap electrodes are right circular cylinders with a 2.0 cm diameter inner radius. They provide a quadratic potential near the trap center. Laser beams pass through gaps in the insulators between the electrodes or vertically along the trap axis. Compensation electrodes (not shown) are mounted on the insulators in the experimental trap. In recent work, B≈4.5 T and V_T ranges from 10 V to about 1.5 kV. For Be⁺ ions this gives a cyclotron frequency of 7.6 MHz and a trap axial frequency ranging from 80 to 980 kHz.

0.01° of the applied magnetic field direction.

The load trap has electrode surfaces machined to lie on hyperboloids of revolution. It is used for producing ions and for ejection of impurity ions. Ions are loaded by ionizing neutral atoms in the trap with a beam of electrons that pass through the endcaps. Small atomic ovens provide the source of neutral atoms. A hyperbolic Penning trap provides a large volume over which the electrostatic potential is quadratic, which improves the resonant ejection of impurity ions using an rf field. The symmetry of the load trap is degraded by the holes and slits required for the atomic ovens and electron guns, and by patch effects due to the atomic beams plating onto the electrodes. Laser cooling and imaging optics (f/7 input) are provided so that crude mass spectrometry can be performed to check for impurity ions.

From the load trap, the collected ions are transferred along the magnetic field direction into a precision cylindrical Penning trap (the experimental trap in Fig. 1). The potentials applied to the electrode assembly are varied slowly to adiabatically shift the axial trapping well from the center of the (hyperbolic) load trap to the center of the (cylindrical) experimental trap. The electrodes of the experimental trap form a 12.7 cm long stack (including gaps) of OFHC copper rings. In each gap between the cylindrical electrodes are compensation electrodes divided into six equal sectors with 0.64 cm separation between neighboring sectors. The compensation electrodes allow rf excitation of plasma modes and minimization of small azimuthal asymmetries. The separation between neighboring compensation electrodes permits optical access to the trap center. For spectroscopy experiments, a nuclear spin flipping rf drive can be coupled to the ions through one of these gaps. A microwave drive can also be applied for measurements of ω_r from the Doppler shift of an electron spin flip transition (8).

A new feature of this apparatus is the ability to simultaneously image the ions in two orthogonal directions. The optical axis of an f/2 system is aligned with the symmetry axis of the trap and provides a top-view image of the ions. Perpendicular to the magnetic field, an f/5 optical system is used to obtain a side-view image of the ion plasma. This side-view image gives the spheroidal boundary of the plasma. The density and rotation frequency of the plasma can be determined from the aspect ratio of such an image.

III. STRONG COUPLING

In a frame of reference rotating with the ions, ions in a Penning trap behave as if they are immersed in a uniform density background of opposite charge (19), where the trapping fields provide the uniform background of opposite charge. A single species of charge immersed in a uniform density background of opposite charge is a one-component plasma (20). The thermodynamic properties of a one-component plasma are determined by the Coulomb-coupling parameter

$$\Gamma = \frac{q^2}{4\pi\epsilon_o a_s k_B T} \, ,$$

(1)

where a_s is the Wigner-Seitz radius given by $4\pi n_0 a_s^3/3 = 1$ and T is the ion temperature. Calculations (20,21) for the infinite one-component plasma predict that for $\Gamma > 2$, the plasma should exhibit liquid-like behavior characterized by short range order, and at around $\Gamma=170$ a liquid-solid phase transition to a body-centered cubic (bcc) lattice should take place. With laser cooling, coupling parameters of several hundred can be routinely obtained with ions in traps (5,13). In principle, with Be$^+$ ions at the Doppler laser-cooling limit of 0.5 mK and the Brillouin density at B=4.5 T of 5.9×10^9 cm^{-3}, the coupling parameter can be as high as 9700.

A. Shell structure

Ion trap experiments until now have laser-cooled small plasmas. Typical plasma dimensions are less than 20 interparticle spacings (7). (In the quadrupole storage ring trap of Ref. 22, the plasmas were dimensionally large along the quadrupole circumference, but less than 8 interparticle spacings in the other directions.) Such small systems should not behave like infinite-volume one-component plasmas. Computer simulations (23,24,25) with less than a few thousand cold ions in a trap found something quite different than bcc order. The ions eventually solidified into concentric, approximately spheroidal shells rather than bcc planes.

Experimentally, shell structure has been observed with up to 20 000 laser-cooled Be^+ ions in a Penning trap (7). [Analogous shell structure has also been observed in a miniature rf storage ring trap (22).] The shells were observed by illuminating a thin cross section of the plasma with laser beams and imaging the laser-induced fluorescence onto a photon-counting imaging tube. Imaging works well for observing shell structure because the structure is preserved by the plasma rotation. Qualitative agreement was observed with the simulations (7), except that, in some cases, open-ended cylindrical shells were observed in the experiments. The cause of these open-ended shells remains an open question. One possibility is that shear in the plasma rotation might produce such a structure.

So far the observed ion correlations (shell structure) are strongly affected by the finite size and boundary of the plasma. How large must the ion plasmas be in order to exhibit infinite-volume behavior? Two different analytical methods (26,27) give similar predictions. The plasmas may have to be greater than 60 interparticle spacings along their smallest dimension in order to exhibit bcc structure. A spherical plasma with 60 shells has about 10^6 trapped ions.

B. Bragg scattering

In the trap shown in Fig. 1, we measured the densities and temperatures of plasmas of several hundred thousand laser-cooled Be^+ ions. Only the temperature parallel to the magnetic field was measured. These measurements indicated that the plasmas were strongly coupled with couplings typically greater than 100 and in some cases greater than 300. Shell structure has been difficult to observe in the images of these larger plasmas. In some cases, shells were observed near the plasma boundary, but not in the plasma interior. As the clarity of the shells decreases with increasing ion number, a different technique is required to obtain information on the ion correlations.

Bragg scattering (28) can be used to obtain this information. The Bragg scattering signal as a function of the scattering angle is just the Fourier transform of the ion pair correlation function. The relative positions and heights of the Bragg scattering peaks therefore give information on the spacings and relative positions of the ions in the strongly coupled plasma. For example, in our experimental setup, a laser beam (with frequency tuned slightly below the cooling transition) can be directed along the symmetry axis of the trap with a beam waist large enough to encompass the ion plasma. Light is approximately elastically scattered from the ions and will interfere constructively in certain directions. Because the 313 nm wavelength of our cooling laser is small compared to the inter-ion spacing (from 15 to 5 μm depending on the density), constructive interference will occur mainly in the forward-scattered light. We estimate that the first and strongest Bragg scattering peak will occur at a

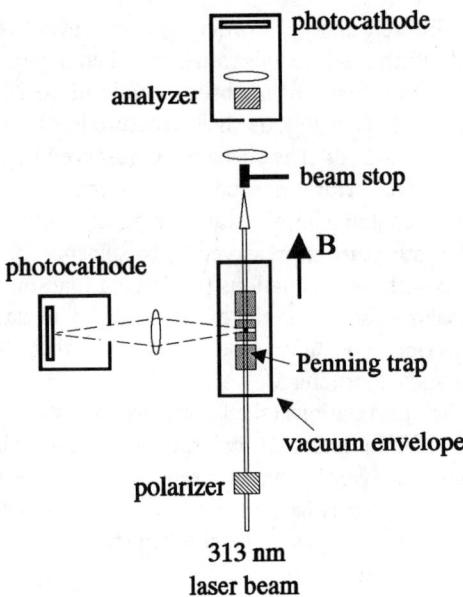

Figure 2. Schematic diagram of the Bragg scattering setup. A 313 nm laser beam is directed through the ions along the magnetic field axis and then blocked by a beam stop. The forward-scattered ion fluorescence is observed as a function of the scattering angle. Light scattered off the laser windows is minimized by an aperture at an image of the ions and with a pair of crossed polarizers (the polarizer and analyzer). An image of the ion plasma is obtained from light scattered at 90° from the magnetic field direction.

small scattering angle between 1 and 3°.

Very recently at NIST, Bragg scattering from Be^+ ions in a Penning trap has been observed. The detection system was designed to separate the forward-scattered ion fluorescesce from the background of forward-scattered light off the vacuum windows. A simple sketch of the experimental set-up is given in Fig. 2. The laser beam is linearly polarized and is blocked by a beam stop after passing through the vacuum chamber. Two steps were taken to minimize the forward-scattered background light. First, because the ion fluorescence is mainly circularly polarized along the direction of observation (close to the $\vec{B}$-field direction) and the forward-scattered light from the windows is mainly linearly polarized, a pair of crossed polarizers can be used to preferentially attenuate the forward-scattered background light. Second, an image of the ion fluorescence was formed in the center of a small aperture. This aperture helped skim off some of the forward-scattered background light. The Bragg scattered light was then observed with a photon-counting camera. With these steps, it was possible to observe Bragg scattering off plasmas of 3×10^4 ions with good signal-to-noise ratio. Figure 3 shows an example of the observed Bragg scattering pattern. The positions of the first two Bragg peaks provide some information on the

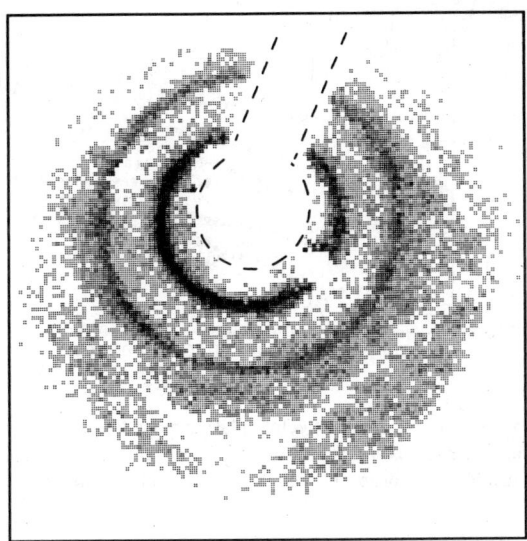

Figure 3. Bragg scattering image observed with approximately 3×10^4 Be$^+$ ions. The first two Bragg peaks are visible. The ion density here was about 9% of the Brillouin density which produced a scattering angle for the first peak of about $1.5°$. The square shadow near the edge of the image is from a wire mesh on the top of the upper experimental trap endcap. The dashed line outlines the beam stop.

type of ion spatial correlations. We are in the process of analyzing this information.

IV. ELECTROSTATIC MODES

A cold nonneutral plasma stored in a Penning trap where the plasma dimensions are small compared to the trap is a constant density spheroid. The electrostatic modes of a spheroidal plasma in a uniform magnetic field can be calculated exactly (29). The modes can be described (9,29) with the use of spheroidal coordinates by two integers (ℓ,m) with $\ell \geq 1$ and $m \geq 0$. (Negative values of m are allowed but do not give rise to new modes.) The index m indicates that the plasma mode displays an $\exp(im\phi)$ dependence where ϕ is the azimuthal angle. The index ℓ describes the mode dependence along a spheroidal surface (for example, the plasma boundary) in a direction perpendicular to $\hat{\phi}$. As an example, the $\ell=1$ modes are the familiar center of mass modes. Plasma modes have been detected in two different experiments at NIST (2,8,9). In one experiment, some of the $\ell=2$ modes were excited on plasmas of laser-cooled Be$^+$ ions (8,9). Laser torques were used to vary the plasma rotation frequency. The rotation frequency could be measured from Doppler shifts induced by the rotation. In this way the theoretical predictions for the mode frequencies were tested at the level of a few percent. The new trap shown in Fig. 1 has segmented compensation electrodes which should permit the excitation of azimuthally

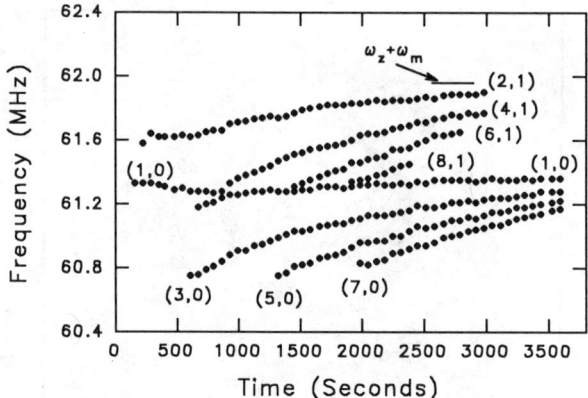

Figure 4. A plot of the measured cryogenic electron plasma mode frequencies vs time after the plasma was loaded into the trap. The modes are labeled by their (ℓ,m) values. The measurement uncertainty in the frequency is less than the symbol size used in the plot and is the same for all points.

asymmetric modes and extend the studies in Refs. 8 and 9. Measurement of the modes can be a useful tool in diagnosing the plasma density and shape. This should be particularly useful for antimatter (30,31,32) and electron plasmas where the standard techniques for obtaining this information involves dumping the plasma out of the trap. Plasma modes have been used as a diagnostic tool in several experiments (2,33,34). At NIST plasma modes have been used to obtain information on the aspect ratio and density of a cryogenic electron plasma (2).

In the NIST experiment the modes were detected by spectrum analyzing the image currents induced in one of the endcap electrodes of a Penning trap at cryogenic (4 K) temperatures. Experimentally, modes with frequencies near the single-particle axial frequency ω_z could be detected. Immediately after a plasma of electrons was loaded into the trap, only the center-of-mass mode was observed. However, after a time that depended on the magnetic field, other modes were observed. The evolution of the mode frequencies was followed as a function of time as shown in Fig. 4. The detected modes were identified within the framework of Dubin's cold fluid theory (29) as drumhead modes of a 2-dimensional disk. In Fig. 4, we identified 8 modes which were used to provide 7 different estimates of the plasma aspect ratio shown in Fig. 5. The discrepancy between the different estimates is on the order of 20 %. Figure 5 shows that the drumhead modes were detected when the plasma aspect ratio α was about 0.02 or smaller. The decrease in α as a function of time could then be followed until $\alpha \approx 0.002$. The number of electrons, determined from the center-of-mass mode signal (35), varied from 20 000 to 100 000. The modes observed here may be related to the features observed in Ref. 36. This experiment also demonstrated compression of the electron plasma by a "sideband cooling drive" at frequency $\omega_z + \omega_m$.

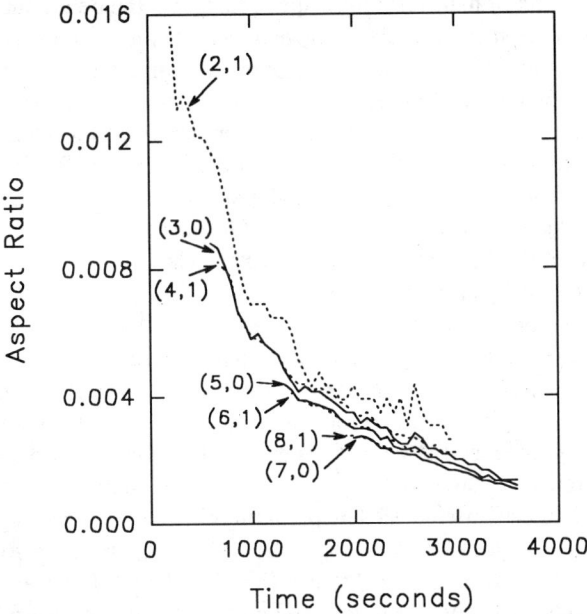

Figure 5. Estimates of the plasma aspect ratio α as a function of the time after the plasma was loaded, for the data of Fig. 4. For each time, the (1,0) mode is used to estimate ω_r. Each (ℓ, m) mode (except the (1,0) mode) is then used to estimate α as described in Ref. 2. The estimated values of α have been connected by straight lines for each mode. The values of α from $m=1$ ($m=0$) modes are connected by dashed (solid) lines.

V. APPLICATIONS OF PENNING TRAP PLASMAS

A. Spectroscopy

Ion traps have several properties which are useful for high-resolution spectroscopy. The fact that the ions are stored means their velocity averages to 0; this virtually eliminates the first order Doppler shift of ion spectra. Long storage times can lead to very high resolution; linewidths smaller than 0.001 Hz have been observed (37). Ions are typically stored under high vacuum conditions; this minimizes perturbations of the ions' internal structure caused by background gas collisions. The Coulomb repulsion between ions has both bad and good effects. It limits the attainable density and therefore the overall numbers of ions; this limits the signal-to-noise ratio, thereby limiting the measurement precision (38). On the other hand, the Coulomb repulsion reduces the perturbing effects of ion-ion collisions; this is particularly true at low temperatures where the ions never get very close together.

One application of high-resolution spectroscopy is the determination of atomic structure and measurement of nuclear moments. Another application is atomic clocks and frequency standards. Atomic clocks generate time intervals by counting the cycles of an oscillator which is made synchronous with the natural oscillations in atoms. The first frequency standard using a Penning trap and the first using laser-cooled atoms was demonstrated with a hyperfine transition in $^9Be^+$ ions (39). It had an accuracy of about 1 part in 10^{13} limited in part by an (unexpected) large pressure shift due to background gas molecules (40). (At present, the most accurate atomic clocks have inaccuracies of about 1 part in 10^{14}.) This shift could reduced by many orders of magnitude by cryogenic cooling.

We anticipate that the largest systematic frequency shift in a clock based on stored ions in a Penning trap will be the time-dilation shift caused by the rotational speed of the ions in the plasma (41). (We assume that the cyclotron and axial temperatures are reduced to low enough values that the contribution to the overall speed of the ions due to these components of the motion is negligible.) Recently we investigated this problem theoretically and found that the time-dilation shift due to the velocity of rotation of the ions can be minimized by preparing the ions in a special spheroid (42). For very low $(\omega_r \rightarrow \omega_m)$ and very high $(\omega_r \rightarrow \Omega - \omega_m)$ rotation frequencies, the spheroid is stretched into a thin circular disk ($\alpha \rightarrow 0$ if the $\vec{v} \times \vec{B}$ radial restoring force is weak or the centrifugal force is large), and consequently, the time-dilation shift for the thin disk of ions is very large. The time dilation shift goes to a minimum at a particular (low) value of ω_r which brings the ions closer to the trap axis. Near this minimum there is only a second-order dependence on rotation frequency, and thus greater frequency stability can be obtained. This is illustrated in Table 1.

Table 1. Expected performance for Penning-trap-based frequency standards which use hyperfine transitions. We assume 10^6 ions, $r_o = 4.2$ mm, and $z_o/r_o = 0.46$. The fractional measurement uncertainty $(\sigma_y(\tau) = \delta v(\text{measurement})/v_o)$ is evaluated as described in Ref. 41. The fractional time dilation shift is $\Delta v/v_o$. If the fluctuations in plasma radius are held to 5% of the optimal radius, the fluctuations in $\Delta v/v_o$ are 1% of the last column. (From Ref. 42)

Ion	v_0 (GHz)	B (T)	σ_y(1 s) $\times 10^{15}$	minimum $\Delta v/v_o \times 10^{15}$
$^9Be^+$	0.303016	0.8194	53	-241
$^{25}Mg^+$	0.291996	1.2398	55	105
$^{67}Zn^+$	≈ 1	≈ 8.0	16	-2.5
$^{199}Hg^+$	20.9	43.9	0.76	-0.84
$^{201}Hg^+$	7.73	3.91	2.1	-11

B. Multiple species plasmas

Two or more ion species (of the same sign of charge) can be stored simultaneously in a Penning trap in thermal equilibrium. At low enough temperatures, the species separate spatially due to differences in centrifugal forces; the ions with the highest charge-to-mass ratio are pushed to the center of the ion cloud and form a column on the axis (43). For densities sufficiently below the Brillouin density, the separation between the ion species can be approximately equal to the mean interparticle spacing so the thermal contact between ions can remain fairly strong. This can be useful in different kinds of experiments. For example, if one of the stored species is an ion which can be laser cooled, this cooling can be transferred to the second species - "sympathetic" laser cooling (10). This technique was useful in the $^9Be^+$ Penning trap clock experiments. There, laser cooling could not be applied directly to the $^9Be^+$ ions when the clock transition was being driven (durations ≈ 100 s) because of the large "light" shifts of the clock levels caused by the laser. On the other hand, if cooling had not been provided continuously, the ions would have heated leading to a relatively large time-dilation shift (39). Therefore, simultaneously stored laser-cooled Mg^+ ions were used to keep the $^9Be^+$ ions cold (48).

Sympathetic laser cooling may also be useful to cool ion species where no other direct means of cooling is available. If the cooling laser is also used to compress the laser-cooled species through laser-induced torque, this torque is also sympathetically transferred to the other species thereby preventing or reversing radial diffusion. The transfer of cooling and torque may be useful, for example, in experiments which capture high-Z ions from other sources such as storage rings or electron beam ion sources (44). At NIST we have recently become interested in experiments which capture and cool positrons in a Penning trap through collisions with a laser-cooled $^9Be^+$ ion plasma. We have analyzed this problem theoretically by making estimates of capture efficiencies under possible experimental conditions (45). (We have refined the basic idea outlined in Ref. 45 and are preparing a manuscript on this). The idea is as follows: positrons from a moderated ^{22}Na source are assumed to be captured in the trap through Coulomb collisions with the $^9Be^+$ ions. Once the positrons are trapped, cooling will occur through sympathetic cooling with the $^9Be^+$ plasma and positron cyclotron radiation. As the positrons cool, a centrifugal separation will occur between the positrons and the $^9Be^+$ ions and force the positrons to coalesce into a cold column along the trap axis. We have analyzed an experiment for a cylindrical Penning trap with B = 6 T. In this field, a laser-cooled $^9Be^+$ plasma can reach a uniform density of up to 10^{10} cm^{-3} based on our previous studies (8). Positrons from a ^{22}Na source will enter the UHV trap region axially through a thin Ti foil. After passing through the plasma, the positrons will strike a Cu(111) single crystal moderator and about 0.1 % will emerge backwards towards the trap as a beam of slow positrons. Some of these positrons will lose enough axial energy through Coulomb collisions with the $^9Be^+$ plasma to remain trapped. A Monte-Carlo simulation indicates a capture efficiency of greater than 10% of the moderated positrons for a $^9Be^+$ density of 10^{10}

atoms/cm^3 and a column length of about 1 cm. Given a 2 mCi positron source, we expect to capture positrons at a rate of 2800 s^{-1} for these conditions. The resulting dense reservoirs of cold positrons may be useful for antihydrogen production (46) and for reaching a plasma state in which the mode dynamics must be treated quantum mechanically.

C. Other studies

Many other uses of Penning traps are described in papers from this workshop (see also Ref. 44). At NIST we have used the high measurement precisions and accuracies obtained from spectra on ions in Penning traps to perform accurate tests of physical laws. For example, Penning traps have been used to make precise tests of Local-Lorentz invariance (47) and the linearity of quantum mechanics (48), and they have allowed searches for anomalous spin interactions (49). We have also used the high detection sensitivity capable of detecting transitions in individual atoms to study various quantum effects in measurements such as "quantum jumps" (50), the "quantum Zeno" effect (51), and quantum "projection" noise (38). Starting with the early Penning trap experiments at NIST (3), we have also performed many experiments and theoretical studies on laser cooling in Penning traps. The NIST work has been summarized in several Technical Notes (52) (available upon request).

ACKNOWLEDGEMENTS

We gratefully acknowledge the support of the U. S. Office of Naval Research. We thank Brana Jelenkovic, John Miller, and Matt Young for their suggestions and careful reading of the manuscript.

REFERENCES

1. H.G. Dehmelt, Adv. At. Mol. Phys. **3**, 53 (1967); **5**, 109 (1969); D.J. Wineland, W.M. Itano, and R.S. Van Dyck, Jr., ibid **19**, 135 (1983); R.C. Thompson, Adv. At. Mol. Opt. Phys. **31**, 63 (1993).
2. C.S. Weimer, J.J. Bollinger, F.L. Moore, and D.J. Wineland, Phys. Rev. A **49**, 3842 (1994).
3. D.J. Wineland, R.E. Drullinger, and F.L. Walls, Phys. Rev. Lett. **40**, 1639 (1978).
4. W.M. Itano and D.J. Wineland, Phys. Rev. A **24**, 1364 (1981).
5. J.J. Bollinger and D.J. Wineland, Phys. Rev. Lett. **53**, 348 (1984).
6. L.R. Brewer, J.D. Prestage, J.J. Bollinger, W.M. Itano, D.J. Larson, and D.J. Wineland, Phys. Rev. A **38**, 859 (1988).
7. S.L. Gilbert, J.J. Bollinger, and D.J. Wineland, Phys. Rev. Lett. **60**, 2022 (1988).
8. D.J. Heinzen, J.J. Bollinger, F.L. Moore, W.M. Itano, and D.J. Wineland, Phys. Rev. Lett. **66**, 2080 (1991).
9. J.J. Bollinger, D.J. Heinzen, F.L. Moore, W.M. Itano, and D.J. Wineland., Phys. Rev. A **48**, 525

(1993).

10. D.J. Larson, J.C. Bergquist, J.J. Bollinger, W.M. Itano, and D.J. Wineland, Phys. Rev. Lett. **57**, 70 (1986).
11. J.J. Bollinger, D.J. Wineland, and D.H.E. Dubin, Phys. Plasmas **1**, 1403 (1994).
12. J.N. Tan, J.J. Bollinger, and D.J. Wineland, Bull. Am. Phys. Soc. **38**, 1972 (1993).
13. R.C. Davidson, "Physics of Non-Neutral Plasmas" (Addison-Wesley, Redwood City, CA, 1990).
14. C.F. Driscoll, J.H. Malmberg, and K.S. Fine, Phys. Rev. Lett. **60**, 1290 (1988).
15. S.A. Prasad and T.M. O'Neil, Phys. Fluids **22**, 278 (1979).
16. D.J. Wineland and W.M. Itano, Phys. Today $\underline{40}$(6), 34 (1987).
17. D.J. Wineland, J.J. Bollinger, W.M. Itano, and J.D. Prestage, J. Opt. Soc. Am. B**2**, 1721 (1985).
18. J.B. Jeffries, S.E. Barlow, and G.H. Dunn, Int. J. Mass Spectrom. Ion Processes **54**, 169 (1983).
19. J.H. Malmberg and T.M. O'Neil, Phys. Rev. Lett. **39**, 1333 (1977).
20. S. Ichimaru, I. Iyetomi, and S. Tanaka, Phys. Rep. **149**, 91 (1987).
21. E. Pollack and J. Hansen, Phys. Rev. A **8**, 3110 (1973); W.L. Slaterly, G.D. Doolen, and H.E. DeWitt, ibid. **21**, 2087 (1980); **26**, 2255 (1982); S. Ogata and S. Ichimaru, ibid. **36**, 5451 (1987); D.H.E. Dubin, ibid. **42**, 4972 (1990).
22. G. Birkl, S. Kassner, and H. Walther, Nature **357**, 310 (1992).
23. D.H.E. Dubin and T.M. O'Neil, Phys. Rev. Lett. **60**, 511 (1988).
24. A. Rahman and J.P. Schiffer, Phys. Rev. Lett. **57**, 1133 (1986).
25. H. Totsuji, in "Strongly Coupled Plasma Physics," edited by F.J. Rogers and H.E. Dewitt (Plenum, New York, 1987) p. 19.
26. D.H.E. Dubin, Phys. Rev. A **40**, 1140 (1989).
27. R.W. Hasse and V.V. Avilov, Phys. Rev. A **44**, 4506 (1991).
28. C. Kittel, "Introduction to Solid State Physics, Fourth Edition," (John Wiley, New York, 1971), Chap. 2.
29. D.H.E. Dubin, Phys. Rev. Lett. **66**, 2076 (1991).
30. T.J. Murphy, and C.M. Surko, Phys. Rev. A **46**, 5696 (1992).
31. G. Gabrielse, L. Haarsma and K. Abdullah, "Trapped positrons for antihydrogen and ion cooling", submitted for publication.
32. G. Gabrielse, X. Fei, L.A. Orozco, R.L. Tjoelker, J. Haas, H. Kalinowsky, T.A. Trainor, and W. Kells, Phys. Rev. Lett. **65**, 1317 (1990).
33. M.D. Tinkle, R.G. Greaves, C.M. Surko, R.L. Spencer, and G.W. Mason, Phys. Rev. Lett. **72**, 352 (1994).
34. R.G. Greaves, M.D. Tinkle, and C.M. Surko, submitted for publication.
35. D.J. Wineland and H.G. Dehmelt, J. Appl. Phys. **46**, 919 (1975).
36. Stephan Barlow, Ph. D. Thesis, Dept. of Physics, University of Colorado, 1984.
37. J. J. Bollinger, D. J. Heinzen, W. M. Itano, S. L. Gilbert, and D. J. Wineland, IEEE Trans. on Instrum. and Measurement **40**, 126 (1991).
38. W. M. Itano, J. C. Bergquist, J. J. Bollinger, J. M. Gilligan, D. J. Heinzen, F. L. Moore, M. G. Raizen, and D. J. Wineland, Phys. Rev. A **47**, 3554 (1993).
39. J.J. Bollinger, J.D. Prestage, W.M. Itano, and D.J. Wineland, Phys. Rev. Lett. **54**, 1000-1003 (1985).
40. D. J. Wineland, J. C. Bergquist, J. J. Bollinger, W. M. Itano, F. L. Moore, J. M. Gilligan, M. G. Raizen, D. J. Heinzen, C. S. Weimer, and C. H. Manney, in "Laser Manipulation of Atoms and Ions", ed. by E. Arimondo, W. D. Phillips, and F. Strumia (North-Holland, Amsterdam, 1992), pp. 553-567.
41. D. J. Wineland, J. C. Bergquist, J. J. Bollinger, W. M. Itano, D. J. Heinzen, S. L. Gilbert, C. H. Manney, and M. G. Raizen, IEEE Trans. on Ultrasonics, Ferroelectrics, and Frequency Control **37**, 515 (1990).
42. J. N. Tan, J. J. Bollinger, and D. J. Wineland, Proc. Conf. Prec. Electromagnetic Meas., Boulder, CO, June, 1994; to be published in special issue of IEEE Trans. Instrum. Meas.
43. T. M. O'Neil, Phys. Fluids **24**, 1477 (1981).

44. For a forthcoming review, see: "Proceedings of the Nobel Symposium on Trapped Charged Particles and Related Fundamental Physics, Lysekil, Sweden, August, '94", to be published in Physica Scripta.

45. D. J. Wineland, C. S. Weimer, and J. J. Bollinger, "Proc. of the Anti Hydrogen workshop, Munich, July 30-31, 1992", ed. by J. Eades, in Hyperfine Interactions 76, 115 (1993).

46. G. Gabrielse, S. L. Rolston, L. Haarsma, and W. Kells, Phys. Lett. A 129, 38 (1988).

47. J.D. Prestage, J.J. Bollinger, W.M. Itano, and D.J. Wineland, Phys. Rev. Lett. 54, 2387-2390 (1985).

48. J. J. Bollinger, D. J. Heinzen, W. M. Itano, S. L. Gilbert and D. J. Wineland, Phys. Rev. Lett. 63, 1031 (1989).

49. D. J. Wineland, J. J. Bollinger, D. J. Heinzen, W. M. Itano, and M. G. Raizen, Phys. Rev. Lett. 67, 1735 (1991).

50. R.G. Hulet, D.J. Wineland, J.C. Bergquist, and W.M. Itano, Phys. Rev. A 37, 4544 (1988).

51. W. M. Itano, D. J. Heinzen, J. J. Bollinger, and D. J. Wineland, Phys. Rev. A 41, 2295 (1990).

52. "Trapped Ions and Laser Cooling", ed. by D.J. Wineland, W.M. Itano, J.C. Bergquist and J.J. Bollinger, NBS Technical Note 1086 (1985); "Trapped Ions and Laser Cooling II", ed. by D. J. Wineland, W. M. Itano, J. C. Bergquist, and J. J. Bollinger, NIST Technical Note 1324, 1988; "Trapped Ions and Laser Cooling III," NIST Technical Note 1353, ed. by J. C. Bergquist, J. J. Bollinger, W. M. Itano, and D. J. Wineland (U. S. Government Printing Office, Washington, 1992).

Modes of spheroidal ion plasmas at the Brillouin limit

M. D. Tinkle, R. G. Greaves, and C. M. Surko

Physics Department, University of California, San Diego,
La Jolla, California 92093-0319

Abstract. Brillouin-density pure ion plasmas have been generated in a quadrupole Penning trap by electron-beam ionization of a low-pressure gas. Large, spheroidal, steady-state plasmas are produced that extend to one of the trap electrodes. With the density fixed at the Brillouin limit by the high ion production rate, the electrode potentials determine the plasma shape. The frequencies of azimuthally propagating cyclotron and diocotron modes are found to vary significantly with the plasma aspect ratio. For oblate plasmas, we are able to test theoretical predictions of a simple fluid model, and the frequencies are in good agreement.

1   INTRODUCTION

A fundamental feature of single-component plasmas is the existence of a maximum achievable density, first predicted by Brillouin (1), below which the repulsive and centrifugal forces on the rotating plasma can be balanced by the force due to the confining magnetic field.

The Brillouin density limit is a fundamental restriction on the maximum achievable density of a single-species plasma, and it is critical to many applications. For example, it represents the most stringent limitation on alternative fusion energy concepts based on Penning traps, and a significant theoretical and experimental effort has recently been directed at developing techniques for overcoming this restriction (2–5). Dense trapped plasmas are now being developed as targets for high-energy beam experiments. Dense ion plasmas have been also proposed as a method of trapping and cooling positrons for the production of low-emittance beams (6). Finally, the Brillouin limit places a practical restriction on the accumulation of antiprotons (7) and positrons (8,9) for antimatter studies (10).

There have been relatively few experiments on confined Brillouin-density plasmas because most nonneutral plasma experiments have been performed with pure electron plasmas, in which the density limit is typically quite high. An experiment by Dimonte (11) demonstrated a Brillouin-density Li^+ ion plasma produced by a relatively simple source. Most ion plasma devices, however, are designed for use with laser diagnostics in high magnetic fields and do not naturally generate plasmas at the Brillouin limit. Through the application of laser torques, microplasmas consisting of about 10^5 ions have been compressed to the Brillouin limit (12), resulting in the discovery of an interesting heating resonance.

Among the predicted modes of nonneutral plasmas are diocotron waves near multiples of the plasma rotation frequency, ω_r, and modes near the cyclotron frequency, $\Omega_c = qB/Mc$, where B is the magnetic field, c is the speed of light, and q and M are the charge and mass of the plasma particles, respectively. While the low-frequency modes such as the diocotron mode (13) and the Trivelpiece-Gould mode (11) have been studied extensively in nonneutral plasmas, there have been few studies of the cyclotron modes (14), which have inconveniently high frequencies in typical electron plasmas. Furthermore, there have been only a few studies of modes in plasmas near the Brillouin limit (15,16).

We have recently developed a new steady-state mode of operation of a quadrupole trap that enables us to explore a range of plasma phenomena including plasma confinement and cyclotron and diocotron modes in pure ion plasmas near the Brillouin density. These experiments complement the plasma mode studies carried out in precision quadrupole traps (15,16), in that a different family of modes is studied in plasmas that are hotter and much larger than the laser-cooled plasmas. Except for minor perturbations which we ascribe to image charge effects, the mode frequencies we observe are well described by the same cold fluid theory (17,18) that was developed for the microplasmas.

Understanding of plasma modes can also provide diagnostic information. For example, low-order modes are being developed as diagnostics of the temperature and aspect ratio of spheroidal plasmas (12, 15, 19, 20). The mode frequencies of the higher-order diocotron and cyclotron branches in spheroidal plasmas have been predicted to depend on aspect ratio (17). Such diagnostics are expected to be particularly important for monitoring the antimatter plasmas now under study (6–8).

2   QUADRUPOLE TRAP CONFINEMENT

2.1   Critical Voltage

An ideal quadrupole Penning trap consists of a uniform magnetic field, B, directed along the z axis and an electrostatic potential of the form $\Phi(\rho, z) = V(z^2 - \rho^2/2)/z_0^2$, where $\rho = (x^2 + y^2)^{1/2}$ is the cylindrical radius and V and z_0 are constants. This potential is often realized by machining electrodes to approximate the hyperboloidal equipotential surfaces defined by the equation $z^2 - \rho^2/2 = \pm z_0^2$, and biasing them to potentials $\pm V$. The result is a pair of mirror-symmetric "endcaps" biased at a potential V and a "ring" electrode

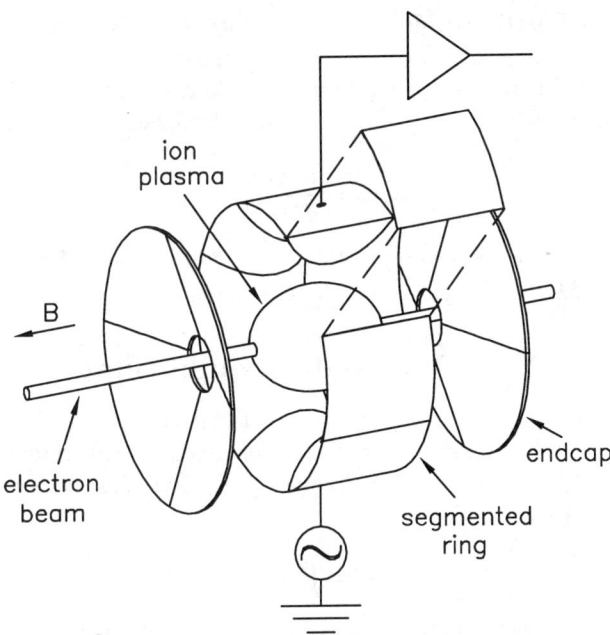

FIGURE 1: Exploded schematic of the ion plasma experiment, showing the electron beam, the spheroidal ion plasma, and the electronics used for mode excitation and detection.

biased at a potential $-V$. The trap used in these experiments, shown schematically in Fig. 1, is an approximation to such a trap, with a scale parameter $z_0 = 6.3$ cm. For this geometry, the distance to the endcaps from the center of the trap is z_0, and $\rho_0 = \sqrt{2}z_0$ is the inner radius of the ring electrode.

A single trapped ion executes simple harmonic oscillations along the z direction at a frequency

$$\omega_z = \left(\frac{2qV}{Mz_0^2}\right)^{1/2}$$

The motion in the x and y coordinates consists of a rapid circular cyclotron motion at a frequency

$$\Omega_c' = \frac{\Omega_c}{2} + \frac{1}{2}\sqrt{\Omega_c^2 - 2\omega_z^2}, \tag{1}$$

accompanied by a slower circular drift around the z axis at the magnetron frequency,

$$\Omega_M = \frac{\Omega_c}{2} - \frac{1}{2}\sqrt{\Omega_c^2 - 2\omega_z^2}, \tag{2}$$

where Ω_c is the cyclotron frequency in the absence of the trap electric field.

The confined particle orbits described above are only possible when $\omega_z^2 < \Omega_c^2/2$; for higher values of ω_z, the particle trajectories in x and y are combinations of hyperbolic sine and cosine functions, and the particle will strike the ring electrode quickly. As a result, V must be kept below a critical value,

$$V_c = \frac{B^2 z_0^2 q}{4Mc^2},$$
(3)

for the trap to work. Conversely, for a particular choice of V and B, only particles with $M/q < B^2 z_0^2 / V c^2$ will be confined.

2.2   Spheroidal Equilibrium

The thermal equilibrium of a large number of particles confined in a quadrupole Penning trap at a low temperature is a uniform-density spheroidal plasma, rotating rigidly (21). The rotation frequency, ω_r, satisfies the same equation found for cylindrical plasmas,

$$\omega_p^2 = 2\omega_r(\Omega_c - \omega_r),$$
(4)

where $\omega_p = (4\pi q^2 n/M)^{1/2}$ is the plasma frequency, and n is the number density of the plasma. The spheroids are biaxial ellipsoids with rotational symmetry about the z axis, so they are completely specified by their length, L, along the z axis and their radius, r_p, at $z = 0$. The ratio of length to diameter,

$$\alpha = \frac{L}{2r_p},$$
(5)

is referred to as the aspect ratio. In equilibrium, α is related to the plasma density by the equation (17)

$$\omega_p^2 = \omega_z^2 \frac{2}{A_3(\alpha)},$$
(6)

where

$$A_3(\alpha) = \frac{2Q_1^0 \left[\alpha(\alpha^2 - 1)^{-1/2}\right]}{\alpha^2 - 1},$$
(7)

and Q_1^0 is a Legendre function of the second kind. The plasma frequency is a monotonically increasing function of α which approaches its minimum value, $\omega_p = \omega_z$, as $\alpha \to 0$.

2.3   Brillouin Limit

As in the cylindrical plasma equilibrium, the radial electric field inside the plasma is proportional to ρ, having the form $E_\rho = M\omega_p^2\rho/2qz_0^2$. Comparing this with the radial electric field of the trap, $E_\rho = M\omega_z^2\rho/2qz_0^2$, which led to the condition $\omega_z^2 < \Omega_c^2/2$, it is clear that a similar condition applies to the plasma frequency. Just as the trap field prevents confinement for $V \geq V_c$, plasma self-fields limit the density of trapped particles to the Brillouin density (1),

$$n_B = \frac{B^2}{8\pi Mc^2},$$
(8)

at which $\omega_p^2 = \Omega_c^2/2$. At this limiting density, the plasma rotation frequency is $\omega_r = \Omega_c/2$, and the plasma has the highest aspect ratio available for the specified values of ω_z and Ω_c.

As the Brillouin limit is approached ($\omega_p \to \Omega_c/\sqrt{2}$), the orbit size of the plasma particles diverges, just as the orbit size of single trapped particles diverges when the confinement limit is approached ($\omega_z \to \Omega_c/\sqrt{2}$). In this case, however, particles may remain confined in the trap as long as the plasma boundary is well separated from the electrode surfaces, because their orbit radius becomes finite once they leave the plasma. In the limit of high magnetic field or low temperature, the particle motions in a plasma at the Brillouin density limit are straight-line trajectories followed by specular reflections from the plasma boundary, when viewed in a frame rotating with the plasma. If the Brillouin limit is exceeded, the plasma will expand across the magnetic field until the density is reduced to $n = n_B$.

2.4   Brillouin Spheroid Model

The Brillouin density for argon ions in a magnetic field of 1 kG is $n_B = 6.63 \times 10^5$ cm^{-3}. Small numbers of ions have a confinement time of the order of 1 s at a neutral gas pressure of 10^{-6} torr, so that with ions being formed in the path of the beam at a rate greater than 10^8 cm^{-3}s^{-1}, the density will approach the Brillouin limit in a matter of milliseconds, unless space charge is sufficient to allow ions to escape through the endcaps. This latter possibility will occur for low confining potentials. We can estimate the minimum value of V required to prevent such a loss by considering a spheroidal equilibrium plasma with the diameter of the electron beam, the length of the trap, and the Brillouin density. The value of V for which this equilibrium is obtained is found from Eq. (6), which may be written

$$2\pi z_0^2 q\frac{n}{V} = \frac{\alpha^2 - 1}{Q_1^0[\alpha(\alpha^2 - 1)^{-1/2}]},$$
(9)

by setting $n = n_B$ and $\alpha = z_0/r_{\text{beam}}$. For argon ions in a magnetic field of 1 kG, the result is $V_{\text{min}} = 0.45$ volts.

With the aspect ratio determined by the Brillouin density and the confining voltage V according to Eq. (9), the remaining factor of the model is the plasma size. With ions being continually produced, the plasma will grow in size (maintaining a fixed aspect ratio) until it contacts one of the electrodes, after which additional ions formed will flow to that electrode, and a steady state will be established. For a fixed value of B, the aspect ratio varies with V, with increasing values of V leading to increasingly oblate spheroids. A plasma with an aspect ratio of $\alpha = 1/\sqrt{2}$ has the right shape to touch both the endcaps and the ring electrode, giving it a volume of about 2×10^3 cm^3, the largest spheroidal plasma that may be confined in this trap. The confinement voltage, V_t, that produces this plasma shape may be found from Eq. (9), and can be shown to be a constant multiple of the critical voltage, V_c, above which confinement is lost:

$$V_t = (2 - \pi/2)V_c. \tag{10}$$

For $V < V_t$, the plasma length is fixed at $L = 2z_0$, and excess ions leave through the endcaps. Due to collisions, some particles will be transported radially out of the Brillouin-density core and will diffuse to the ring electrode. Our inference is that they form a tenuous halo plasma surrounding the core. For $V > V_t$, the plasma radius is fixed at $r_p = \rho_0$, and there is no halo plasma. The plasmas described by this model for various values of V are shown schematically in Fig. 2.

3   EXPERIMENT

3.1   Plasma Production

The experiments were performed in a Penning trap designed to accumulate and store low-energy positrons $(8, 9)$. A magnetic field of up to 1.3 kG is aligned with the electrode structure shown in Fig. 1, which approximates the hyperboloidal electrodes of a precision quadrupole trap. The ring electrode is split azimuthally into eight sectors.

Ions are formed by passing an electron beam along the axis of the trap, as shown schematically in Fig. 1. The density of ion production is

$$\frac{dn}{dt} \simeq \frac{n_n \sigma_i I_{\text{beam}}}{\pi e r_b^2}, \tag{11}$$

where σ_i is the ionization cross-section, n_n is the density of neutral gas atoms, I_{beam} is the beam current, and $r_b \simeq 3$ mm is the beam radius. A typical value is about 4×10^8 cm^{-3}s^{-1} for a one microamp beam passing through 10^{-6} torr of argon in a magnetic field of one kiloGauss. As with other sources of pure ion plasmas, production of doubly ionized particles is a concern. To avoid these complications, the beam energy is kept below the second ionization energy of the gas.

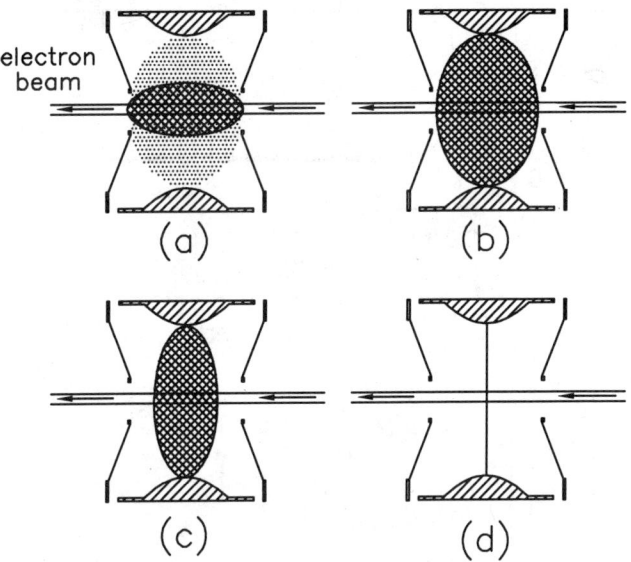

FIGURE 2: Model of the steady-state ion plasma for (a): $V_{min} < V < V_t$, endcap-limited plasma with halo, $\alpha > 1/\sqrt{2}$; (b): $V = V_t$, transition plasma, $\alpha = 1/\sqrt{2}$; (c): $V_t < V < V_c$, ring-limited plasma, $\alpha < 1/\sqrt{2}$ and (d); $V = V_c$, thin disk-shaped plasma, $\alpha = 0$. For $V > V_c$, there is no appreciable plasma.

3.2 Total Charge Measurement

It would be desirable to measure the plasma profiles directly by dumping the plasma onto the collector plates. There are two difficulties with this approach. One problem is that if the plasmas are indeed near the Brillouin density, then the particles are not following magnetic field lines in the tight helices familiar from low-density electron plasmas. A second problem—specific to this experiment—is that these large plasmas must be dumped through the moderate-sized hole in one of the endcaps.

Under some conditions, a measurement of the total plasma charge can be obtained by measuring the image charge flowing onto the confining electrodes as the trap fills. Because the filling is rapid (about 100 ms), a current spike is produced on the ring electrode when the electron beam is first switched on. For $V < V_t$, this spike is clearly distinguishable from the dc current which is established once the trap has been filled. Typical current traces, obtained using an electrometer, are shown in Fig. 3(a). The total charge is obtained by integrating the current spikes from all electrodes and correcting for the contribution from the image charge of the beam, which causes the small negative current spike seen at $t = 0$. The measurement is unambiguous for

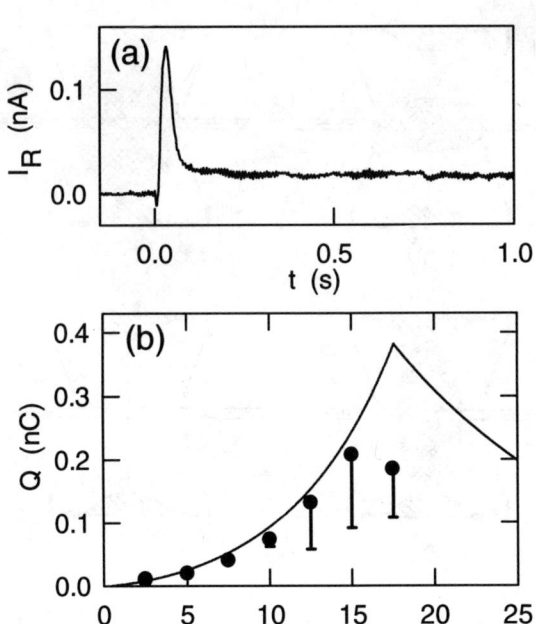

FIGURE 3: (a) Current transient to the ring electrode when electron beam is switched on at $t = 0$, for an argon plasma with $B = 1.31$ kG and $V = 5$ volts. (b) Total charge of argon ion plasma with $B = 1.31$ kG and various values of V. The solid line is from the steady-state model.

small values of V, but as V approaches V_t, the current traces become more complicated, making it difficult to decide when to stop the integration. For $V > V_t$, there is no clear separation between image charge current and the arrival of the steady state current, and the technique is not applicable.

Figure 3(b) shows how the measured stored charge increases as V is increased. The error bars represent the uncertainty introduced by transient features described above. The solid symbols represent the most plausible analysis. Using the model described in Sec. 2.4, the total charge in the Brillouin core is given by

$$Q = 4\pi r_p^3 \alpha n_B/3, \tag{12}$$

which is shown by the solid line in Fig. 3. The agreement between the data and the model is reasonably good, and at least verifies that the total charge is within about a factor of two of the model for $V < V_t$.

3.3 Transition in Ring Current

Less direct, but more convincing, evidence for the Brillouin spheroid model can be obtained by monitoring the steady-state current to the ring electrode. In Fig. 4(a), the steady-state current to the ring is plotted as a function of V for various values of B and I_{beam}. In each case, the ring current is small for low values of V, corresponding in the model to the condition [Fig. 2(a)] in which the core plasma is not in contact with the ring, and most excess ions leave via the endcaps. The current rises gradually with V until an abrupt increase occurs, presumably when $V \simeq V_t$ and the plasma contacts the ring [Fig. 2(b)]. Further increases in V result in only slight change in the ring current, which should now be the entire ion formation current. Comparing the different cases, we note that the value of B affects the value of V_t, but not the current for $V > V_t$.

Figure 4(b) shows the measured value of the transition voltage as a function of magnetic field for four ion species and the expected dependences, $V_t = (2 - \pi/2)V_c \sim B^2/M$, from the Brillouin spheroid model. No adjustable parameters were used.

3.4 Plasma Modes

Plasma modes are excited by applying a signal to one of the sectors of the ring electrode and detecting the plasma response on the opposite sector, as shown in Fig. 1. Frequency spectra are measured using a spectrum analyzer and a tracking generator. A typical spectrum is shown in Fig. 5(a), and two families of modes are discernible. One family has frequencies near Ω_c and has some similarity to the cyclotron modes studied by Gould and LaPointe (14) in electron plasmas. The lower-frequency family are the diocotron modes (22). A series of spectra like that shown in Fig. 5(a) were taken for steady-state argon plasmas for various values of V in the range $V_{\text{min}} < V \leq V_c$. Figure 5(b) displays the frequencies of the resonances observed as functions of V. It is clearly evident that the character of the modes changes near $V = V_t$.

The cold fluid theory of spheroidal nonneutral plasma modes (17) predicts that modes of the form $\delta\Phi \sim e^{i(m\phi - \omega t)}$, which are expected to be excited by this technique, will have frequencies

$$\omega_m = (m - 1)\omega_r + \frac{\Omega_c}{2} \pm \frac{1}{2}\left[(\Omega_c - 2\omega_r)^2 + 2\omega_p^2 G_m(\alpha)\right]^{1/2}, \qquad (13)$$

for positive m. The geometrical factor, G_m, is given by

$$G_m(\alpha) = 2\left[m - \frac{Q_m^{m\,\prime}(k_2)}{\alpha(\alpha^2 - 1)^{1/2}Q_m^m(k_2)}\right]^{-1}, \qquad (14)$$

where $k_2 = \alpha(\alpha^2 - 1)^{-1/2}$ and the primes indicate differentiation with respect to the entire argument. The plus and minus signs in Eq. (13) give solutions

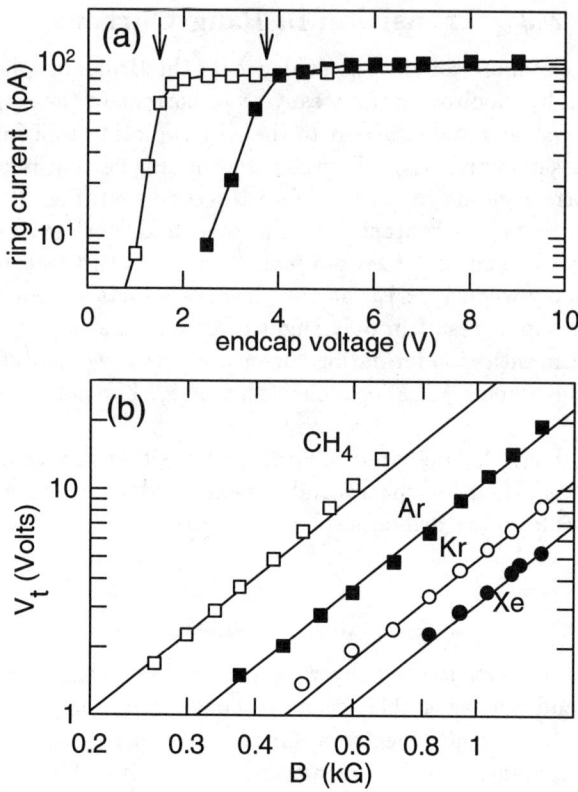

FIGURE 4: (a) Steady-state current to the ring electrode as a function of V for an argon plasma at a pressure of 2.4×10^{-8} torr and with ($\square$): $B = 290$ G, $I_{\text{beam}} = 10\mu$A; ($\blacksquare$): $B = 580$ G, $I_{\text{beam}} = 10\mu$A. Arrows show transition voltages. (b) Measured transition voltage as a function of B for four different ion species. Solid lines are from the model.

ω_m^+ and ω_m^- corresponding to the cyclotron and diocotron branches of the dispersion relation, respectively. The plasma rotation frequency may be found from ω_m^+ and ω_m^- for any $m > 1$:

$$\omega_r = \frac{\omega_m^+ + \omega_m^- - \Omega_c}{2m - 2}. \tag{15}$$

These azimuthal "flute" modes are very similar to the analogous modes in a cylindrical single-component plasma. In the cylindrical case (22), it is possible to include the effects of the image charge of the plasma. The mode frequencies again satisfy Eq. (13), with geometrical factors now depending on the ratio of the plasma radius, r_p, to r_w, the radius of the cylindrical wall, according to the relation $G_m(r_p) = 1 - (r_p/r_w)^{2m}$. If image charges are neglected in

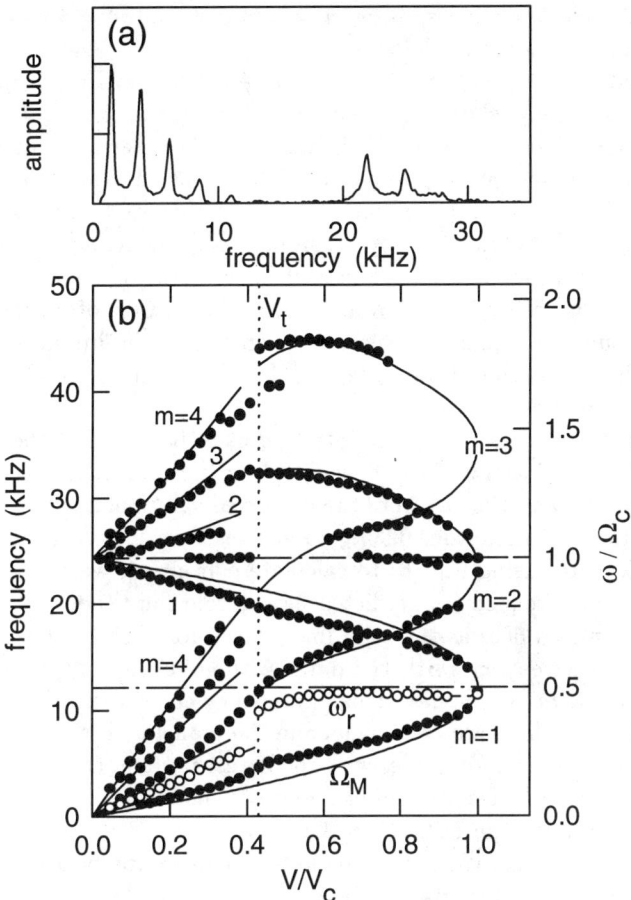

FIGURE 5: (a) Spectrum of azimuthal modes of an argon plasma. (b) Azimuthal mode frequencies of argon plasmas with $B = 640$ G as functions of V. (○) ω_r calculated from mode frequencies using Eq. (15). Solid lines are described in the text.

the cylindrical case by letting $r_w \to \infty$, the geometrical factors reduce to $G_m(r_p) = 1$, and the mode frequencies match those of the $\alpha \to \infty$ limit of the spheroidal case, which did not include image charge effects.

The values of m indicated in Fig. 5(b) were determined experimentally by the phase differences between the signals detected on different sectors. Because pairs of cyclotron and diocotron modes with the same values of m are detected, the rotation frequency may be deduced using Eq. (15). The open circles in Fig. 5(b) are the values of ω_r obtained by applying this procedure to the observed mode frequencies, using $m = 2$, $m = 3$, and $m = 4$ modes for $V < V_t$ and the $m = 2$ modes for $V > V_t$.

Consider now the modes for $V < V_t$. As shown in Fig. 5(b), the inferred rotation frequency (shown by open circles) is close to $2\Omega_M$, rather than $\Omega_c/2$, which would characterize the Brillouin core. We interpret this to mean that the observed modes are supported in the halo plasma, which effectively shields the core. Using these inferred rotation frequencies, we find that the frequency data may be fit to Eq. (13) by assuming $\omega_p^2 G_m = K_m \omega_z^2$, where $K_1 = 0.69$ and $K_m = 1.24$ for $m > 1$. These fit curves are the solid lines plotted for $V < V_t$ in Fig. 5(b). Also shown, for the whole range of V, are Ω_M and Ω_c', which should not depend on plasma parameters. The implications of this fit and of the observation that $\omega_r \simeq 2\Omega_M$ for the nature of the halo plasma are not understood at present.

For $V > V_t$, the ring current data lead us to believe that the entire plasma can be described as a uniform-density spheroid with $n \simeq n_B$. Using Eq. (4) to determine ω_p^2 (and thus n) from the measured values of ω_r shown in Fig. 5(b), we find that n varies from $0.96 n_B$ to greater than $0.99 n_B$ for all values of V above V_t. After using Eq. (6) to calculate α from ω_p^2, we can use Eq. (13) to find the mode frequencies predicted by the cold fluid theory. The results are displayed in Fig. 5(b) as the solid lines shown for $V > V_t$. These calculations are in good agreement with the data, without using any fitted parameters, except for the $m = 1$ modes. This discrepancy could result from the effects of image charge, which are not included in the model and which are expected to be most pronounced for modes with the lowest m-numbers. We note, however, that the sum of the frequencies of the $m = 1$ modes is quite close to Ω_c, which could account for the excitation of the weak mode observed near Ω_c, via a nonlinear coupling. We note also that a mode coupling occurs between the $m = 1$ cyclotron mode and the $m = 2$ diocotron mode near $V = 0.75 V_c$.

4   CONCLUDING REMARKS

In conclusion, we have demonstrated that plasmas near the Brillouin limit can be produced and studied for a variety of ion species, using an unusual steady-state mode of operation of a Penning trap. The technique produces spheroidal plasmas with density $n \simeq n_B$, and aspect ratio varying with V. We have studied azimuthally propagating modes, which can be described by a simple fluid model for V greater than some transition voltage, corresponding to a condition in which the Brillouin plasma extends radially to the confining electrode. It appears that creating and studying plasmas at the Brillouin limit is both interesting and relatively straightforward. This technique could be easily implemented in other devices, opening up the possibility of new insights into the unusual physics of the regime near the Brillouin limit.

ACKNOWLEDGMENTS

We are pleased to acknowledge helpful conversations with D. H. E. Dubin and the technical assistance of E. A. Jerzewski. This work is supported by the Office of Naval Research.

REFERENCES

1. Brillouin, L., *Physical Review* **67**, 260–266 (1945).
2. Barnes, D. C. and Turner, L., *Physics of Fluids B* **4**, 3890–901 (1992).
3. Barnes, D. C., Nebel, R. A., Turner, L., and Tiouririne, T. N., *Plasma Physics and Controlled Fusion* **35**, 929–40 (1993).
4. Tiouririne, T. N., Turner, L., and Lau, A. W. C., *Physical Review Letters* **72**, 1204–7 (1994).
5. Turner, L. and Barnes, D. C., *Physical Review Letters* **70**, 798–801 (1993).
6. Wineland, D. J., Weimer, C. S., and Bollinger, J. J., *Hyperfine Interactions* **76**, 115–25 (1993).
7. Gabrielse, G., Fei, X., Orozco, L. A., Tjoelker, R. L., Haas, J., Kalinowsky, H., Trainor, T. A., and Kells, W., *Physical Review Letters* **63**, 1360–3 (1989).
8. Surko, C. M., Leventhal, M., and Passner, A., *Physical Review Letters* **62**, 901–4 (1989).
9. Greaves, R. G., Tinkle, M. D., and Surko, C. M., *Physics of Plasmas* **1**, 1439–1446 (1994).
10. Charlton, M., Eades, J., Horvath, D., Hughes, R. J., and Zimmermann, C., *Physics Reports* **241**, 65–117 (1994).
11. Dimonte, G., *Physical Review Letters* **46**, 26–29 (1981).
12. Bollinger, J. J., Wineland, D. J., and Dubin, D. H. E., *Physics of Plasmas* **1**, 1403 (1994).
13. Fine, K. S., Driscoll, C. F., and Malmberg, J. H., *Physical Review Letters* **63**, 2232–4 (1989).
14. Gould, R. W. and LaPointe, M. A., *Physical Review Letters* **67**, 3685–8 (1991).
15. Heinzen, D. J., Bollinger, J. J., Moore, F. L., Itano, W. M., and Wineland, D. J., *Physical Review Letters* **66**, 2080–3 (1991).
16. Bollinger, J. J., Heinzen, D. J., Moore, F. L., Itano, W. M., Wineland, D. J., and Dubin, D. H. E., *Physical Review A* **48**, 525–545 (1993).
17. Dubin, D. H. E., *Physical Review Letters* **66**, 2076–9 (1991).
18. Dubin, D. H. E., *Physics of Fluids B* **5**, 295–324 (1993).
19. Weimer, C. S., Bollinger, J. J., Moore, F. L., and Wineland, D. J., *Physical Review A* **49**, 3842 (1994).
20. Tinkle, M. D., Greaves, R. G., Surko, C. M., Spencer, R. L., and Mason, G. W., *Physical Review Letters* **72**, 352–5 (1994).
21. Turner, L., *Physics of Fluids* **30**, 3196–3203 (1987).
22. Davidson, R. C., (1990) *Physics of Nonneutral Plasmas* Addison-Wesley, Redwood City, California.

One-Dimensional Classical Plasmas

Hiroo Totsuji

Department of Electrical and Electronic Engineering
Okayama University, Tsushimanaka 3-1-1, Okayama 700, Japan

Abstract. As plasmas with extremely reduced dimensionality, properties of one-dimensional classical plasmas are analyzed in the domain of strong coupling. These plasmas may be realized in Penning traps with sufficiently strong confinement and also in semiconductor quantum wires under appropriate conditions.

INTRODUCTION

One of key observations to understand various phenomena in Penning traps may be that charged particles in the trap can be regarded as embedded in the uniform background of opposite charges.[1] When we neglect correlations between particles, we thus have a charged fluid of finite extension. Normal modes of oscillations of these finite plasmas in the fluid limit and related properties enable us to diagnose physical parameters of these plasmas.[2] Finiteness of extension also leads to novel phenomena not observed in infinite plasmas.

When the correlations or the discreteness of charges is taken into account, we have shell structures, two-dimensional lattices on shells, and inter-shell correlation of positions of charges.[3, 4, 5, 6] The inter-shell correlation may be neglected as a first approximation and we can reproduce shell structure observed in real and numerical experiments by considering the intra-shell cohesive energy and inter-shell electrostatic energy.[7] In these cases, the aspect of plasmas with limited dimensionality becomes clear.

By introducing an extra electrode on the central line of the trap, we can control the structure of confined charges. Especially, we can produce purely two-dimensional system of charges by such a method.[8]

In this paper, we consider classical plasmas with smallest degrees of freedom, the one-dimensional plasmas. As one-dimensional plasmas, the system composed of charged sheets has been extensively investigated.[9] The potential in this case is proportional to mutual distances. In our one-dimensional plasmas, particles interact through potential proportional to the inverse of the mutual distance. We have this kind of one-dimensional plasmas in the central part of Penning (and Paul) traps with cylindrical symmetry, when the effect of confining force is strong. We also note that in quantum wires in semiconductors we have a system of classical charges with one degree of freedom interacting via the inverse r potential under appropriate conditions.

CHARACTERISTIC PARAMETERS
Ions in Traps

Ions in Penning (and Paul) traps have a variety of ordered structures at low temperatures.[3, 4, 5, 6, 7] These structures are determined as a result of the competition between the confining force and mutual Coulomb repulsion.[7] When the former is sufficiently strong, ions are aligned on a line, and with its decrease, we have one-dimensional lattices on a cylinder, two concentric cylinders (layers), three concentric cylinders, and so on.

We first consider the ions on a line: The cases of ions on cylindrical surfaces will be discussed in the last section. As characteristic energies of this system, we have, in addition to the thermal energy $k_B T$, the Coulomb interaction at mean distance E_C and the Fermi energy E_F given by

$$E_C = nq^2, \qquad E_F = (\pi^2/8)(\hbar^2 n^2/m). \qquad (1)$$

By q, n, T, and m, we denote the charge, the line density, the temperature, and the ionic mass, respectively. (The energy E_F characterizes the energy for which we have to take the effect of quantum statistics into account even if the ion is a boson.) These parameters are compared for Be_9^+ ion in Fig.1.

We observe that these plasmas are usually in the classical domain where $k_B T > E_F$ due to large ionic mass. We there define the Coulomb coupling parameter γ by

$$\gamma = E_C/k_B T = nq^2/k_B T. \qquad (2)$$

As is shown in Fig.1, we can be in either the weakly coupled domain $\gamma < 1$ (A) or the strongly coupled domain $\gamma > 1$ (B) according to experimental conditions.

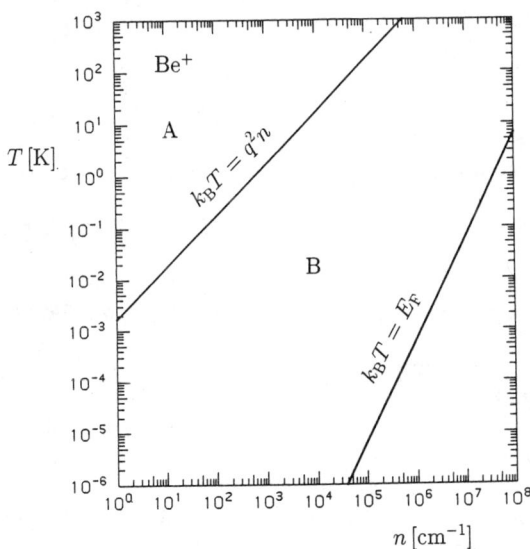

FIGURE 1. Characteristic parameters of ions in traps.

Ions in Storage Rings

In ion storage rings, alternating quadrupole magnetic fields serve as confining force which focuses ions in directions perpendicular to the beam. Sufficiently low parallel temperatures of the order of meV are attained by employing the electron and laser coolings, while the perpendicular temperatures are much higher and of the order of eV.

Ions in storage rings can be regarded as classical one-dimensional plasmas along the direction of the beam.[10] The effective interactions in this system are modified by the finite perpendicular extensions and the existence of grounded wall surrounding the beam. Correlations between ions can be observed in the spectrum of Schotkky noise picked up by electrodes placed near the beam: In the domains of strong coupling we have a reduction of integrated spectrum and the appearance of collective mode.[11] These effects can be accounted for by regarding the ion beam as one-dimensional plasma with an effective interaction.[10]

Electrons in Quantum Wires

In microstructures such as quantum wires in semiconductors, electrons (and holes) are confined by the potential due to the band offset between different semiconductors. Typical potential depth in structures composed of GaAs and $Al_xGa_{1-x}As$ is of the order of 100meV. For electrons in quantum wires, the mass is to be replaced by the effective mass and the characteristic Coulomb energy is screened by the dielectric constant ε as $E_C = q^2 n/\varepsilon$. (We assume that the difference in energies of the ground and the first excited states in the plane perpendicular to the wire ΔE satisfies $\Delta E > Max(k_B T, E_F)$, and only the ground state is occupied.)

In the classical domain where $k_B T > E_F$, the parameter $\gamma = E_C/k_B T = q^2 n/\varepsilon k_B T$ characterizes the importance of Coulomb interactions;

$$\gamma < 1 \quad \text{weakly coupled (A)}, \quad \gamma > 1 \quad \text{strongly coupled (B)}. \tag{3}$$

In the degenerate case $k_B T < E_F$, the ratio $E_C/E_F = (8/\pi^2)R_s$ characterizes the importance of Coulomb interactions;

$$R_s < 1 \quad \text{weakly coupled (C)}, \quad R_s > 1 \quad \text{strongly coupled (D)}. \tag{4}$$

These three energies are compared with one another in Fig.2 for electrons in quantum wires composed of GaAs with $m = 0.067 m_e$ and $\varepsilon = 13.1$, m_e being the electronic mass. In the domain A, our system is classical and weakly coupled, in B, classical and strongly coupled, in C, degenerate and weakly coupled, and in D, degenerate and strongly coupled.

We note that, except for very low temperatures and high densities, electrons in quantum wires are classical with respect to the motion along the wire. In other words, we have one-dimensional classical plasmas in quantum wires under a wide range of conditions.

We keep the dielectric constant ε in following expressions: $\varepsilon = 1$ for ions in traps and storage rings.

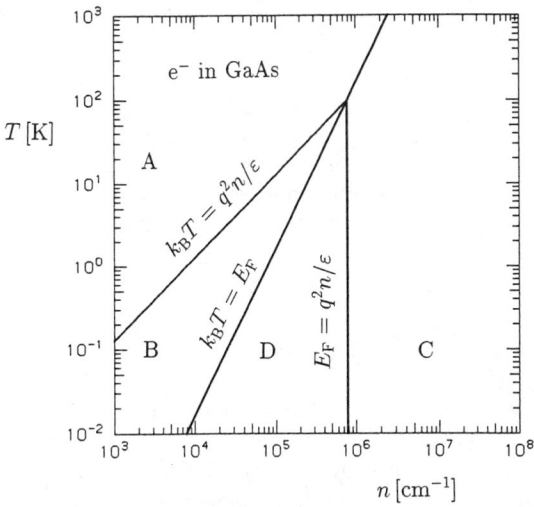

FIGURE 2. Characteristic parameters of electrons in quantum wires.

RANDOM PHASE APPROXIMATION

In the domain of weak coupling, we may apply the random phase approximation to our one-dimensional plasmas. When we cutoff the interaction $q^2/\varepsilon|z|$ at δ as $q^2/\sqrt{z^2 + \delta^2}$, its Fourier component is given by $2K_0(k\delta) \sim 2q^2(-\ln k)$ ($K_0(x)$ is the modified Bessel function) and the plasmon dispersion $\omega(k)$ and the static form factor $S(k)$ are calculated as

$$\omega^2(k) \sim \frac{2q^2 n}{\varepsilon m} k^2(-\ln k),\tag{5}$$

$$S(k) \sim \left(1 + \frac{2nq^2}{\varepsilon k_B T}(-\ln k)\right)^{-1},\tag{6}$$

respectively.[10]

STRONGLY COUPLED DOMAIN

The Hamiltonian of our system may be written as

$$H = \sum_i \frac{p_i^2}{2m} + \frac{q^2}{2\varepsilon}\sum_{i \neq j}\frac{1}{|z_i - z_j|} + H_b.\tag{7}$$

Here z_i denotes the position of i-th electron and H_b, the effect of neutralizing background. Let us note that, since our system is one-dimensional and composed of the same particles, we can always number our particles in the order of their distance from some arbitrary taken origin. We thus have

$$z_i \geq z_{i-1}. \tag{8}$$

When the couplings between classical charges are sufficiently strong, it is natural to expect those charges have correlations which resemble those in a solid phase. For our system in the strongly coupled regime, we therefore imagine a lattice of particles with positions $\{z_i^0\}$ and express the positions of particles $\{z_i\}$ by deviations from those in the lattice $\{\zeta_i\}$ as

$$z_i = z_i^0 + \zeta_i \tag{9}$$

with

$$z_i^0 = \frac{1}{n}i. \tag{10}$$

It is well known that we do not have solids in one dimension due to enhanced effect of thermal fluctuations: The mean square displacement calculated in the harmonic approximation diverges and the Debye-Waller factor reduces to zero due to the contribution of acoustic modes with long wavelengths.[12] In this respect, we emphasize that, in our calculation, it is *not* necessary to assume the existence of such a lattice.

In terms of ζ_i, $|z_i - z_j| = |(i-j)/n + (\zeta_i - \zeta_j)|$, and the Hamiltonian is rewritten as

$$H = \sum_i \frac{p_i^2}{2m} + \frac{q^2}{2\varepsilon} \sum_{i \neq j} \frac{1}{|(i-j)/n + (\zeta_i - \zeta_j)|} + H_b, \tag{11}$$

where $p_i = m(d\zeta_i/dt)$ conjugate to ζ_i. We now expand each term in the potential with respect to $\zeta_i - \zeta_j$. Within the harmonic approximation, we have

$$H = \sum_i \frac{p_i^2}{2m} + \frac{q^2 n}{2\varepsilon} \sum_{i \neq j} \frac{1}{|i-j|} + H_b + \frac{q^2 n^3}{2\varepsilon} \sum_{i \neq j} \frac{1}{|i-j|^3}(\zeta_i - \zeta_j)^2. \tag{12}$$

The second and third terms on the right hand side give the Madelung energy of linear chain of charges. When the effect of neutralizing background charge is properly taken into account, they reduce to

$$U_M/2 + \ln(Rn), \tag{13}$$

where $U_M = -0.231803 \cdots$ and R is the radius of cylinder containing neutralizing background.[13]

The rest of the Hamiltonian (12) gives a collection of harmonic oscillations. Their dispersion relation is calculated as

$$\omega^2(k) = \frac{4q^2 n^3}{\varepsilon m} F\left(\frac{k}{n}\right) \qquad |k| < \frac{\pi}{n}, \tag{14}$$

where

$$F(x) = \sum_{j=1}^{\infty} \frac{1}{j^3}(1 - \cos jx). \tag{15}$$

For long wavelengths, (14) reduces to

$$\omega^2(k) \sim \frac{2q^2n}{\varepsilon m} k^2(-\ln k). \tag{16}$$

We show the dispersion relation in Fig.3.

To confirm the applicability of the harmonic approximation, we now calculate $< (\zeta_i - \zeta_j)^2 >$. Here $< >$ denotes the thermal average. At the temperature T, we have

$$< (\zeta_i - \zeta_j)^2 > n^2 = \frac{1}{\pi\gamma} \int_0^\pi dx \frac{\sin^2\left(\frac{(i-j)}{2}x\right)}{F(x)} \tag{17}$$

in the harmonic approximation. We plot the ratio of $< (\zeta_i - \zeta_j)^2 >$ to $(i-j)^2/n^2$ normalized by $1/\gamma$ in Fig.4. We note that the average $< (\zeta_i - \zeta_j)^2 >$ is finite even if the mean square displacement $< \zeta_i^2 >$ diverges and the ratio plotted in Fig.4 is a decreasing function of $|i-j|$ with the maximum of $0.4/\gamma$ at $|i-j| = 1$. When $|i - j| \gg 1$, we have

$$\frac{< (\zeta_i - \zeta_j)^2 >}{(i - j)^2/n^2} \sim \frac{1}{2\gamma} \frac{1}{|i - j| \ln |i - j|}. \tag{18}$$

We also note that this asymptotic expression works as a good approximation for $|i - j| \geq 2$.

The approximation we applied to the Hamiltonian (12) is the expansion of $1/|z_i - z_j| = 1/|(i - j)/n + (\zeta_i - \zeta_j)|$ with respect to $\zeta_i - \zeta_j$. Our harmonic approximation is thus applicable at least as a first approximation when $\gamma > 1$ or in the domain of strong coupling.

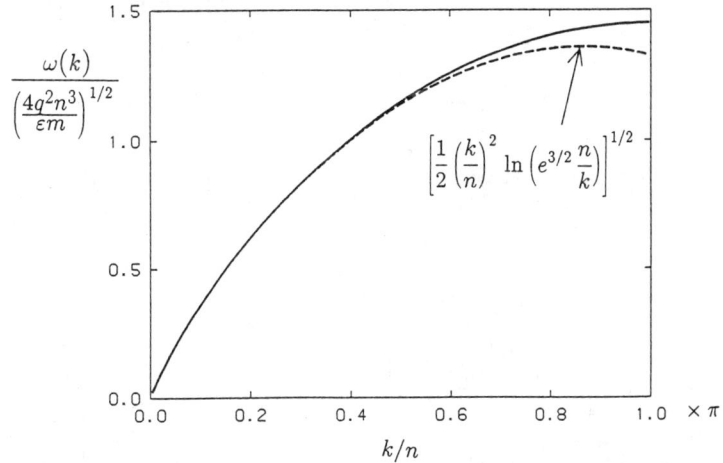

FIGURE 3. Dispersion relation of one-dimensional plasmon.

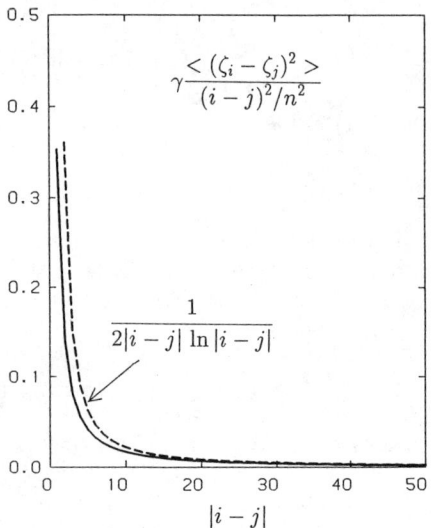

FIGURE 4. Relative displacement from (fictitious) lattice points.

The static correlation between charges is characterized by the structure factor $S(k)$ and the pair distribution function $g(z)$ given by

$$S(k) = \frac{1}{N} < \rho_k \rho_{-k} >, \qquad (19)$$

$$g(z) - 1 = \frac{1}{2\pi} \int_{-\infty}^{\infty} dk \exp(ikz)[S(k) - 1]. \qquad (20)$$

Here $\rho_k = \sum_j \exp(-ikz_j)$ is the density fluctuation and N is the total number of charges. Within the harmonic approximation, the structure factor and the pair distribution function are calculated as

$$S(k) = 1 + 2\sum_{j=1}^{\infty} \cos\left(j\frac{k}{n}\right) \exp\left[-\frac{k^2}{2} < (\zeta_0 - \zeta_j)^2 >\right]$$

$$= 1 + 2\sum_{j=1}^{\infty} \cos\left(j\frac{k}{n}\right) \exp\left[-\frac{1}{4\pi}\frac{\epsilon k_B T}{q^2 n}\frac{k^2}{n^2}\int_0^{\pi} dx \frac{1}{F(x)}[1 - \cos(jx)]\right], \quad (21)$$

$$g(z) = \frac{1}{(2\pi)^{1/2}} \sum_{j=1}^{\infty} \frac{1}{n < (\zeta_0 - \zeta_j)^2 >^{1/2}}$$

$$\times \left\{\exp\left[-\frac{(z - nj)^2}{2 < (\zeta_0 - \zeta_j)^2 >}\right] + \exp\left[-\frac{(z + nj)^2}{2 < (\zeta_0 - \zeta_j)^2 >}\right]\right\}. \qquad (22)$$

We show the structure factor and the correlation function in Fig.5 for several values of coupling constant γ.

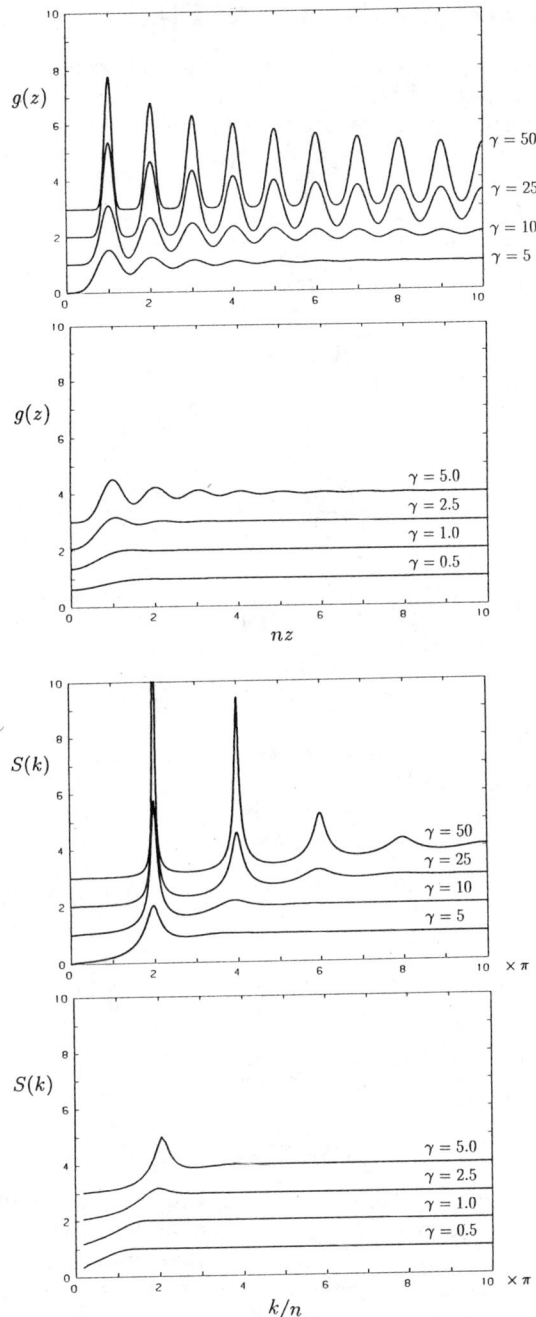

FIGURE 5. Pair distribution function and static form factor.

ONE-DIMENSIONAL LATTICE ON CYLINDERS

We here consider one-dimensional lattices of charges on a cylinder which are realized in Penning and Paul traps.[7, 14] We write the position of i-th charge $\mathbf{r}_i$ referring to the lattice (denoted by the superscript 0) as

$$\mathbf{r}_i = (\mathbf{R}_i^0 + \mathbf{R}_i, z_i^0 + \zeta_i). \tag{23}$$

Here $\mathbf{R}_i^0$ and $\mathbf{R}_i$ have x and y components. Expanding the mutual interaction $q^2/|\mathbf{r}_i - \mathbf{r}_j|$ with respect to $\zeta_{ij} = \zeta_i - \zeta_j$ and $\mathbf{R}_{ij} = \mathbf{R}_i - \mathbf{R}_j$, we have the deviation of the interaction potential from the Madelung energy to the second order as

$$-\frac{1}{2}\frac{q^2}{[(z_{ij}^0)^2 + (\mathbf{R}_{ij}^0)^2]^{3/2}}(\zeta_{ij}^2 + \mathbf{R}_{ij}^2) + \frac{3}{2}\frac{q^2}{[(z_{ij}^0)^2 + (\mathbf{R}_{ij}^0)^2]^{5/2}}(z_{ij}^0\zeta_{ij} + \mathbf{R}_{ij}^0\cdot\mathbf{R}_{ij})^2, \tag{24}$$

where $z_{ij}^0 = z_i^0 - z_j^0$ and $\mathbf{R}_{ij}^0 = \mathbf{R}_i^0 - \mathbf{R}_j^0$. The equations of motion for ζ_i and $\mathbf{R}_i$ are given by

$$m\frac{d^2\zeta_i}{dt^2} = 2\sum_{j(\neq i)}\frac{2(z_{ij}^0)^2 + (\mathbf{R}_{ij})^2}{[(z_{ij}^0)^2 + (\mathbf{R}_{ij})^2]^{5/2}}\zeta_{ij} + 6\sum_{j(\neq i)}\frac{z_{ij}^0}{[(z_{ij}^0)^2 + (\mathbf{R}_{ij})^2]^{5/2}}\mathbf{R}_{ij}^0\cdot\mathbf{R}_{ij}, \tag{25}$$

$$m\frac{d^2\mathbf{R}_i}{dt^2} = -k\mathbf{R}_i + 2\sum_{j(\neq i)}\frac{1}{[(z_{ij}^0)^2 + (\mathbf{R}_{ij})^2]^{5/2}}[((z_{ij}^0)^2 + (\mathbf{R}_{ij})^2)\mathbf{R}_{ij} - 3\mathbf{R}_{ij}^0\mathbf{R}_{ij}^0\cdot\mathbf{R}_{ij}]$$

$$+ 6\sum_{j(\neq i)}\frac{z_{ij}^0}{[(z_{ij}^0)^2 + (\mathbf{R}_{ij})^2]^{5/2}}\mathbf{R}_{ij}^0\zeta_{ij}, \tag{26}$$

respectively. Here k in the equation for $\mathbf{R}_i$ is the confining force constant. When we note that $|\mathbf{R}_{ij}^0| \leq$ (diameter of cylinder) while $|z_{ij}^0| \propto |i-j|$ and the convergence of summations are relatively slow, the equations of motion for ζ_i and $\mathbf{R}_i$ are approximately decoupled as

$$m\frac{d^2}{dt^2}\zeta_i \sim 4\sum_{j(\neq i)}\frac{\zeta_{ij}}{|z_{ij}^0|^3}, \tag{27}$$

$$m\frac{d^2}{dt^2}\mathbf{R}_i \sim -k\mathbf{R}_i + 2\sum_{j(\neq i)}\frac{1}{|z_{ij}^0|^3}\mathbf{R}_{ij}. \tag{28}$$

The longitudinal fluctuations in these lattices are thus given by the dispersion relation (14) and density fluctuation is given by (21).

CONCLUSION

We have shown that the strongly coupled classical one-dimensional plasma with $1/r$ interaction can be described by the harmonic approximation. The harmonic approximation which is originally devised for crystalline state does apply to one-dimensional system where crystalline order cannot exist due to enhanced effect of long wavelength fluctuations. These results apply to one-dimensional string of charges in long traps and also to electrons in quantum wires under appropriate conditions.

References

[1] J. M. Malmberg and T. M. O'Neil, Phys. Rev. Lett. **39**, 1333(1977)

[2] D. H. E. Dubin, Phys. Rev. Lett. **66**, 2076(1991); D. J. Heinzen, J. J. Bollinger, F. L. Moore, W. M. Itano, and D. J. Wineland, Phys. Rev. Lett. **66**, 2080(1991); J. J. Bollinger, D. J. Heinzen, F. L. Moore, W. M. Itano, D. J. Wineland, and D. H. E. Dubin, Phys. Rev. A**48**, 525(1993).

[3] A. Rahman and J. P. Schiffer, Phys. Rev. Lett. **57**, 1133(1986).

[4] D. H. E. Dubin and T. M. O'Neil, Phys. Rev. Lett. **60**, 511(1988).

[5] J. P. Schiffer, Phys. Rev. Lett. **61**, 1843(1988).

[6] S. L. Gilbert, J. J. Bollinger, and D. J. Wineland, Phys. Rev. Lett. **60**, 2022(1988).

[7] H. Totsuji and J.-L. Barrat, Phys. Rev. Lett. **60**, 2484(1988).

[8] H. Totsuji, Phys. Rev. E**47**, 3784(1993).

[9] For example, Ph. Choquard, H. Kunz, Ph. A. Martin, and M. Navet, *Physics in One Dimension*, edited by J. Bernasconi and T. Schneider (Springer, New York, 1981), 335.

[10] H. Totsuji, Phys. Rev. A**46**, 2106(1992).

[11] See, for example, *Proceedings of the Workshop on Electron Cooling and New Cooling Techniques, Legnaro, Italy, 1990*, edited by R. Calabrese and L. Tecchio (World Scientific, Singapore, 1991).

[12] For example, let us fix both ends of a system of size L. In this case, the mean square displacement of charges at finite temperatures increases with the distance from the fixed end z as $z/\gamma \ln z$ and diverges as $L/\gamma \ln L$ when $L \to \infty$.

[13] H. Totsuji, Phys. Rev. A**38**, 5444(1988).

[14] M. G. Raizen, J. M. Gilligan, J. C. Bergquist, W. M. Itano, and D. J. Wineland, Phys. Rev. A**45**, 6493(1992).

Dense Plasma as a Statistical Ensemble of Coulomb Clusters: A New Paradigm

Kenji Tsuruta and Setsuo Ichimaru

Department of Physics, University of Tokyo, Bunkyo, Tokyo 113, Japan

Abstract: A novel framework for the description of thermodynamic and correlation properties in a dense one-component plasma (OCP) is presented in a representation based on an ensemble of Coulomb clusters. The properties of spherical and non-spherical Coulomb clusters with number of particles N in the ground state as well as in a thermally excited state are investigated for $N=2$–13 by computer simulations. It is shown that a Coulomb cluster with a magic-number structure, such as an octahedron at $N=6$ and an icosahedron at $N=12$, is conspicuously stable even near to the melting temperature of a bulk OCP. Thermodynamic states of the OCP in such a Coulomb-cluster model are obtained by minimizing the model free energy with respect to fractional population of the clusters with different sizes. In light of such a cluster-model description, we investigate the issues of cluster formation and possible local order in a dense Coulomb matter.

I. INTRODUCTION

The one-component plasma (OCP) is a fundamental representation for a dense Coulombic system found in nature.[1] It is a classical system consisting of identical point particles with electric charge Ze immersed in a uniform background of neutralizing charges with density $-\rho_e$. A theoretical method in approaching the interparticle correlation in such an OCP depends on a hierarchical description in terms of the multiparticle distribution functions.[2] Thus for a tenuous high-temperature plasma, a perturbation-theoretic analysis, based on a giant-cluster expansion[3] or on Bogoliubov-Born-Green-Kirkwood-Yvon hierarchy,[4] has been successful for elucidation of the correlation properties. For dense plasmas, a liquid-theoretic approach such as a hypernetted-chain scheme has been employed, coupled with computer-simulation analyses.[5]

As the density is increased, the plasma starts to exhibit features characteristic of a condensed matter, in which both short-range directional and long-range translational orders conspire to give the plasma a character of a strongly coupled many-particle system. Modeling of a dense plasma as a statistical ensemble of the Coulomb clusters, a paradigm which we wish to explore in this paper, provides a natural framework in which such an interplay between the directional and translational orders may be accounted for in the process of melting transition in a Coulomb solid.

The *Coulomb cluster* is defined as a system of N charged particles confined in a uniform neutralizing charges of a volume Ω_N, with the relation

$$\rho_e \Omega_N = Ne. \tag{1}$$

Spherical and non-spherical Coulomb clusters are illustrated in Fig. 1. Here

$$a \equiv [3Ze/(4\pi\rho_e)]^{1/3} \tag{2}$$

denotes the *ion-sphere radius*, which is approximately a half of the average distance between neighboring particles.

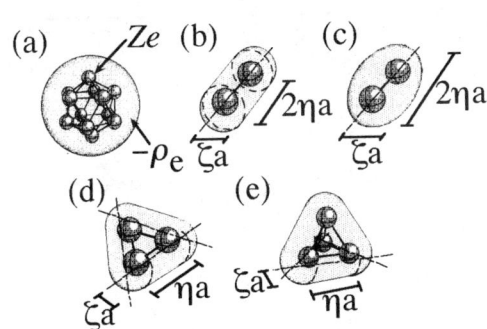

FIGURE 1. Schematic view of Coulomb clusters. (a) A spherical cluster with an icosahedral structure at $N=12$; (b) a non-spherical, two-particle cluster with a spherocylindrical (SC) structure; (c) a non-spherical, two-particle cluster with a spheroidal (SD) structure; (d) a non-spherical, three-particle cluster with a parallel-body structure of equilateral triangle (PT3); (e) a non-spherical, four-particle cluster with a parallel-body structure of tetrahedron (PT4).

In introducing a statistical ensemble of Coulomb clusters as a model representation of dense plasmas, we assume that interaction between the clusters is weak and negligible; thus Coulomb clusters constitute good independent units in an OCP. It is also assumed that the thermal state of the ensemble, obtained through minimization of its free energy, may correspond to the state of a thermodynamic equilibrium for the OCP.

II. COULOMB CLUSTERS IN THE GROUND STATE

The ground-state (i e., $T=0$) properties of the spherical Coulomb clusters have been investigated in Ref. 6; their geometrical structures are summarized in Table I. Ground-state energy per particle of such spherical clusters at $N=2$–13 are plotted in Fig. 2. The lowest energy configurations of the cluster particles were approached by a molecular dynamics (MD) simulation; in such an MD run, the "temperature" of the cluster was gradually lowered by imposing an artificial friction force at each particle. The ground-state energy of a cluster is then calculated according to

$$E(N;\{\mathbf{r}_i\}) = \frac{1}{2}\int\int \frac{d\mathbf{r}d\mathbf{r}'}{|\mathbf{r}-\mathbf{r}'|}\left[\rho(\mathbf{r})\rho(\mathbf{r}') - (Ze)^2\sum_{i=1}^{N}\delta(\mathbf{r}-\mathbf{r}_i)\delta(\mathbf{r}'-\mathbf{r}_i)\right],$$
(3)

where

$$\rho(\mathbf{r}) = Ze\sum_{i=1}^{N}\delta(\mathbf{r}-\mathbf{r}_i) - \rho_e\theta(\mathbf{r}\in\Omega_N)$$
(4)

refers to distribution of the electric charges, the particle positions $\{\mathbf{r}_i\}$ ($i=1$–N) are those of the lowest energy configurations, and $\theta(\mathbf{r}\in\Omega_N)$ is the unit step function that is set to be unity only when the coordinates $\mathbf{r}$ are placed in the volume Ω_N, and zero otherwise. In Ref. 6, it has been shown that the Coulomb clusters assume particularly stable configurations at certain "magic numbers" such as $N=6$ with an octahedral structure and $N=12$ with an icosahedral structure.

In addition to these cases of spherical clusters, the configurations for non-spherical clusters at $N=2$–4, shown in Figs. 1(b)–1(e), may be calculated by

TABLE I. Ground-State Structures of Coulomb Clusters with Spherical Background.

N	Geometrical Structure
2	Dumbbell
3	Equilateral triangle
4	Tetrahedron
5	Triangular bipyramid
6	Octahedron
7	Pentagonal bipyramid
8	Square skew-prism
9	3-Capped triangular prism
10	2-Capped square skew-prism
11	Polyhedron
12	Icosahedron
13	Icosahedron+1(center)

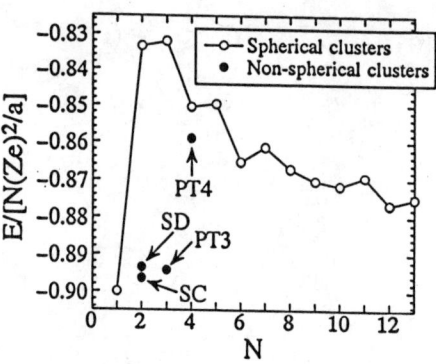

FIGURE 2. Normalized ground-state energy for the spherical (open circles) and the non-spherical (solid circles) Coulomb clusters with sizes $N=2$–13.

minimizing the electrostatic energy (3) with respect to the aspect ratio of the cluster (i e., the ratio of η to ζ in Fig. 1) and the interparticle distance.[7] The ground-state energy of these non-spherical clusters are likewise plotted in Fig. 2. It is observed that a deformation of the cluster from a spherical shape for $N=2$ and 3 may significantly lower the ground state energy.

III. THERMAL STATES: EOS AND CORRELATION

The thermodynamic properties of a classical OCP at temperature T depend on the dimensionless *Coulomb coupling parameter*[1]

$$\Gamma = (Ze)^2/(ak_BT), \qquad (5)$$

where k_B denotes the Boltzmann constant. Here and hereafter, we shall call $1/\Gamma$ "temperature", which is an average kinetic energy in units of $(Ze)^2/a$.

Equation of state (EOS) for a bulk OCP has been investigated accurately in the fluid[8] and crystalline[9] phases mainly by computer-simulation methods. Comparing the Helmholtz free energy in a fluid phase with that in a crystalline phase, it has been found that a Coulomb solid may melt at $\Gamma_m \approx 172$. It is therefore significant to investigate thermal properties of Coulomb clusters at temperatures near $1/172$. Thus we consider a range of temperature at $\Gamma=80$–2000 for Coulomb clusters with $N=2$–13; Monte Carlo (MC) simulations with the Metropolis algorithm[10] have been used for this study. The excess internal energy per particle for a cluster of size N, defined by

$$u_N(\Gamma) \equiv \int_{\Omega_N} \frac{E}{N(Ze)^2/a} e^{-\beta E} d\mathbf{r}_1 d\mathbf{r}_2 \cdots d\mathbf{r}_N \Big/ \int_{\Omega_N} e^{-\beta E} d\mathbf{r}_1 d\mathbf{r}_2 \cdots d\mathbf{r}_N \qquad (6)$$

has been sampled by the MC method at each Γ. Setting the positions of N

particles initially at those obtained in the ground state (Sec. II), we started the simulations at a lowest temperature corresponding to $\Gamma=2000$. After an equilibration has been confirmed, the temperature was raised stepwise, with equilibration being secured in each step, until a highest temperature in the present study corresponding to $\Gamma=80$ has been reached.

Temperature dependence of the excess internal energy per particle is plotted in Fig. 3a for the spherical clusters and in Fig. 3b for the non-spherical clusters. The statistical errors in the samplings are smaller than sizes of the points in the figures, except in the cases where error bars are explicitly shown. The values at $1/\Gamma=0$ correspond to the ground-state energies per particle in Fig. 2. The solid lines depict those derived from a formula in the harmonic approximation,

$$u_N^{\text{harm}}(\Gamma) = a_N + \frac{M}{2N} \cdot \frac{1}{\Gamma}, \qquad (7)$$

where a_N is the ground-state energy per particle. With the value of M chosen as

$$M = \begin{cases} 4 & \text{for a spherical cluster with } N=2, \\ 3N-3 & \text{for a spherical clusters with } N \geq 3, \\ 3N & \text{for a non-spherical cluster,} \end{cases} \qquad (8)$$

we find a good agreement between the MC results and Eq. (7).

Such a linear dependence of the second term in Eq. (7) on $1/\Gamma$ stems from the equipartition theorem[11] in the harmonic-lattice approximation, where M corresponds to the *vibrational degrees of freedom* in particle motions. Since the potential field acting on the particles in a spherical cluster is point-symmetric with respect to the center of the cluster, the rotational degrees of freedom do not contribute to the internal energy and hence are subtracted in Eq. (8). For the non-spherical clusters, on the other hand, the potential fields for the particles are not rotationally symmetric; the rotational degrees of freedom thus contribute in the resultant EOS.

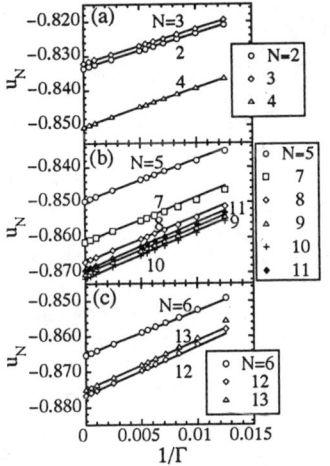

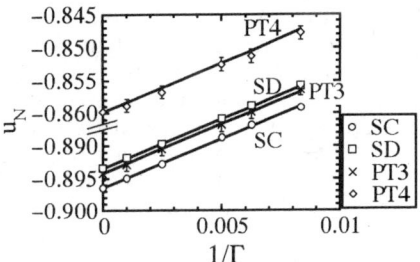

FIGURE 3a. Normalized excess internal energy of the spherical clusters with sizes $N=2$–13. The solid lines depict the equipartition formula Eq. (7).

FIGURE 3b. Same as Fig. 3a, but for the non-spherical clusters of $N=2$–4.

To elucidate the internal structures of Coulomb clusters in a thermal state, we investigate the pair distribution function $g(r)$, the bond-angle distribution function $p(\theta, \phi)$, and the bond-length fluctuation function Δ, calculated as

$$g(r) = \frac{1}{4\pi r^2 N(N-1)} \left\langle \sum_{i \neq j} \delta(|\mathbf{r}_{ij}| - r) \right\rangle, \tag{9}$$

$$p(\theta, \phi) = \frac{1}{N(N-1)(N-2)} \left\langle \sum_i \sum_{j \neq i} \sum_{k \neq i,j} \delta(\theta - \theta_{ij}) \delta(\phi - \phi_{ijk}) \right\rangle, \tag{10}$$

$$\Delta = \frac{1}{N(N-1)} \sum_{i \neq j} \sqrt{\frac{\langle |\mathbf{r}_{ij}|^2 \rangle - \langle \mathbf{r}_{ij} \rangle^2}{\langle \mathbf{r}_{ij} \rangle^2}}, \tag{11}$$

where $\mathbf{r}_{ij} = \mathbf{r}_j - \mathbf{r}_i$ denotes the relative position between the ith and jth particles, and

$$\theta_{ij} = \cos^{-1}\left(\frac{\mathbf{r}_i \cdot \mathbf{r}_j}{|\mathbf{r}_i||\mathbf{r}_j|}\right) \quad \text{and} \quad \phi_{ijk} = \cos^{-1}\left(\frac{\mathbf{r}_i \times \mathbf{r}_j}{|\mathbf{r}_i \times \mathbf{r}_j|} \cdot \frac{\mathbf{r}_i \times \mathbf{r}_k}{|\mathbf{r}_i \times \mathbf{r}_k|}\right) \tag{12}$$

define the relative latitude and longitude, respectively, of the jth and kth particles with respect to the ith particle. The brackets $\langle \cdots \rangle$ imply a thermal average over the MC-generated configurations.

The pair distribution functions $g(r)$ for the spherical clusters with $N=2$–4, 6, 12, 13 and $N=5$, 7–11 are plotted in Figs. 4a and 4b. The bond-angle distributions $p(\phi, \theta)$ for the spherical clusters with $N=3$, 4, 6, 12 and $N=9$–11 are likewise plotted in Figs 5a and 5b. In the cases of $N=2$–4, 6, 12, 13, which assume regular polyhedrons in the ground state, we find conspicuous peaks in $g(r)$ and in $p(\phi, \theta)$ for larger Γ in Figs. 4a and 5a, and simple broadening of the peaks for smaller Γ. The regular structures in the clusters with $N=2$–4, 6, 12, 13 are thus maintained in a stable manner even at these high temperatures.

In the cases of $N=5$, and 7–11 in Figs. 4b and 5b, which do not have a regular structure, $g(r)$ and the contour lines of $p(\phi, \theta)$ exhibit complicated shapes at low temperatures. These entangled structures then vanish by thermal broadening at high temperatures. It may thus be concluded that cluster structures for $N=5$, 7–11 are thermally unstable.

The quantity Δ defined by Eq. (11) may be regarded as a *Lindemann ratio* appropriate to a finite system; it is a ratio of fluctuation in the particle positions to an average nearest-neighbor distance.

Figure 6 shows temperature dependence of Δ for the spherical clusters of $N=2$–13. Analogous to the cases of $g(r)$ and $p(\theta, \phi)$, we clearly observe differences in the temperature dependence between clusters with $N=2$–4, 6, 12, 13 and with $N=5$, 7–11. In the clusters with $N=2$–4, 6, 12, 13, Δ depends linearly on $1/\Gamma^{1/2}$ for $\Gamma \geq 140$, a feature typical in an ion-sphere description,[1] while clusters with $N=5$, 7–11 exhibit deviations from such a linear dependence.

At still higher temperatures, Δ starts to depart away from the linear dependence in the cases of $N=6$, 12, 13. This implies an onset of itinerant behaviors for the particles and resultant cluster melting. The critical values of Δ for such a cluster melting are here observed close to 0.1, a melting criterion found empirically in various substances such as crystalline metals.[12]

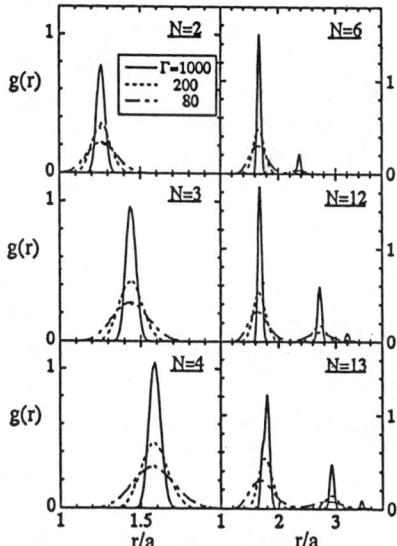

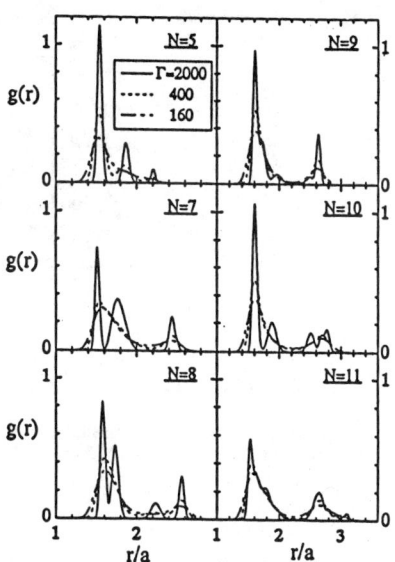

FIGURE 4a. Pair-distribution functions for the spherical clusters with sizes $N=2-4$, 6, 12, and 13.

FIGURE 4b. Same as Fig. 4a, but for the spherical clusters of $N=5$, and 7–11.

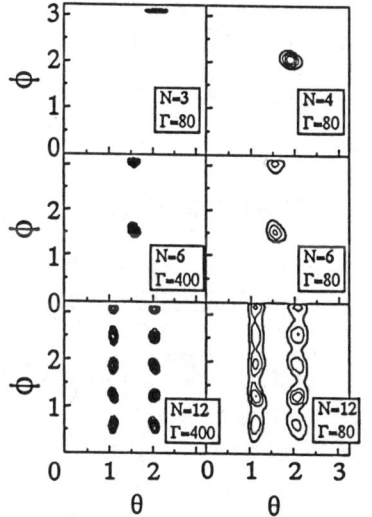

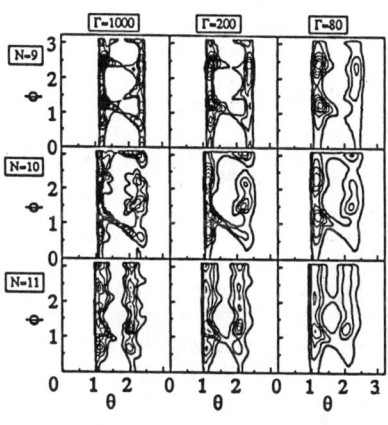

FIGURE 5a. Bond-angle distributions for the spherical clusters of $N=3$, 4, 6, and 12.

FIGURE 5b. Same as Fig. 5a, but for the spherical clusters of $N=9-11$.

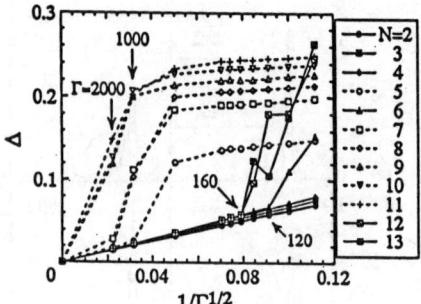

FIGURE 6. Bond-length fluctuation functions for the spherical clusters of $N=2$–13.

IV. CONSTRUCTION OF A STATISTICAL ENSEMBLE REPRESENTING A DENSE PLASMA

We now proceed to construct a statistical ensemble with c_{tot} Coulomb clusters in a volume Ω_{tot}, representing a dense OCP. Denoting the number of N-particle clusters by c_N, we have the total number of particles as given by $N_{tot} \equiv \sum_N N c_N$.

In setting the model, we assume that interaction between clusters is negligible and take into account a surface energy per unit area of the cluster boundary, which is estimated as

$$\sigma \equiv \frac{a_{bcc} - a_1}{4\pi a^2} \cdot \frac{(Ze)^2}{a} = 0.004071(Ze)\rho_e. \tag{13}$$

Here $a_1 = -0.9$ and $a_{bcc} = -0.895929$ are the ground-state energy of an ion sphere in units of $(Ze)^2/a$ and Madelung energy of the bcc crystalline OCP, respectively. For an N-particle cluster, the surface energy is added to the ground-state energy per particle in accordance with

$$a'_N \equiv a_N + \sigma \cdot S_N/[(Ze)^2/a], \tag{14}$$

where S_N is a surface area of an N-particle cluster divided by N, which varies approximately as $S_N \approx N^{-1/3}$.

In Sec. III, we have shown that the internal energy is well represented by the harmonic approximation Eq. (7) at low temperatures. In the limit of high temperatures, the internal energy may be given by that of an ideal-gas system. In these circumstances, particles in a cluster tend to escape away from its volume and therefore the excess internal energy of a cluster at $\Gamma=0$ should be $u_N(0)-0$. We thus adopt

$$u_N(\Gamma) = a'_N + \frac{M}{2N} \cdot \frac{1}{\Gamma - M/(2Na'_N)}, \tag{15}$$

which satisfies the conditions,

$$u_N(\Gamma) \begin{cases} \to a'_N + \frac{M}{2N} \cdot \frac{1}{\Gamma} & \text{for } \Gamma \gg 1, \\ \to 0 & \text{for } \Gamma \ll 1. \end{cases} \tag{16}$$

The Helmholtz free energy F_N for an N-particle OCP system may be expressed as

$$f_N \equiv \frac{F_N}{Nk_{\mathrm{B}}T} = \ln(\Lambda^3) + f_N^{(1)}(\Gamma), \tag{17}$$

where $\Lambda \equiv \sqrt{2\pi\hbar^2/(mk_{\mathrm{B}}T)}/a$ indicates the ratio of the thermal de Broglie length to the ion-sphere radius, and m is the mass of a particle. The second term on the right-hand side of Eq. (17), the interaction free energy, may be calculated by the coupling constant integration as[1]

$$f_N^{(1)}(\Gamma) \equiv -\frac{1}{N}\ln\left[\frac{1}{N!}\int \exp(-\Gamma V_N)d\mathbf{x}_1 d\mathbf{x}_2\cdots d\mathbf{x}_N\right], \tag{18}$$

$$= \int_0^\Gamma u_N(\Gamma')d\Gamma' + f_N^{(1)}(0), \tag{19}$$

where $V_N \equiv E/[(Ze)^2/a]$, $\mathbf{x}_i \equiv \mathbf{r}_i/a$, and $u_N(\Gamma)$ denotes the normalized excess internal energy defined by Eq. (6). Since the particles can move freely in the bulk OCP at $\Gamma=0$, it is not appropriate to presume confinement of the particles in a cluster; the possibility of interchanging particles between clusters should be taken into account in the evaluation of the free energy. We may thus take the free energy as that of an infinite ideal-gas system, i e.,

$$f_N^{(1)}(0) = \ln[3/(4\pi e)]. \tag{20}$$

In Fig. 7, we plot the interaction free energies $f_N^{(1)}(\Gamma)$ for $N=1$–13 calculated by substituting Eqs. (15) and (20) into Eq. (19). Here and hereafter we shall adopt the cluster-energy value of an SC cluster for $N=2$, that of a PT3 cluster for $N=3$, and that of a PT4 cluster for $N=4$. The bcc and fluid lines in Fig. 7 depict the free energy of a bulk OCP in the bcc crystalline phase[9] and that in the fluid phase;[8] $\Gamma_\mathrm{m}=172$ represents the melting condition for a bulk Coulomb solid.

In terms of the cluster free energy $f_N^{(1)}$, we formulate the free energy of an ensemble of Coulomb clusters with the set of variables, $\{c_1, c_2, \cdots, c_N, \cdots\}$, as

$$f_{\mathrm{c}}(\Gamma) \equiv \sum_N \frac{Nc_N}{N_{\mathrm{tot}}}f_N^{(1)}(\Gamma). \tag{21}$$

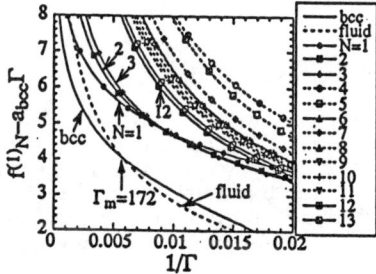

FIGURE 7. Normalized interaction free energies of the Coulomb clusters with sizes $N=1$–13. "bcc" and "fluid" indicate free energies of a bulk OCP in bcc crystalline and fluid phases, respectively.

The *ideal entropy of mixing* for the model system is then given by

$$S = k_B \ln \left[\frac{N_{\text{tot}}!}{(c_1!)(2c_2!) \cdots (Nc_N!) \cdots} \right], \tag{22}$$

which contributes to the free energy in the amount,

$$f_s = -S/(k_B N_{\text{tot}}) \approx \sum_N \left(\frac{Nc_N}{N_{\text{tot}}} \right) \ln \left(\frac{Nc_N}{N_{\text{tot}}} \right) \quad \text{(for } c_N \gg 1\text{)}. \tag{23}$$

Equation (22) implies that interchanges of particles situated in clusters with different sizes may be accounted for in counting the number of states, in the way as considered in the evaluation of the cluster free energy (20).

Summation of the energies, Eqs. (21) and (23), yields the interaction free energy for the statistical ensemble of the Coulomb clusters,

$$f_{\text{tot}}^{(1)}(\Gamma; \{x_N\}) = \sum_N x_N [f_N^{(1)}(\Gamma) + \ln(x_N)], \tag{24}$$

where $x_N \equiv Nc_N/N_{\text{tot}}$ defines the fractional number of N-particle clusters. By minimizing $f_{\text{tot}}^{(1)}$ with respect to $\{x_N\}$ ($N=1$–13) under the constraint $\sum_N x_N = 1$, we obtain the free energy of the OCP in thermodynamic equilibrium as

$$f_{\text{model}}^{(1)}(\Gamma) = -\ln \left[\sum_N \exp(-f_N^{(1)}) \right], \tag{25}$$

and the fractional distribution of N-particle clusters as

$$x_N = \exp(-f_N^{(1)}) / \sum_N \exp(-f_N^{(1)}). \tag{26}$$

In Fig. 8, we plot the model free energy Eq. (25), the bcc and fluid free energies of the bulk OCP, f_c, and f_s as functions of the temperature. It should be noted that the entropy of mixing significantly lowers the model free energy, so that a melting transition is predicted at $\Gamma_m \approx 70$, a value rather close to $\Gamma_m = 172$.

The fractional numbers x_N of the clusters are exhibited in Fig. 9. As temperature is raised and approaches the melting condition, cluster populations with $N=2$ and 3 grow drastically. We may regard such clusters with $N=2$ and 3 as representing "dislocations" in a crystalline solid. In the fluid phase (i e., $\Gamma < \Gamma_m$), we find gradual increase in the probabilities for larger regular clusters such as $N=12$ and 13 to emerge in the system. We thus expect that clusters with such an icosahedral symmetry may be observed in a fluid OCP with a probability of approximately 10%.

V. CONCLUDING REMARKS

We have investigated a Coulomb-cluster model for a description of thermodynamics and short-range order in a dense OCP. The study of individual Coulomb clusters has shown that the symmetric structures inherent in the clusters with $N=2$–4, 6, 12, 13 are conspicuously stable against thermal melting. A statistical ensemble of the Coulomb clusters and its free energy have been formulated on the basis of EOS for these individual clusters. Minimizing

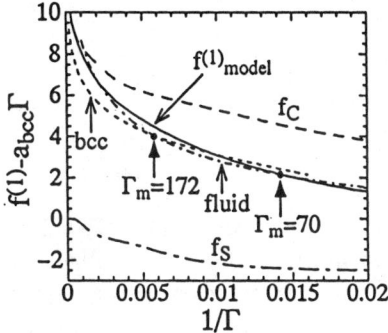

FIGURE 8. Normalized interaction free energies for the statistical ensemble of Coulomb clusters; see the text for the meaning of f_c and f_s.

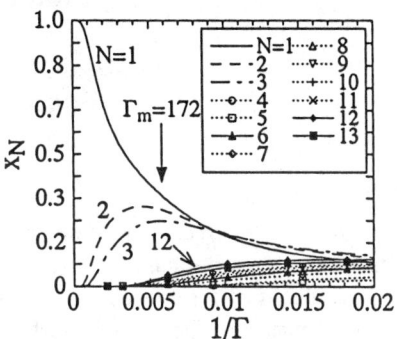

FIGURE 9. Fractional distributions of N-size clusters for the statistical ensemble of Coulomb clusters.

the free energy with respect to fractional distribution of clusters, we have obtained a statistical description of a dense OCP system, which exhibits an overall agreement with some of the known properties of the bulk OCP. The roles that clusters with $N=1$–3 play in the vicinity of the melting transition and the growth in fractional numbers of larger icosahedral clusters in fluid phase have been particularly remarked.

A major shortcoming of the present analysis lies in its neglect of interaction between clusters. Dipole-dipole interaction being attractive, we expect a possibility that an inclusion of such an interaction may lower the model free energy and thereby decrease the predicted melting temperature. A study in these directions is in progress.

ACKNOWLEDGEMENTS

One of the authors (K.T.) gratefully acknowledges the support of the Japan Society for the Promotion of Science through a Post-Doctoral Fellowship and Research Grant No. 3005.

REFERENCES

1. Ichimaru, S., *Statistical Plasma Physics Vol. II: Condensed Plasmas*, Reading, MA: Addison-Wesley, 1994.
2. Ichimaru, S., *Statistical Plasma Physics Vol. I: Basic Principles*, Redwood City, CA: Addison-Wesley, 1992.
3. Abe, R., *Progr. Theor. Phys.* **21**, 475 (1959).
4. O'Neil, T.M., and Rostoker, N., *Phys. Fluids* **8**, 1109 (1965).
5. Baus, M., and Hansen, J.-P., *Phys. Rep.* **59**, 1 (1980).
6. Tsuruta, K., and Ichimaru, S., *Phys. Rev. A* **48**, 1339 (1993).
7. Ichimaru, S., Ogata, S., and Tsuruta, K., *Phys. Rev. E* **50**, No. 4 (1994).
8. Ogata, S., and Ichimaru, S., *Phys. Rev. A* **36**, 5451 (1987).
9. Dubin, D. H. E., *Phys. Rev. A* **42**, 4972 (1990).
10. Metropolis, N., Rosenbluth, A. W., Rosenbluth, M. N., Teller, A. H., and Teller, E., *J. Chem. Phys.* **21**, 1087 (1953).
11. For example, Landau, L. D., and Lifshitz, E. M., *Statistical Physics*, New York: Pergamon Press, 1980.
12. Ashcroft, N. W., and Stroud, D., in *Solid State Phys.* **33**, edited by Ehrenreich, H., Seitz, F., and Turnbull, D., New York: Academic, 1978, 1.

Open Questions in Non-Neutral Plasma Physics
Panel Discussion

Moderated by Charles Roberson
Edited by Dan Dubin

Charles Roberson: Let me explain how we will run the session. First, I will make a few remarks and so I have a few viewgraphs here to do. Then I will also use the privileges of the moderator by asking the first question of our panel members to get that part of the session started. After that we will open the questions up to the audience. The other thing I want to try to do here is to encourage the students and recent graduate students to ask questions. So the last half hour or so will be devoted just to the questions from students or recent graduates. You can ask questions at any other time but at that time, if you don't ask questions we're just going to sit here and wait. With those ground rules, let's begin. Now the title of this session is "Open Questions." For me, one of the open questions is, "Where are we, and where do we go in this kind of field in general?" One of the things that I did was look back in the proceedings of the Non-neutral Symposium we ran in 1988. I've always found a lot of quotes in there that are quite interesting. Here is a paraphrased quote from our colleague Dave Crandall of the Department of Energy. He told me that he looked upon this field as the equivalent of rediscovering the hydrogen atom, and I think there's some truth to this, in that we are re-discovering or in some sense reinventing plasma physics. My colleague Dr. Junker in the Office of Naval Research very much likes the work of Wineland and Bollinger. Bruce Robinson in his welcoming remarks commented on what an exciting area of science this is and its potential impact on technologies of interest to the Navy. It was very interesting to see a number of people at NIST and in other labs using Dubin's theory to get at modes and there's really starting to be a coming together that was very much evident in the Irvine workshop 2 years ago, and I think we are finding a lot of common ground with the area of atomic and molecular physics. Of our colleagues at the Department of Energy, for example, Dave Sutter who oversees the accelerator research there, says that they are finally admitting that plasmas do exist. In the high energy or accelerator community I think that is tantamount to trying to introduce the uncertainty principle into classical mechanics. Davidson also says that the connection to accelerators and beam physics is one direction where we ought to be going. Of course, John Malmberg put it very simply, that the most important problem in all of plasma physics was how particles cross field lines, and that neutral plasmas are just too complicated to get at this problem and so we're driven to non-neutral plasmas. Roy Gould saw an opportunity to go from the essentially linear phase of plasma evolution to the whole turbulent phase in confined nonneutral plasmas, and I think we've seen a lot of progress in that arena in this workshop. My own perspective at the time of the '88 conference was summarized by the diagram used as a 'logo' for the conference - that non-neutral plasmas had a potential of having an impact on a lot of fields, including which were some of my interests at the time, high-power microwaves, FELs and accelerators, anti-matter, atomic, molecular, and so on. At the Irvine workshop two years ago it was essentially decided to concentrate on the trap physics, that would limit the main interest of this workshop and focus it sufficiently so that we could make a lot of progress. I think that was a wise choice, and I think this meeting confirms it. As I say, I wanted to be able to ask the first question and that's for a deliberate reason. Two years ago, you know we did a workshop at Univ. of California at Irvine at the Beckman Center of the National Academy, and we did that in conjunction with the National Academy In fact, in this program from its very beginning we have tried to work closely with the new National Academy Plasma Science Committee. The committee is writing a report on opportunities in plasma physics. Cliff, who is chairman of the committee probably can't say too much, but we certainly can ask him what impact this field has had on that study.

Cliff Surko: Yes, there was a panel put in place by the Academy two years ago to make a broad-based study of opportunities in plasma science and technology. It should come as no surprise that I can't tell you what the recommendations or the findings are in detail because they've not been approved and there's a very lengthy and extensive review process for any Academy report. But dating from the days of the Brinkman report, one of the key observations was, and continues to be, that basic plasma science is in terrible shape, and the seeds of that started perhaps in the 70's, and it's been on a downward trend ever since. If that's gone on so long, one might well lose one's confidence that there can be a field of basic plasma science or even know what such a field should look like. Well, what many of us have realized is that if you look around the scene, perhaps the healthiest corner is this enterprise of non-neutral plasma research. And so if one wants a model for basic neutral plasma research, this provides one. Some answers to key questions such as how basic plasma science can stay healthy, what scale of enterprise it takes, and what style works, may be found here in this nonneutral plasma effort. I think that's my answer to Chuck's question. Having said that, I would like to say just a couple of other things. The fact that nonneutral plasmas is successful has raised the visibility of this program, and with that increase in visibility, then there comes more scrutiny, and my own opinion is that it's important for us to be clear about the impact that we have and can have on other fields. Chuck has mentioned atomic physics in the viewgraph as well as "fluid physics" - we can't be glib or cavalier or about that; we have to be very clear about the real issues and see that a reasonable expert in those fields will agree that we can make a contribution or at least listen to the argument. So in some sense the success also comes with a responsibility to be clear about articulating what we're really doing. And there's another issue that personally I would like to have some answer to, and I'll throw it out as a question. I don't think it's necessary that one has an answer to this, but I suspect that one does exist. It would be useful to have a clear picture of what impact non-neutral plasma research has had and can have on the larger field of neutral plasmas. Being clear on that then helps in arguing for what fraction of national resources should be devoted to basic research in this area as opposed to other areas. .

Roberson: OK, I would like questions and comments from the audience.

Fred Driscoll: This relates to the question that Cliff asked about the importance of nonneutral plasmas to neutral plasma physics. I've always thought that the name non-neutral plasmas was a little odd. Actually, we work on un-neutralized plasmas. I would allege that neutral plasmas are in general non-neutral plasmas and that their non-neutrality is one of the things that is least understood about them. For example, a tokamak plasma is non-neutral to the extent of perhaps one particle out of a million and that generates electric fields which can dominate the motion of the plasma; this is not news. The hard question that nobody can answer is, how do you predict the non-neutrality that arises in such non-neutral plasmas? Now whether we can contribute to that, I don't know. It seems that perhaps we're contributing something in our studies of convective transport in very simple systems. Unfortunately, most of the neutral plasmas such as tokamaks are such horrendously complicated systems that our examples appear too simplified to be directly relevant. On the other hand, if there were a more healthy neutral basic plasma physics program, just a basic research effort, then there would be more connection between that research effort and our area.

Tom O'Neil: Cliff's question is one that Cliff and Dan and I have talked about before, and I think it's important. It seems to me that for a long time, much of laboratory neutral plasma physics consisted of a search for a quiescent well-confined plasma sample. Non-neutral plasmas are a perfect example of that. It's exactly what you want; it's long-lived, it's quiescent, and it's easy to confine. It makes possible high precision experiments. It pushes plasma physics into whole new parameter regimes; for example, a regime of strong correlation, and a regime of strong magnetization. There are other more direct connections to neutral plasmas. For example, we've cleared up misunderstandings about like-particle transport. Furthermore, transport in rotating plasma systems such as tandem mirrors was once thought to be dominated by the input of angular momentum through resonances with individual particles. We now know that there can be resonances with collective modes of oscillation that dominate the transport in such systems.

Dan Dubin: I'd also like to make a comment about the connection between non-neutral and neutral plasmas. This also relates to possible future directions for our research area. We've seen in the talks that people are starting to think about what happens when you take positrons, for example, and mix them with a non-neutral plasma consisting of electrons. So it seems that one interesting new direction might be to find out what happens to a non-neutral plasma when it starts to become neutral. When one starts to add an oppositely-charged species to the plasma, at what point does one start to lose the good confinement characteristics which we all understand in a non-neutral plasma, and how is that confinement lost? Perhaps consideration of such questions might be a way to make our field more directly relevant to neutral plasma physics.

Joel Fajans: One of the obvious areas of overlap is plasma processing, because there are necessarily sheaths at the edges of plasmas which are non-neutral, and I think, some plasma processing is done with unneutralized beams directly. I don't think this community has made much attempt really to reach out to the plasma processing community. However, we do happen to have a representative of that community here, who I'm sure is dying to make some remarks about this, Ned.

Ned Birdsall: I think that part of what you are getting at is that plasmas do have edges, and edges tend to have non-neutral parts. Our particular effort has been to try to understand the sheath, and the pre-sheath. All the crazy myths (and I call them myths) that exist about what goes on at the edge of the plasma are far from solved. But I think we can learn a lot from the effort that has gone on for many years at San Diego because the things that you learn in a somewhat less complex system with only one species lead us to think more about what we're doing with two species. For example, if we use a high-power electron beam in a microwave tube, a hollow beam for example, it's likely to break up through the diocotron instability; so everything that you've discovered about diocotron modes is helpful to people trying to build such high power devices. What I'm trying to say is that a lot of things that go on in non-neutral plasmas have been an immense help in dealing with multispecies plasmas. I would also challenge you to think of not only having ions and electrons around but think of having negative ions around, and think of the chaos that leads to and so forth.

Roberson: If I can make one comment about the plasma processing connection. The Plasma Science Committee did do a study on plasma processing, and that came about mainly because the industry people came to this committee and asked them to do something, and I participated in many of those workshops. It was absolutely remarkable to see this, because you realized that they had built up a multi-billion dollar industry and they didn't talk to any theorists. They really just did these things by making discharges, and there was no serious theory involved. The reactors they are building are complicated, even by neutral plasma standards, involving chemical reactions and surface physics that is not well understood theoretically. This is in contrast to this program which is driving towards the simplest and most predictable sorts of plasma effects.

John Bollinger: Let me first say at least from the perspective of the time and frequency standards group at NIST, we've benefited greatly from what we have learned from this community of non-neutral plasma physicists. I don't think we would be as far along in understanding our small non-neutral plasmas in our ion traps if it wouldn't have been for the work of this community. Let me give an example of how plasma physics can help us design a better clock. I came prepared here with one transparency I'd like to put up. This is something Joseph Tan showed during his talk. This is a plot of the second-order Doppler shift in the radiation emitted by the ions due to the rigid rotation which the ion clouds undergo when they are in thermal equilibrium in the trap. One can see that the second order Doppler shift goes through a minimum as a function of the rotation frequency, so in order to reduce fluctuations in the emitted radiation due to fluctuations in the cloud's rotation frequency clearly that's where you want to operate. The key limitation to the frequency stability of a clock based on a Penning trap frequency standard will be how well can we stabilize the rotation frequency to this minimum. Now, there's an obvious way to control the rotation frequency right now using lasers; we can set the rotation frequency wherever we want by varying the laser torque on the ion cloud, and so in principle then by imaging our cloud and looking at the radius we can feedback and move our laser beam so that we stabilize the radius or rotation frequency where we want, but that doesn't sound like a very promising approach to me, just because it's a very long feedback loop, and I'm not sure how sensitive that would be. So any ideas people would have on this would be very helpful. One thing in particular struck me in looking over the program here. There's been a number of talks in using a rotating wall to apply a torque to a plasma, and so that's a way of applying a torque which can increase the rotation frequency. I assume that if the rotating wall were going slower than the plasma was rotating, then that would apply a drag, so in principle just by applying a rotating wall to our traps at this rotation frequency would provide a torque which would tend to stabilize our rotation frequency at that value. But my guess is that, it's going to be a very weak torque; are there ways to enhance it somehow? But basically we'd like to set up a trap where the torque goes through zero at the rotation frequency corresponding to the minimum which we see here.

Let me mention another area where I think plasma physics might make a contribution to ion traps. There are cooling mechanisms known as resistive cooling, which typically take energy out of clouds through the center of mass mode, and I don't think a lot is known how that energy eventually gets taken out of internal modes,. The problem of how internal modes in a quadratic Penning trap couple to the center of mass modes is important. So let me stop there and let Gerry continue.

Gerry Gabrielse: I may not be an ideal representative of atomic-molecular-optical physics, because for the last 10 years until this past year I didn't use an atom; I used only elementary particles. I guess I stumbled into this field almost by accident. I was mostly in the camp that thought collisions were something to be avoided like the plague. In those days I thought transport was so complicated that I used only one particle. I guess what got me interested in plasma physics issues was partly that we needed to know some things in order to do some of our experiments, and then the attractiveness of a small contained system which could be understood took over in its own right. For example, we used electron cooling to cool anti-protons. I didn't talk about that today but you know it well. One is interested in increasing the density of the cold particles. Now that's very much like electron cooling in an accelerator. At LEAR, we deal with an electron cooling machine which seems to work very effectively at low particle density. When you start to succeed, your success does you in. That is to say, you get collective behaviors that start happening, and frankly for a while we used plasma as a dirty word. Whenever something happened that we couldn't understand at all we said, Oh, there's a collective plasma mode that happened. If we now would like to take our electron cooling techniques and push them even further to make anti-hydrogen, I think it's going to be important to understand some of these collective modes.

Now the other thing I think I'd just like to emphasize is John Bollinger's point, that frequently we cool trapped particles by coupling to resistors or cavities, and cooling the center-of-mass motions. We know very little about how this cooling really works.

The last thing I'd like to say is this: we started doing plasma experiments for their own sake once we started seeing simple features emerge, simple features which we thought had no right to occur, based on the seemingly messy set of coupled equations of motion of the trapped particles. Now whenever that happens there's a better way to look at things, and even though it is complicated collective behavior, there must be a simple way to intuitively understand it. I think that's the charm of this particular interface field. These are simple enough systems where I think one can really make progress.

Roberson: Perhaps one of the answers to Cliff's comments about why things have improved so much is that I noticed here a fairly large group of graduate students, larger than in the past. So the field has been infused with a lot of new talent, and of course the experiments have progressed as well. Unless somebody has something else to say, I think we want to now try to encourage the students and recent graduates to make some comments.

Jay Hartley: My name is Jay Hartley, I'm with the UC Davis Department of Applied Science which is basically centered at Livermore National Laboratory, and I'm working with Tom Cowan on the accelerator project. The question I have to all of you is, Imagine you have a graduate student or post-doc that you are sending out to the wild world to start their own program, and they are going to a place where they have a complete blank slate to begin with. What would you recommend they do as an upgrade to your own apparatus or as an extension of your work, or maybe something completely new, that would result in them being able to study some new physics that you couldn't simply do by twiddling a knob on your own devices?

Dubin: I'm glad you asked that question! I've been thinking about something for a little while, and I ran it through the group meeting a couple of weeks ago in San Diego; I call it the plasma wind tunnel. However, I don't know that I would recommend this for the first project of a newly minted Ph.D. since it is rather speculative. I don't even know whether it would work; but this session is supposed to be in part about new directions for the field and this might be something to pursue.

Now, we've seen some nice experiments by Fred Driscoll on 2-dimensional fluid dynamics where one confines an electron plasma in 2-dimensional geometry. This turns out to be a very good approximation of an ideal incompressible 2-D Euler fluid. However, there's only one sign of charge in the electron system, which means that there is one sign of vorticity in the Euler fluid analog. However, if you could set up a system where you have both electrons and ions, then you could start thinking about doing more realistic fluid dynamics problems - both signs of vorticity would be present. One way to set up such a system, and probably not the only way, is to place surfaces in the confinement region which have been chemically treated in order to emit both electron and ions when electric fields are applied to the surfaces. For example, consider a treated wire oriented parallel to the magnetic field, and set between and parallel to the two plates of a parallel plate capacitor which is at sufficient voltage to induce field emission of electrons and ions from the wire. It's easy to show that the shielding of the electric field due to charged particle emission from the wire is equivalent to a no-slip boundary condition at the wire; the $\mathbf{E} \times \mathbf{B}$ drift velocity approaches a small value at the surface due to the cold emission. The emitted particle density would then mimic the vorticity that is created by the flow of a 2-d fluid past the wire. That's why I call this system a plasma wind tunnel.

Surko: Here's a problem that I've always felt that someone smart enough in the plasma community could make a great contribution to. You've heard all the litanies about trouble slowing down fast positrons from radioactive sources in order to trap them - the efficiency we can get is rather low, at best in the half a percent range. If there were someone smart enough to be able to take a small radioactive source emitting positrons with a broad spectrum of energies from a couple of hundred keV and ranging up to 500 keV, and efficiently trap them, then it would make a great impact on many people's research.

David Moore: I'm a lonely researcher at Princeton. I wanted to make a few comments on a couple of things. First, about the need for an introduction to the field. As I have tried to begin research in this field from a sort of satellite position, I found that having the UCSD reprints is very useful. But I did invest a lot of time going over those, and at times I have run into situations in which I have missed things by not being tightly knit into this community. So I think if someone generates a good new review or a simple introduction to the field, it should be stored next to Joel's bibliography, so that whenever you FTP it you can get that at the same time. I wanted to mention something else. We're forgetting that we're using the name Penning here, and there's a long history of crossed field discharges like our Russian colleague is working on, and there are a lot of very old results and very old questions that have been unanswered for a long time. Most of these questions involve adding gas into your systems, which very few of you do, but that is why I started doing my experiments.

Here's another question. There's been this idea that you can make a more quiescent plasma by adding an azimuthally symmetric voltage perturbation to one of your cylinders, and that the mode growth of a lot of modes seems to be less strong under such a perturbation. I didn't hear anything about this effect here, and I didn't see it much about it in the literature when I was first looking, but it seems very interesting.

Fajans: If you've got a multi-segment trap, and you put a few volts bias on one of the cylinders so that the line density is a little bit different in that area than in the rest, it seems that you get less growth of modes. Occasionally Bret Beck and I, working on the cryogenic machine down in San Diego, noticed that if we did bias one of the cylinders upwards or downwards the confinement properties changed dramatically. I think Kevin Fine should address this because he systematically investigated tilts of the magnetic field and also changing gate potentials.

Kevin Fine: We did do a series of experiments where we induced am $m = 1$ diocotron mode and then looked at the amplitude as a function of time after you increase the voltage on a cylinder for a while, and while the voltage was on the mode was damped. At the same time, there was strong cross-field particle transport. I don't think that the mechanism for either the damping or the transport is understood.

Driscoll: I think that what you're talking about is a transport process, the understanding of which is certainly lacking, not only in these plasmas but in all plasmas. However, we are beginning to make some headway in understanding transport in one example. The induced damping of the $m = 1$ diocotron mode has been thoroughly studied experimentally on the CV apparatus by Brian Cluggish and is pretty well understood theoretically by Crooks and O'Neil. We take that as significant because it's the first transport process that we actually have a really good handle on, and the hope is that it can be generalized to the transport induced by asymmetries, including the addition of cylindrically symmetric voltages used to squeeze part of the plasma.

Closing remarks by Joel Fajans: I would say from an experimental point of view that the really big change in the last 6 years in this field has been the improvement in the diagnostics. I think that particularly the laser diagnostics for ion plasmas have produced some really spectacular results, but even in the electron plasmas the diagnostics are a lot better - the use of fluorescent screens in order to observe the plasma density being a good example.

A final and very cheerful note: one member of this community has recently been recognized—Roy Gould has won this year's Maxwell Prize. APPLAUSE So, thank you all for coming, and I hope you had a great time.

Single Component Plasma Bibliography

Compiled by J. Fajans

Papers in the bibliography are categorized as follows:

A: Equilibrium and stability.
B: Transport and Kinetic Effects.
C: Waves.
D: 2-d fluid effects.
E: Cyclotron motion effects.
F: High density plasmas and energetic plasmas.
G: Correlations and microplasmas.
H: Ion plasmas.
I: Antimatter plasmas.
J: Non-neutral plasmas and atomic physics.
K: Exotic traps or particles.
L: Numerical techniques.
M: Applications.
N: Other.
O: Experimental
P: Theoretical

Entries were submitted by the workshop participants and are limited to published papers and thesis. This bibliography can be obtained by emailing a request to fajans@physics.berkeley.edu. The bibliography is available in LaTeX format and as a BiBTeX database.

Cross-Reference

A: Equilibrium and stability:
avin:91, bhat:92, bhat:92a, bhat:92b, boge:70, boll:93, boll:94a, brew:88, brow:86, chan:90, chen:90, chen:90a, chen:91a, chen:94a, chen:94b, chu:93, chu:93a, clar:76, daug:67, daug:69, davi:69, davi:70, davi:71, davi:72, davi:72a, davi:73, davi:74, davi:75, davi:77, davi:79, davi:82, davi:84, davi:84a, davi:84b, davi:84c, davi:85, davi:85a, davi:85b, davi:86, davi:86a, davi:87, davi:88, davi:88a, davi:90, davi:90a, davi:91, davi:93, davi:94, davi:94a, degr:77, dris:76, dris:85, dris:86b, dris:86c, dris:89, dris:92, dris:94, dubi:86, dubi:86b, dubi:88, dubi:90, dubi:91, dubi:92, dubi:93a, dubi:93b, dubi:94, fang93a, fine:89, gabr:92, glis:94, hein:91, holl:93, huan:93, itan:82, itan:88, iwat:94, kerv:86, kerv:89a, kerv:91, kerv:94, khir:93, kwon:83, kwon:92, lars:86, lund:93, lund:93a, malm:82, malm:84, malm:92, mitc:93, miti:93a, miti:93b, mora:88, murp:92, nott:93b, onei:80b, onei:80c, onei:81, onei:88, onei:94, onei:94a, pass:89, peur:90, peur:92d, pras:79, pras:81, pras:86, pras:87a, pras:88, pras:89,

raiz:921, robe:88, smit:89, smit:90, smit:90a, smit:91, smit:92, smit:92a, smit:92b, spen:92, spen:93a, tan:94, tots:81a, tots:88, tots:88a, tots:93, turn:87, turn:90, turn:91, turn:92, turn:93, turn:93a, turn:93b, turn:94, turn:94a, turn:94b, uhm:78, uhm:80, uhm:80a, uhm:82, uhm:83, weim:94, zave:92

B: Transport and Kinetic Effects:

avin:92, avin:92a, beck:90, beck:92, boll:91, brig:70, brow:86, chan:90, chen:90, chen:90a, chen:91a, chen:93, chen:93a, chen:94, chen:94a, chen:94b, corn:93, craw:85, craw:87, croo:86, davi:69, davi:70, davi:71, davi:72, davi:72a, davi:73, davi:74, davi:75, davi:77, davi:79, davi:82, davi:85, davi:85a, davi:85b, davi:86, davi:86a, davi:88, davi:90, davi:90a, davi:91, davi:93, davi:94, davi:94a, degr:77, degr:77a, degr:80, doug:78, dris:82, dris:83, dris:85, dris:86a, dris:86b, dris:86c, dris:92, dubi:86a, dubi:88a, dubi:94, eggl:87a, faja:93, gabr:92, glin:91, glin:91a, glin:92, hart:91, hjor:86, hjor:87, hjor:88, holl:93, huan:93, huan:94, hyat:87, hyat:88, kape:73, kein:84, kerv:89a, kriv:93, lamb:83, levy:69, malm:82, malm:92, miti:94, mood:92, murp:90, onei:80a, onei:88, onei:90, onei:94, onel:85a, peti:87, peur:90, peur:92d, peur:93e, rama:93, rasb:93, robe:88, spen:90, spen:92, spen:93a, spen:93b, tots:80, turn:93a, turn:93b, uhm:78, uhm:80, uhm:80a, uhm:82, uhm:83, wine:85a

C: Waves:

boll:92, boll:93, boll:94a, boll:94b, brig:70, chan:90, chen:90, chen:90a, chen:91a, craw:85, craw:86, craw:87, croo:94, davi:69, davi:70, davi:73, davi:74, davi:77, davi:82, davi:84, davi:84a, davi:84b, davi:84c, davi:85, davi:85a, davi:85b, davi:86, davi:86a, davi:87, davi:88, davi:88a, davi:90, davi:90a, davi:91, davi:93, degr:77, degr:77a, degr:80, dimo:81, dris:85, dris:86b, dris:90b, dris:92, dris:94, dubi:86a, dubi:91, dubi:91a, dubi:93, dubi:93a, eggl:87, eggl:87a, faja:93, fine:88, fine:89, goul:91, goul:92, grea:94b, hein:91, huan:93, kape:73, kein:81, kein:84, kerv:89b, lamb:83, levy:65, levy:69, malm:82, malm:92, mitc:93, miti:93b, miti:94, mood:92, mora:88, nott:93a, nott:93b, onei:80a, onei:94, peti:87, peur:92d, peur:93b, peur:93c, peur:93e, pill:94, prad:93, pras:81, pras:83, pras:84, pras:86, pras:87a, pras:88, pras:89, rama:93, robe:88, rose:87, rose:90, smit:90, smit:92a, smit:92b, spen:90, spen:93b, tink:94, tink:95, tots:80a, tots:81a, tots:81b, tots:92a, turn:94b, uhm:78, uhm:80, uhm:80a, uhm:82, uhm:83, weim:94, whit:82, wine:87

D: 2-d fluid effects:

brig:70, davi:74, davi:90, dris:90b, dris:92, dris:94, dubi:90, fine:89, hart:91, huan:94, kadt:94, kerv:89b, kerv:89c, kerv:91, malm:92, miti:93a, miti:94, onei:94, peur:92d, peur:93a, peur:93b, peur:93c, peur:93f, pill:94, pras:86, robe:88, rose:87, rose:90, smit:89, smit:90, smit:90a, smit:91, smit:92, smit:92a, smit:92b, spen:90, spen:92, spen:93b, tots:75, tots:76, tots:78

E: Cyclotron motion effects:

avin:92, beck:92, beu:91, boll:92, boll:93, boll:94a, boll:94b, chen:91, chen:93, chen:94a, davi:90, glin:92, gors:93, goul:91, goul:92, hans:89a, hans:89b, hein:91, hjor:86, hjor:87, hyat:87, jeff:83, kerv:85, kerv:89a, kerv:91, lauk:86, onei:90, peur:94a, pras:85, pras:87, robe:88, tots:93, xian:93

F: High density plasmas and energetic plasmas:

boll:92, boll:93, boll:94a, boll:94b, davi:74, davi:90, hein:91, mich:89, robe:88

G: Correlations and microplasmas:

alex:84, boll:84, boll:90, boll:94a, boll:94b, chen:93a, chen:94, davi:74, davi:90, dubi:86, dubi:86b, dubi:88, dubi:89, dubi:90, dubi:90a, dubi:91, dubi:92, dubi:93b, dubi:94, gilb:88, hang:91:, hang:94, hass:90, itan:82, itan:88, krau:94, lars:86, malm:84, niel:94, onei:80b, onei:88, onei:94, rafa:91, rahm:86a, rahm:88, raiz:92, raiz:921, rama:93, robe:88, schi:85, schi:86, schi:88, schi:88a, schi:89, schi:91, schi:93, schi:93a, schi:93b, schi:94, schi:94a, schu:88, tots:75, tots:76, tots:78, tots:79, tots:80, tots:80a, tots:81, tots:81a, tots:81b, tots:82, tots:84, tots:84a, tots:86, tots:87, tots:88, tots:88a, tots:92, tots:92a, whit:82, wine:87

H: Ion plasmas:
barl:86, barn:93, boll:84, boll:85, boll:90, boll:91, boll:92, boll:93, boll:94a, boll:94b, brew:88, brow:86, daug:68, davi:74, davi:90, dimo:81, dris:85, dris:86b, dubi:88, dubi:90, dubi:92, dubi:93b, gilb:88, hang:91:, hang:94, hass:90, hein:91, itan:88, kerv:85, kerv:89a, kerv:91, krau:94, lars:86, mich:89, mora:88, niel:94, pras:87a, pras:88, pras:89, rafa:91, rahm:86a, rahm:88, raiz:92, raiz:921, robe:88, rose:87, rose:90, schi:85, schi:86, schi:88, schi:88a, schi:89, schi:91, schi:93, schi:93a, schi:93b, schi:94, schi:94a, schu:88, tan:94, tots:84a, tots:88, tots:88a, tots:92a, tots:93, whit:82, wine:85a, wine:87, wine:90

I: Antimatter plasmas:
cowa:91, cowa:93, davi:74, davi:90, fang93a, gabr:92, glis:94, grea:94a, hang:91:, hang:94, hass:90, iwat:93, iwat:94, krau:94, kwon:83, murp:90, murp:91, murp:92, niel:94, onei:80b, rafa:91, rahm:86a, rahm:88, robe:88, schi:85, schi:86, schi:88, schi:88a, schi:89, schi:91, schi:93, schi:93a, schi:93b, schi:94, schi:94a, surk:86a, surk:86b, surk:86c, surk:88, surk:89, surk:90, surk:93, tang:92, tink:94, turn:87, wine:93, wyso:88

J: Non-neutral plasmas and atomic physics:
barn:93, boll:84, boll:85, boll:90, boll:91, boll:93, boll:94a, boll:94b, brew:88, davi:74, davi:90, fang93a, fang93b, fang94, gilb:88, glis:94, itan:82, itan:88, iwat:93, iwat:94, kerv:85, kwon:83, kwon:89a, kwon:89b, kwon:90, kwon:92, kwon:93a, kwon:93b, lars:86, murp:90, murp:91, pass:89, raiz:92, raiz:921, robe:88, schu:88, surk:88, surk:90, surk:93, tan:94, tang:92, weim:94, whit:82, wine:85, wine:85a, wine:87, wine:90

K: Exotic traps or particles:
avin:91, avin:92, barl:86, bhat:92a, bhat:92b, brow:86, chen:93a, chen:94, clar:76, daug:67, daug:68, daug:69, davi:74, davi:90, gabr:92, glin:91, gors:92, hang:91:, hang:94, hass:90, khir:93, krau:94, mich:89, mora:88, niel:94, onei:94a, prad:93, pras:87a, pras:88, pras:89, rafa:91, rahm:86a, rahm:88, robe:88, schi:85, schi:86, schi:88, schi:88a, schi:89, schi:91, schi:93, schi:93a, schi:93b, schi:94, schi:94a, schu:88, tots:93, turn:90, turn:91, turn:92, turn:93, turn:93a, turn:93b, turn:94, turn:94b, wang:89, yin:92, zave:92

L: Numerical techniques:
davi:74, davi:90, fang93b, robe:88, spen:93a, spen:93b, tots:84a, tots:92

M: Applications:
chan:90, chen:91a, daug:68, davi:72, davi:74, davi:82, davi:84a, davi:84b, davi:85a, davi:85b, davi:88, davi:88a, davi:90, davi:93, glin:91, peti:87, robe:88, schu:88, turn:92, turn:93a, turn:93b, turn:94b, uhm:80, uhm:82, uhm:83

N: Other:
cowa:91, cowa:93, davi:74, davi:90, eggl:92, mich:89, robe:88

O: Experimental:
beck:90, beck:92, brow:86, chu:93, clar:76, cowa:91, cowa:93, daug:67, daug:68, daug:69, davi:74, davi:90, degr:77, degr:77a, degr:80, dimo:81, dris:83, dris:85,

dris:86a, dris:86b, dris:88, dris:90, dris:90a, dris:90b, dris:92, dris:94, dubi:86a,
eggl:84, eggl:87, eggl:87a, eggl:92, fine:88, fine:89, fine:91, gabr:92, glis:94, goul:91,
goul:92, grea:94a, grea:94b, huan:93, huan:94, hyat:87, hyat:88, iwat:93, iwat:94,
kadt:94, malm:75, malm:80, malm:82, malm:84, malm:88, malm:92, mitc:93, miti:93a,
miti:93b, miti:94, mood:92, murp:90, murp:91, murp:92, nott:92, nott:93a, nott:93b,
nott:94, onei:90, onei:94, pass:89, peur:92d, peur:93a, peur:93b, peur:93c, peur:93e,
peur:93f, pill:94, robe:88, rose:87, rose:90, schu:88, surk:86a, surk:86b, surk:86c,
surk:88, surk:89, surk:90, surk:93, tang:92, tink:95, whit:82, wyso:88

P: Theoretical.
avin:92a, barn:93, bhat:92b, boge:70, boll:92, brow:86, chen:93, chen:93a, chen:94,
chu:93, chu:93a, corn:93, craw:85, craw:86, craw:87, croo:94, davi:69, davi:70, davi:71,
davi:73, davi:74, davi:90, doug:78, dris:76, dris:82, dris:86c, dris:89, dubi:86, dubi:86a,
dubi:86b, dubi:88, dubi:88a, dubi:89, dubi:90a, dubi:91, dubi:91a, dubi:92, dubi:93,
dubi:93a, dubi:93b, dubi:94, faja:93, fine:92, gabr:92, glin:91, glin:91a, glin:92,
hart:91, hjor:86, hjor:87, hjor:88, itan:82, kein:84, levy:65, levy:69, malm:77, nott:92,
nott:93a, onei:79, onei:80, onei:80a, onei:80b, onei:80c, onei:81, onei:83, onei:85,
onei:87, onei:88, onei:90, onei:92, onei:94, onei:94a, onel:85a, peur:90, peur:93f,
pras:79, pras:81, pras:83, pras:84, pras:86, rahm:86a, raiz:921, schu:88, smit:89,
smit:90, smit:90a, smit:91, smit:92, smit:92a, smit:92b, tan:94, tots:75, tots:76,
tots:78, tots:79, tots:80, tots:80a, tots:81, tots:81a, tots:81b, tots:82, tots:84, tots:84a,
tots:86, tots:87, tots:88, tots:88a, tots:92, tots:92a, tots:93, turn:87, turn:90, turn:91,
turn:92, turn:93a, turn:93b, turn:94, turn:94a, turn:94b, weim:94

References

[1] **alex:84**   S. Alexander, P. M. Chaikin, P. Grant, G. J. Morales, P. Pincus, and D. Hone, Charge renormalization, osmotic pressure, and bulk modulus of colloidal crystals: Theory *J. Chem. Phys.* **80**, 5776 (1984), Cat G.

[2] **avin:91**   K. Avinash, On toroidal equilibrium of non-neutral plasma *Phys. Fluids B* **3**, 3226 (1991), Cat A, K.

[3] **avin:92**   K. Avinash and S. N. Bhattacharyya, Brillouin limit in cylinders and torus with shaped cross sections *Phys. Fluids B* **4**, 3863 (1992), Cat B, E, K.

[4] **avin:92a**   K. Avinash, The evolution of slightly nonideal, non-neutral plasma *Phys. Fluids B* **4**, 2658 (1992), Cat B, P.

[5] **barl:86**   S. E. Barlow, J. A. Luine, and G. H. Dunn, Measurement of ion/molecule reactions between 10 and 20 k *Int. J. Mass Spectrom. Ion Processes* **74**, 97 (1986), Cat H.

[6] **barn:93**   Paul N. Barnes and Grant W. Hart, Precision spectroscopy using the lamb dip in a pure ion plasma *Rev. Sci. Instrum.* **64**, 579 (1993), Cat H, J, P.

[7] **beck:90**   B. R. Beck, *Measurement of the Magnetic and Temperature Dependence of the Electron-Electron Anisotropic Temperature Relaxation Rate*, PhD thesis, University of California, San Diego (1993), Cat B, O

[8] **beck:92**   B. R. Beck, J. Fajans, and J. H. Malmberg, Measurement of collisional anisotropic temperature relaxation in a strongly magnetized pure electron plasma *Phys. Rev. Lett.* **68**, 317 (1992), Cat B, E, O.

[9] **beu:91**   S. C. Beu and D. A. Laude, Radial ion transport dur to resistive-wall destabilization in fourier transform mass spectrometery *Int. J. Mass Spectrom. Ion Processes* **108**, 255 (1991), Cat E.

[10] **bhat:92**   S. N. Bhattacharyya and K. Avinash, Equilibrium of non-neutral plasma in toroidal geometry with applied electric field *Phys. Fluids B* **4**, 1702 (1992), Cat A, K.

[11] **bhat:92a**   S. N. Bhattacharyya and K. Avinash, Stability of a toroidal non-neutral plasma with elongated cross-section to rigid displacements *Physics Letters A* **171**, 367 (1992), Cat A, K.

[12] **bhat:92b**   S. N. Bhattacharyya and K. Avinash, Toroidal equilibrium of a non-neutral plasma with toroidal current, inertia and pressure *Journal of Plasma Physics* **47**, 349 (1992), Cat A, K, P.

[13] **boge:70**   Jr. B. L. Bogema and R. C. Davidson, Rotor equilibria of non-neutral plasmas *Phys. Fluids* **13**, 2772 (1970), Cat A, P.

[14] **boll:84**   J. J. Bollinger and D. J. Wineland, Strongly coupled non-neutral ion plasma *Phys. Rev. Lett.* **53**, 348–351 (1984), Cat G, H, J.

[15] **boll:85**   J. J. Bollinger, J. D. Prestage, W. M. Itano, and D. J. Wineland, Laser cooled atomic frequency standard *Phys. Rev. Lett.* **4**, 1000–1003 (1985), Cat H, J.

[16] **boll:90**   J. J. Bollinger and D. J. Wineland, Microplasmas *Sci. Am.* **262**(1), 124–130 (1990), Cat G, H, J.

[17] **boll:91**   J. J. Bollinger, D. J. Heinzen, W. M. Itano, S. L. Gilbert, and D. J. Wineland, A 303-MHz frequency standard based on trapped Be^+ ions *IEEE Trans. Instrum. Meas.* **40**, 126–128 (1991), Cat H, J.

[18] **boll:92**   J. J. Bollinger, D. J. Heinzen, F. L. Moore, W. M. Itano, and D. J. Wineland, Low order modes of an ion cloud in a Penning trap *Physica Scripta* **46**, 282–284 (1992), Cat C, E, F, H.

[19] **boll:93**   J. J. Bollinger, D. J. Heinzen, F. L. Moore, W. M. Itano, D. J. Wineland, and D. H. E. Dubin, Electrostatic modes of an ion trap plasma *Phys. Rev. A* **48**, 525–545 (1993), Cat A, C, E, F, H, J.

[20] **boll:94a**   J. J. Bollinger, D. J. Wineland, and D. H. E. Dubin, Non-neutral ion plasmas and crystals, laser cooling, and atomic clocks *Phys. Plasmas* **1**, 1403–1414 (1994), Cat A, C, E, F, G, H, J.

[21] **boll:94b**   J. J. Bollinger, J. N. Tan, W. M. Itano, D. J. Wineland, and D. H. E. Dubin, Non-neutral ion plasmas and crystals in Penning traps *Physica Scripta (to be published)* (1994), Cat C, E, F, G, H, J.

[22] **brew:88**   L. R. Brewer, J. D. Prestage, J. J. Bollinger, W. M.Itano, D. J. Larson, and D. J. Wineland, Static properties of a non-neutral Be^+ ion plasma *Phys. Rev. A* **38**, 859–873 (1988), Cat A, H, J.

[23] **brig:70**   R. J. Briggs, J. D. Daugherty, and R. H. Levy, Role of landau damping in crossed-field electron beams and inviscid shear flow *Phys. Fluids* **13**, 421 (1970), Cat B, C, D.

[24] **brow:86**   L. S. Brown and G. Gabrielse, Geonium theory: physics of a single electron or ion in a penning trap *Rev. Modern Phys.* **58**, 233 (1986), Cat A, B, H, K, O, P.

[25] **chan:90**   H.-W. Chan, C. Chen, and R.C. Davidson, Computer simulation of relativistic multiresonator cylindrical magnetrons *Appl. Phys Lett* **57**, 1271 (1990), Cat A, B,,C, M.

[26] **chen:90**   C. Chen and R.C. Davidson, Chaotic electron dynamics for electron beam propagation through a planar-wiggler magnetic field *Phys. Rev. A* **A42**, 5041 (1990), Cat A, B, C.

[27] **chen:90a**   C. Chen and R.C. Davidson, Self-field induced chaoticity in the electron orbits in a helical wiggler free electron laser with axial guide field *Phys. Fluids B* **B2**, 171 (1990), Cat A, B, C.

[28] **chen:91**   S. P. Chen and M. B. Comisarow, Simple physical models for coulomb-induced frequency shifts and coulomb-induced inhomogeneous broadening for like and unlike ions in fourier transform ion cyclotron resonance mass spectrometry *Rapid Commun. Mass Spectrom.* **5**, 450 (1991), Cat E.

[29] **chen:91a**   C. Chen and R.C. Davidson, Chaotic particle dynamics in free electron lasers *Phys. Rev. A* **A43**, 5541 (1991), Cat A, B, C, M.

[30] **chen:93**   S. J. Chen and D. H. E. Dubin, Equilibration rate of spin temperature in a strongly magnetized pure electron plasma *Phys. Fluids B* **5**, 691 (1993), Cat B, E, P.

[31] **chen:93a**   S.-J. Chen and D. H. E. Dubin, Temperature equilibration of a 1d coulomb chain and a many-particle adiabatic invariant *Phys. Rev. Lett.* **71**, 2721 (1993), Cat B, G, K, P.

[32] **chen:94**   Chen S.-J., *Temperature Equilibration and Many-Body Adiabatic Invariants*, PhD thesis, University of California, San Diego (1994), Cat B, G, K, P.

[33] **chen:94a**   C. Chen and R.C. Davidson, Properties of the kapchinskij-vladimirskij equilibrium and envelope equation for an intense charged particle beam in a periodic focussing field *Phys. Rev. E* **E49**, 5679 (1994), Cat A, B.

[34] **chen:94b**   C. Chen and R.C. Davidson, Nonlinear resonances and chaotic behavior in a periodically focused intense charged particle beam *Phys. Rev. Lett.* **72**, 2195 (1994), Cat A, B.

[35] **chu:93**   R. Chu, J. S. Wurtele, J. Notte, A. J. Peurrung, and J. Fajans, Pure electron plasmas in asymmetric traps *Phys. Fluids B* **5**, 2378 (1993), Cat A, O, P.

[36] **chu:93a**   R. Chu, *Theoretical Studies of Pure Electron Plasmas in Asymmetric Traps*, PhD thesis, Massachusetts Institute of Technology (1993), Cat A, P.

[37] **clar:76**   W. Clark, P. Korn, A. Mondelli, and N. Rostoker, Experiments on electron injection into a toroidal magnetic field *Phys. Rev. Lett.* **37**, 592 (1976), Cat A, K, O.

[38] **corn:93**   N. R. Corngold, Virial equation for the two-dimensional pure electron plasma *Phys. Fluids B* **5**, 3847 (1993), Cat B, P.

[39] **cowa:91**   T. E. Cowan, R. H. Howell, R. R. Rohatgi, and J. Fajans, Proposed search for resonant states in positron-electron scattering using a positron gas target *Nucl. Instrum. and Methods B* **56**, 599 (1991), Cat I, N, O.

[40] **cowa:93**   T. E. Cowan, B. R. Beck, J.H. Hartley, R. H. Howell, R. R. Rohatgi, J. Fajans, and R. Gopalan, Development of a pure cryogenic positron plasma using a linac positron source *Hyperfine Interactions* **172**, 1 (1993), Cat I, N, O.

[41] **craw:85**   J. D. Crawford, T. M. O'Neil, and J. H. Malmberg, Effect of nonlinear collective processes on the confinement of a pure-electron plasma *Phys. Rev. Lett.* **54**, 697 (1985), Cat B, C, P.

[42] **craw:86**   J. D. Crawford, S. Johnston, A. N. Kaufman, and C. Oberman, Theory of beat-resonant coupling of electrostatic modes *Phys. Fluids* **29**, 3219 (1986), Cat C, P.

[43] **craw:87**   J. D. Crawford and T. M. O'Neil, Nonlinear collective processes and the confinement of a pure-electron plasma *Phys. Fluids* **30**, 2076 (1987), Cat B, C, P.

[44] **croo:94**   S. Crooks and T.M. O'Neil, Rotational pumping and damping of the $m = 1$ diocotron mode *Phys. Plasmas* (1994), Cat B, C, P.

[45] **daug:67**   J. D. Daugherty, J. E. Eninger, and G. S. Janes, Equilibrium of electron clouds in toroidal magnetic fields *Phys. Fluids* **10**, 155 (1967), Cat A, K, O.

[46] **daug:68**   J. D. Daugherty, L. Grodzins, G. S. Janes, and R. H. Levy, New source of highly stripped ions *Phys. Rev. Lett.* **20**, 369 (1968), Cat H, K, M, O.

[47] **daug:69**   J. D. Daugherty, J. E. Eninger, and G. S. Janes, Experiments on the injection and containment of electron clouds in a toroidal apparatus *Phys. Fluids* **12**, 2677 (1969), Cat A, K, O.

[48] **davi:69**   R. C. Davidson and N. A. Krall, Vlasov description of an electron gas in a magnetic field *Phys. Rev. Lett.* **22**, 833 (1969), Cat A, B, C, P.

[49] **davi:70**   R. C. Davidson and N. A. Krall, Vlasov equilibria and stability of an electron gas *Phys. Fluids* **13**, 1543 (1970), Cat A, B, C, P.

[50] **davi:71**   R. C. Davidson, Electrostatic shielding of a test charge in a non-neutral plasma *J. Plasma Phys.* **6**, 229 (1971), Cat A, B, P.

[51] **davi:72**   R.C. Davidson and S.M. Mahajan, A relativistic electron ring equilibrium with thermal energy spread *Particle Accelerators* **4**, 53 (1972), Cat,A, B,M.

[52] **davi:72a**   R.C. Davidson and J.D. Lawson, Self-consistent vlasov description of relativistic electron ring equilibria *Particle Accelerators* **4**, 1 (1972), Cat,A, B.

[53] **davi:73**   R. C. Davidson, A. T. Drobot, and C. A. Kapetanakos, Equilibrium and stability of mirror-confined e-layers *Phys. Fluids* **16**, 2199 (1973), Cat A, B, C, P.

[54] **davi:74**   R. C. Davidson, *Theory of Non-Neutral Plasmas*, Benjamin, Reading, MA, (1974), Cat A, B, C, D, E, F, G, H, I, J, K, L, M, N, O, P.

[55] **davi:75**   R.C. Davidson and B.H. Hui, Influence of self fields on the equilibrium and stability properties of relativistic electron beam-plasma systems *Annals of Physics* **94**, 209 (1975), Cat A,B, C.

[56] **davi:77**   R.C. Davidson and H.S. Uhm, Influence of strong self-electric fields on the ion resonance instability in a nonneutral plasma column *Phys. Fluids* **20**, 1938 (1977), Cat A, B, C.

[57] **davi:79**   R.C. Davidson and H.S. Uhm, Thermal equilibrium properties of an intense relativistic electron beam *Phys. Fluids* **22**, 1375 (1979), Cat A, B.

[58] **davi:82**   R.C. Davidson and H.S. Uhm, Stability properties of an intense relativistic nonneutral electron ring in a modified betatron accelerator *Phys. Fluids* **25**, 2089 (1982), Cat A, B, C, M.

[59] **davi:84**   R.C. Davidson, Macroscopic guiding-center stability theorem for nonrelativistic nonneutral electron flow in a cylindrical diode with applied magnetic field *Phys. Fluids* **27**, 1804 (1984), Cat A, C.

[60] **davi:84a**   R.C. Davidson, K.T. Tsang, and J.A. Swegle,   Macroscopic extraordinary-mode stability properties of relativistic nonneutral electron flow in a planar diode with applied magnetic field *Phys. Fluids* **27**, 2332 (1984), Cat A, C, M.

[61] **davi:84b**   R.C. Davidson and K.T. Tsang, Macroscopic electrostatic stability properties of nonrelativistic nonneutral electron flow in a cylindrical diode with applied magnetic field *Phys. Rev. A* **A30**, 488 (1984), Cat A, C, M.

[62] **davi:84c**   R.C. Davidson and W.A. McMullin, Influence of intense equilibrium self fields on the spontaneous emission from a test electron in a relativistic nonneutral electron beam *Phys. Fluids* **27**, 1268 (1984), Cat A, C.

[63] **davi:85**   R.C. Davidson, Quasilinear theory of the diocotron instability for nonrelativistic nonneutral electron flow in planar geometry *Phys. Fluids* **28**, 1937 (1985), Cat A, B, C.

[64] **davi:85a**   R.C. Davidson, Nonlinear bound on unstable electrostatic fluctuation energy for nonrelativistic nonneutral electron flow in a planar diode with applied magnetic field *J. Plasma Phys.* **33**, 157 (1985), Cat A, B, C, M.

[65] **davi:85b**   R.C. Davidson, Kinetic stability theorem for relativistic nonneutral electron flow in a planar diode with applied magnetic field *Phys. Fluids* **28**, 377 (1985), Cat A, B, C, M.

[66] **davi:86**   R.C. Davidson and K.T. Tsang, Nonlinear bound on unstable fluctuation level in low-density nonneutral plasma *J. Plasma Phys.* **36**, 329 (1986), Cat,A, B, C.

[67] **davi:86a**   R.C. Davidson and H.S. Uhm, Kinetic description of betatron oscillation instability for nonrelativistic nonneutral electron flow *Phys. Fluids* **29**, 2273 (1986), Cat A, B, C.

[68] **davi:87**   R.C. Davidson, K.T. Tsang, and H.S. Uhm, Stabilization of diocotron instability by relativistic and electromagnetic effects for intense nonneutral electron flow *Phys. Lett. A* **A125**, 61 (1987), Cat A, C.

[69] **davi:88**   R.C. Davidson, Waves and instabilities in nonneutral plasmas, in C.W. Roberson and C.F. Driscoll, editors, *Non-Neutral Plasma Physics*, volume AIP 175 p. 139, New York (1988), American Institute of Physics, Cat A, B, C, M.

[70] **davi:88a**   R.C. Davidson and K.T. Tsang, Analysis of magnetron instability for relativistic nonneutral electron flow in cylindrical high-voltage diodes *Laser and Particle Beams* **6**, 661 (1988), Cat A, C, M.

[71] **davi:90**   R.C. Davidson, *Physics of Nonneutral Plasmas*, Frontiers in Physics Series, Addison Wesley, Reading, Massachusetts, (1990), Cat A, B, C, D, E, F, G, H, I, J, K, L, M, N, O, P.

[72] **davi:90a**   R.C. Davidson and C. Chen, Self-field-induced chaotic motion in free electron lasers *Nucl. Instrum. and Methods A* **A296**, 471 (1990), Cat A, B, C.

[73] **davi:91**   R.C. Davidson, H.-W. Chan, C. Chen, and S. Lund, Equilibrium and stability properties of intense nonneutral electron flow *Rev. Modern Phys.* **63**, 341 (1991), Cat A, B, C.

[74] **davi:93**   R.C. Davidson, H.-W.Chan, C. Chen, and S.M. Lund, Numerical study of relativistic magnetrons *J. Appl. Phys.* **73**, 7053 (1993), Cat A, B, C, M.

[75] **davi:94**   R.C. Davidson and Q. Qian, Phase advance for an intense charged particle beam propagating through a periodic quadrupole focussing field in the smooth-beam approximation *Phys. Plasmas* **1**, 3104 (1994), Cat A, B.

[76] **davi:94a**   R.C. Davidson and S.M. Lund, Thermal equilibrium properties of non-neutral plasma in the weak coupling approximation, in *Advances in Plasma Physics. Thomas H. Stix Symposium*, volume AIP 314 p. 1, New York (1994), American Institute of Physics, Cat A, B.

[77] **degr:77**   deGrassie J.S., *Equilibrium, Waves and Transport in the Pure Electron Plasma*, PhD thesis, University of California, San Diego (1977), Cat A, B, C, O.

[78] **degr:77a**   J. S. deGrassie and J. H. Malmberg, Wave-induced transport in the pure electron plasma *Phys. Rev. Lett.* **39**, 1077 (1977), Cat B, C, O.

[79] **degr:80**   J. S. DeGrassie and J. H. Malmberg, Waves and transport in the pure electron plasma *Phys. Fluids* **23**, 63 (1980), Cat B, C, O.

[80] **dimo:81**   Guy Dimonte, Ion langmuir waves in a non-neutral plasma *Phys. Rev. Lett.* **46**, 26 (1981), Cat C, H, O.

[81] **doug:78**   M. H. Douglas and T. M. O'Neil, Transport of a non-neutral electron plasma due to electron collisions with neutral atoms *Phys. Fluids* **21**, 920 (1978), Cat B, P.

[82] **dris:76**   C. F. Driscoll and J. H. Malmberg, Hollow electron column from an equipotential cathode *Phys. Fluids* **19**, 760 (1976), Cat A, P.

[83] **dris:82**   C. F. Driscoll, Wall losses for a single species plasma near thermal equilibrium *Phys. Fluids* **25**, 97 (1982), Cat B, P.

[84] **dris:83**   C. F. Driscoll and J. H. Malmberg, Length-dependent containment of a pure electron plasma *Phys. Rev. Lett.* **50**, 167 (1983), Cat B, O.

[85] **dris:85**   C. F. Driscoll, Pure electron plasma experiments, in *Proc. of 3rd Workshop on EBIS Sources and Their Applications*, (1985), Cat A, B, C, H, O.

[86] **dris:86a**   C. F. Driscoll, K. S. Fine, and J. H. Malmberg, Reduction of radial losses in a pure electron plasma *Phys. Fluids* **29**, 2015 (1986), Cat B, O.

[87] **dris:86b**   C. F. Driscoll, Containment of single-species plasmas at low energies, in D. Cline, editor, *Low Energy Anti-Matter* p. 184, World Scientific, (1986), Cat A, B, C, H, O.

[88] **dris:86c**   C. F. Driscoll and T. M. O'Neil, Equilibrium of totally unneutralized plasmas *Physics Today* **39**, S–62 (1986), Cat A, B, P.

[89] **dris:88**   C. F. Driscoll, J. H. Malmberg, and K. S. Fine, Observation of transport to thermal equilibrium in pure electron plasmas *Phys. Rev. Lett.* **60**, 1290 (1988), Cat B O.

[90] **dris:89**   C. F. Driscoll, J. H. Malmberg, K. S. Fine, R. A. Smith, X.-P. Huang, and R. W. Gould, Growth and decay of turbulent vortex structures in pure electron plasmas, in *Plasma Physics and Controlled Nuclear Fusion Research 1988*, volume 3 pp. 507–514, IAEA, Vienna (1989), Cat A C D O P.

[91] **dris:90**   C. F. Driscoll and K. S. Fine, Experiments on vortex dynamics in pure electron plasmas *Phys. Fluids B* **2**, 1359 (1990), Cat A B C D O.

[92] **dris:90a**   C. F. Driscoll, Observation of an unstable $l = 1$ diocotron mode on a hollow electron column *Phys. Rev. Lett.* **64**, 645 (1990), Cat A B C D O.

[93] **dris:90b**   X.-P. Huang C. F. Driscoll, R. A. Smith and J. H. Malmberg, Growth and decay of vortex structures in pure electron plasmas, in *Structures in Confined Plasmas - Proc. of Workshop of U.S.-Japan Joint Institute for Fusion Theory Program* p. 69, Nagoya (1990), National Institute for Fusion Science, Cat A, C, D, O.

[94] **dris:92**   C. F. Driscoll, Wave and vortex dynamics in pure electron plasmas, in V. Stefan, editor, *Research Trends in Physics: Nonlinear and Relativistic Effects in Plasmas* p. 454, New York (1992), American Institute of Physics, Cat A, B, C, D, O.

[95] **dris:94**   F. Driscoll C., K. S. Fine, X.-P. Huang, T. B. Mitchell, A. C. Cass, and T. M. O'Neil, Vortices, holes, and turbulent relaxation in sheared electron columns, in *Proc. of 1994 IAEA Intl. Conf. on Plasma Physics and Controlled Nuclear Fusion*, (1994), Cat A, C, D, O.

[96] **dubi:86**   D. H. E. Dubin and T. M. O'Neil, Adiabatic expansion of a strongly correlated pure electron plasma *Phys. Rev. Lett.* **56**, 728 (1986), Cat A, G, P.

[97] **dubi:86a**   D. H. E. Dubin, T. M. O'Neil, and C. F. Driscoll, Transport toward thermal equilibrium in a pure electron plasma, in *Proc. of Workshop of US-Japan Joint Institute for Fusion Theory Program* pp. 265–279, (1986), Cat B, C, O, P.

[98] **dubi:86b**   D. H. E. Dubin and T. M. O'Neil, Thermal equilibrium of a cryogenic magnetized pure electron plasma *Phys. Fluids* **29**, 11 (1986), Cat A, G, P.

[99] **dubi:88**   D. H. E. Dubin and T. M. O'Neil, Computer simulation of ion clouds in a penning trap *Phys. Rev. Lett.* **60**, 511 (1988), Cat A, G, H, P.

[100] **dubi:88a**   D. H. E. Dubin and T. M. O'Neil, Two-dimensional guiding-center transport of a pure electron plasma *Phys. Rev. Lett.* **60**, 1286 (1988), Cat B, P.

[101] **dubi:89**   D. H. E. Dubin, Correlation energies of simple bounded coulomb lattices *Phys. Rev. A* **40**, 1140 (1989), Cat G, P.

[102] **dubi:90**   D. H. E. Dubin and T. M. O'Neil, Theory of strongly-correlated pure ion plasma in penning traps, in S. Ichimaru, editor, *Strongly Coupled Plasma Physics* p. 189, Elsevier Science Pub. B.V./Yamada Science Foundation, (1990), Cat A, G, H, D.

[103] **dubi:90a**   D. H. E. Dubin, First-order anharmonic correction to the free energy of a coulomb crystal in periodic boundary conditions *Phys. Rev. A* **42**, 4972 (1990), Cat G, P.

[104] **dubi:91**   D. H. E. Dubin and T. M. O'Neil, Pure ion plasmas, liquids and crystals, in W. Rozmus and J.A. Tuszynski, editors, *Nonlinear and Chaotic Phenomena in Plasmas, Solids and Fluids* p. 211, Singapore (1991), World Scientific, Cat A, C, G, P.

[105] **dubi:91a**   D. H. E. Dubin, Theory of electrostatic fluid modes in a cold spheroidal non-neutral plasma *Phys. Rev. Lett.* **66**, 2076 (1991), Cat C, P.

[106] **dubi:92**   D. H. E. Dubin, Pure ion plasmas, liquids and crystals, in V. Stefan, editor, *Research Trends in Physics: Nonlinear and Relativistic Effects in Plasmas* p. 460, New York (1992), American Institute of Physics, Cat A, G, H, P.

[107] **dubi:93**   D. H. E. Dubin, Normal modes in a cryogenic pure ion plasma, in H.M. Van Horn and S. Ichimaru, editors, *Strongly Coupled Plasma Physics* p. 399, Univ. of Rochester Press, (1993), Cat C, P.

[108] **dubi:93a**   D. H. E. Dubin, Equilibrium and dynamics of uniform density ellipsoidal non-neutral plasmas *Phys. Fluids B* **5**, 295 (1993), Cat A, C, P.

[109] **dubi:93b**   D. H. E. Dubin, Theory of structural phase transitions in a trapped coulomb crystal *Phys. Rev. Lett.* **71**, 2753 (1993), Cat A, G, H, P.

[110] **dubi:94**   D. H. E. Dubin and H. Dewitt, Polymorphic phase transition for inverse-power-potential crystals keeping the first-order anharmonic correction to the free energy *Phys. Rev. B* **49**, 3043 (1994), Cat A, G, P.

[111] **eggl:84**   D. L. Eggleston, T. M. O'Neil, and J. H. Malmberg, Collective enhancement of radial transport in a non-neutral plasma *Phys. Rev. Lett.* **53**, 982 (1984), Cat B C O.

[112] **eggl:87**   D. L. Eggleston, J. D. Crawford, T. M. O'Neil, and J. H. Malmberg, Enhancement of radial transport by collective processes, in *Plasma Physics and Controlled Nuclear Fusion Research 1986* pp. 337–342, (1987), Cat B, C, O.

[113] **eggl:87a**   D. L. Eggleston and J. H. Malmberg, Observation of an induced-scattering instability driven by static field asymmetries in a pure electron plasma *Phys. Rev. Lett.* **59**, 1675 (1987), Cat B, C, O.

[114] **eggl:92**   D. L. Eggleston, C. F. Driscoll, B. R. Beck, A. W. Hyatt, and T. H. Malmberg, Parallel energy analyzer for pure electron plasma devices *Phys. Fluids B* **4**, 3432 (1992), Cat N, O.

[115] **faja:93**   J. Fajans, Transient ion resonance instability *Phys. Fluids B* **5**, 3127 (1993), Cat B, C, P.

[116] **fang93a**   Z. Fang, Victor H. S. Kwong, Jiebing Wang, and W. H. Parkinson, Measurements of radiative-decay rates of the $2s^22p(^2P^o)$-$2s2p^2(^4P)$ intersystem transitions of C II *Phys. Rev. A* **48**(2), 1114 (1993), Cat J.

[117] **fang93b**   Z. Fang, Victor H. S. Kwong, and W. H. Parkinson, Radiative lifetimes of the $2s2p^2$ (^4P) metastable levels of N III *Astrophys. J.* **413**(2), L141 (1993), Cat J.

[118] **fang94**   Z. Fang and Vicor H. S. Kwong, Production and storage of $2p^2$ ^3P ground state O^{2+} ions from iron oxide targets *Rev. Sci. Instrum.* **65**(6), 2143 (1994), Cat J.

[119] **fine:88**   K. S. Fine, *Experiments with the l=1 Diocotron Mode*, PhD thesis, University of California, San Diego (1988), Cat A, C, O.

[120] **fine:89**   K. S. Fine, C. F. Driscoll, and J. H. Malmberg, Measurements of a non-linear diocotron mode in pure electron plasmas *Phys. Rev. Lett.* **63**, 2232 (1989), Cat A, C, D, O.

[121] **fine:91**   K. S. Fine, C. F. Driscoll, J. H. Malmberg, and T. B. Mitchell, Measurements of symmetric vortex merger *Phys. Rev. Lett.* **91**, 588 (1991), Cat D O.

[122] **fine:92**   K. S. Fine, Simple theory of a nonlinear diocotron mode *Phys. Fluids B* **4**, 3981 (1992), Cat C D P.

[123] **gabr:92**   G. Gabrielse, Extremely cold antiprotons *Sci. Amer.* **267**, 78 (1992), Cat A, B, I, K, O, P.

[124] **gilb:88**   S. L. Gilbert, J. J. Bollinger, and D. J. Wineland, Shell-structure phase of magnetically confined strongly coupled plasmas *Phys. Rev. Lett.* **60**, 2022–2025 (1988), Cat G, H, J.

[125] **glin:91**   M. E. Glinsky and T. M. O'Neil, Guiding center atoms: Three-body recombination in a strongly magnetized plasma *Phys. Fluids B* **3**, 1279 (1991), Cat B, K, M, P.

[126] **glin:91a**   E. Glinsky M., *Temperature Equilibration and Three-body Recombination in Strongly Magnetized Pure Electron Plasmas*, PhD thesis, University of California, San Diego (1991), Cat B, P.

[127] **glin:92**   M. E. Glinsky, T. M. O'Neil, M. N. Rosenbluth, K. Tsuruta, and S. Ichimaru, Collisional equipartition rate for a magnetized pure electron plasma *Phys. Fluids B* **4**, 1156 (1992), Cat B, E, P.

[128] **glis:94**   G. L. Glish, R. G. Greaves, S. A. McLuckey, L. D. Hulett, C. M. Surko, J. Xu, and D. L. Donohue, Ion production by positron-molecule resonances *Physical Review A* **49**, 2389–93 (1994), Cat I, J, O.

[129] **gors:92**   M. V. Gorshkov, S. Guan, and A. G. Marshall, Dynamic ion trapping for fourier-transform ion cyclotron resonance mass spectrometry: simultaneous positive- and negative-ion detection *Rapid Commun. Mass Spectrom.* **6**, 166 (1992), Cat K.

[130] **gors:93**   M. V. Gorshkov, A. G. Marshall, and E. N. Nikolaev, Analysis and elimination of systematic errors originating from coulomb mutual interaction and image charge in fourier transform ion cyclotron resonance precise mass difference measurements *J. Am. Soc. Mass Spectrom.* **4**, 855 (1993), Cat E.

[131] **goul:91**   R. W. Gould and M. A. LaPointe, Cyclotron resonance in a pure electron plasma *Phys. Rev. Lett.* **67**, 3685 (1991), Cat C, E, O.

[132] **goul:92**   R. W. Gould and M. A. LaPointe, Cyclotron resonance phenomena in a pure electron plasma *Phys. Fluids B* **4**, 2038 (1992), Cat C, E, O.

[133] **grea:94a**   R. G. Greaves, M. D. Tinkle, and C. M. Surko, Creation and uses of positron plasmas *Physics of Plasmas* **1**, 1439–1446 (1994), Cat I, O.

[134] **grea:94b**   R. G. Greaves, M. D. Tinkle, and C. M. Surko, Modes in a pure ion plasma at the brillouin limit, to be published in Physical Review Letters, (1994), Cat C, O.

[135] **hang:91:**   J. S. Hangst, Kristensen, J. S. Nielsen, O. Poulsen, J. P. Schiffer, and P. Shi, Laser cooling of a stored ion beam to 1 mk *Phys. Rev. Lett.* **67**, 1238 (1991), Cat G, H, I, K.

[136] **hang:94**   J. S. Hangst, J. S. Nielsen, O. Poulsen, J. P. Schiffer, P. Shi, and B. Wanner, Laser cooling of a bunched beam in astrid, in J. Bosser, editor, *Proceedings of the Workshop on Beam Cooling and Related Topics, Montreux, Switzerland, October 4-8, 1993* p. 343, CERN 94-03, (1994), Cat G, H, I, K.

[137] **hans:89a**   C. D. Hanson, E. L. Kerley, M. E. Castro, and D. H. Russell, Ion detection by fourier transform ion cycotron resonance: the effect of initial radial velocity on the coherent ion packet *Anal. Chem.* **61**, 2040 (1989), Cat E.

[138] **hans:89b**   C. D. Hanson, M. E. Castro, and D. H. Russell, Phase synchronization of an ion ensemble by frequency sweep excitation in fourier transform ion cyclotron resonance *Anal. Chem.* **61**, 2130 (1989), Cat E.

[139] **hart:91**   Grant W. Hart, The effect of a tilted magnetic field on the equilibrium of a pure electron plasma *Phys. Fluids B* **3**, 2987 (1991), Cat A, D, P.

[140] **hass:90**   R. W. Hasse and J. P. Schiffer, The structure of the cylindrically confined coulomb lattice *Annals of Phys.* **203**, 419 (1990), Cat G, H, I, K.

[141] **hein:91**   D. J. Heinzen, J. J. Bollinger, F. L. Moore, W. M. Itano, and D. J. Wineland, Rotational equilibria and low-order modes of a non-neutral ion plasma *Phys. Rev. Lett.* **66**, 2080–2083, 3087E (1991), Cat A, C, F, H.

[142] **hend:93b**   C. L. Hendrickson, S. C. Beu, and D. A. Laude, Two-dimensional coulomb-induced frequency modulation in fourier transform ion cyclotron resonance: A mechanism for line broadening at high mass and for large ion populations *J. Am. Soc. Mass Spectrom.* **4**, 909 (1993), Cat E.

[143] **hjor:86**   P. G. Hjorth and T. M. O'Neil, Temperature equilibration in a strongly magnetized pure electron plasma, in *Nonequilibrium Statistical Mechanics Session of the VIII IAMP Conference, Marseilles*, (1986), Cat B, E, P.

[144] **hjor:87**   P. G. Hjorth and T. M. O'Neil, Numerical study of a many particle adiabatic invariant *Phys. Fluids* **30**, 2613 (1987), Cat B, E, P.

[145] **hjor:88**   G. Hjorth P., *A Many Pphdthesis Adiabatic Invariant of Strongly Magnetized Pure Electron Plasmas*, PhD thesis, University of California, San Diego (1988), Cat B, P.

[146] **holl:93**   D. L. Holland, G. J. Morales, and B. D. Fried, Effect of particle losses on the equilibrium profiles of a non-neutral plasma *Phys. Fluids B* **5**, 1398 (1993), Cat A, B.

[147] **huan:93**   Huang X.-P., *Experimental Studies of Relaxation of Two-Dimensional Turbulence in Magnetized Electron Plasma Columns*, PhD thesis, University of California, San Diego (1993), Cat A, B, C, O.

[148] **huan:94**   X.-P. Huang and C. F. Driscoll, Relaxation of 2d turbulence to a metaequilibrium near the minimum enstrophy state *Phys. Rev. Lett.* **72**, 2187 (1994), Cat B, D, O.

[149] **hyat:87**   A. W. Hyatt, C. F. Driscoll, and J. H. Malmberg, Measurements of the anisotropic temperature relaxation rate in a pure electron plasma *Phys. Rev. Lett.* **59**, 2975 (1987), Cat B, E, O.

[150] **hyat:88**   A. W. Hyatt, *Measurements of the Anisotropic Temperature Relaxation Rate in a Magnetized Pure Electron Plasma*, PhD thesis, University of California, San Diego (1988), Cat B, O.

[151] **itan:82**   W. M . Itano and D. J. Wineland, Laser cooling of ions stored in harmonic and Penning traps *Phys. Rev. A* **25**, 35–54 (1982), Cat G, J.

[152] **itan:88**   W. M. Itano, L. R. Brewer, D. J. Larson, and D. J. Wineland, Perpendicular laser cooling of a rotating ion plasma in a penning trap *Phys. Rev. A* **38**, 5698–5706 (1988), Cat A, G, H, J.

[153] **iwat:93**   K. Iwata, R. G. Greaves, and C. M. Surko, Annihilation rates of positrons on aromatic molecules, (1994), Cat I, J, O.

[154] **iwat:94**   K. Iwata, R. G. Greaves, T. J. Murphy, D. Tinkle, and C. M. Surko, Measurements of positron annihilation rates on molecules, to be published in Physical Review A, (1994), Cat I, J, O.

[155] **jeff:83**   J. B. Jeffries, S. E. Barlow, and G. H. Dunn, Theory of space-charge shift of ion cyclotron resonance frequencies *Int. J. Mass Spectrom. Ion Processes* **54**, 169 (1983), Cat E.

[156] **kadt:94**   J. B. Kadtke, T. B. Mitchell, C. F. Driscoll, K. S., and Fine, Reconstructing chaotic vortex trajectories from plasma data, in *Current Topics in Astrophysical and Fusion Plasma Research: Proceedings of the International Workshop on Plasma Physics* p. 1, dbv-Verlag Graz, (1994), Cat D, O.

[157] **kape:73**   C.A. Kapetanakos, D.A. Hammer, C.D. Striffler, and R.C. Davidson, Destructive instabilities in hollow intense relativistic electron beams *Phys. Rev. Lett.* **30**, 1303 (1973), Cat A, B, C.

[158] **kein:81**   R. Keinigs, The effect of magnetic field errors on low frequency waves in a pure electron plasma *Phys. Fluids* **24**, 860 (1981), Cat C, P.

[159] **kein:84**   R. Keinigs, Field-error induced transport in a pure electron column *Phys. Fluids* **27**, 206 (1984), Cat B, C, P.

[160] **kerv:85**   N. A. Kervalishvili and V. P. Kortthondzhiya, Rate of electron-impact ionization in the charged plasma of an anode sheath in crossed fields e × h *Sov. J. Plasma Phys.* **11**, 74 (1985), Cat J.

[161] **kerv:86**   N. A. Kervalishvili and V. P. Kortthondzhiya, Rotating instability of the charged plasma of an anode sheath in crossed fields e × h *Sov. J. Plasma Phys.* **12**, 503 (1986), Cat A.

[162] **kerv:89a**   N. A. Kervalishvili, Rotating instability of a charged plasma in crossed fields e × h and generation of electrons of anomalously high energy *Sov. J. Plasma Phys.* **15**, 98 (1989), Cat A, B.

[163] **kerv:89b**   N. A. Kervalishvili, Rotating regular structures in a charged plasma in crossed electric and magnetic fields *Sov. J. Plasma Phys.* **15**, 211 (1989), Cat C, D.

[164] **kerv:89c**   N. A. Kervalishvili, Evolution of nonlinear structures in crossed fields e × h *Sov. J. Plasma Phys.* **15**, 436 (1989), Cat D.

[165] **kerv:91**   N. A. Kervalishvili, Electron vortices in a non-neutral plasma in crossed e × h *fields Phys. Lett. A* **157**, 391 (1991), Cat D.

[166] **kerv:94**   N. A. Kervalishvili, Formation of equilibrium density profile in a non-neutral electron plasma in crossed **e** × **h** fields *Phys. Lett. A* **188**, 170 (1994), Cat A.

[167] **khir:93**   S. S. Khirwadkar, P. I. John, K. Avinash, A. K. Agarwal, and P. K. Kaw, Steady sate formation of a toroidal electron cloud *Phys. Rev. Lett.* **71**, 4334 (1993), Cat A, K.

[168] **krau:94**   G. Kraus, P. Egelhof, C. Fischer, H. Geissel, A. Himmler, F. Nickel, G. Munzenberg, W. Schwab, A. Weiss, J. Friese, A. Gillitzer, H. J. Korner, M. Peter, W. F. Henning, J. P. Schiffer, J. V. Kratz, L. Chulkov, M. Golovkov, A. Ogloblin, and B. A. Brown, Proton inelastic scattering on 56ni in inverse kinematics *Phys. Rev. Lett.* **73**, 1773 (1994), Cat G, H, I, K.

[169] **kriv:93**   S. M. Krivoruchko and I. K. Tarasov, Effect of external perturbations on the expansion of a non-neutral electron plasma in a magnetic field *Plasma Phys. Repts.* **9** (1993), Cat B.

[170] **kwon:83**   H. S. Kwong, B. Carol Johnson, Peter L. Smith, and W. H. Parkinson, Transition probability of the Si III 189.2-nm intersystem line *Phys. Rev. A* **27**(6), 3040 (1983), Cat J.

[171] **kwon:89a**   Victor H. S. Kwong, Production and storage of low-energy highly charged ions by laser ablation and an ion trap *Phys. Rev. A* **39**(9), 4451 (1989), Cat J.

[172] **kwon:89b**   V. H. S. Kwong, Cooling and trapping of laser induced multiply charged ions of molybdenum *J. de Physique* **C1**, 413 (1989), Cat J.

[173] **kwon:90**   V. H. S. Kwong, T. T. Gibbons, Z. Fang, J. Jiang, H. Knocke, Y. Jiang, B. Ruger, S. Huang, E. Braganza, W. Clark, and L. D. Gardner, Experimental apparatus for production, cooling, and storing multiply charged ions for charge-transfer measurements *Rev. Sci. Instrum.* **61**(7), 1931 (1990), Cat J.

[174] **kwon:92**   Victor H. S. Kwong, Z. Fang, Y. Jiang, T. T. Gibbons, and L. D. Gardner, Measurement of thermal-energy charge-transfer rate coefficient of Mo^{6+} and argon *Phys. Rev. A* **46**(1), 201 (1992), Cat J.

[175] **kwon:93a**   Victor H. S. Kwong, Z. Fang, T. T. Gibbons, W. H. Parkinson, and Peter L. Smith, Measurement of the transition probability of the C III 190.9 nanometer intersystem line *Astrophys. J.* **411**(1), 431 (1993), Cat J.

[176] **kwon:93b**   Victor H. S. Kwong and Z. Fang, Charge transfer between O^{2+} ion and helium at electrovolt energy *Phys. Rev. Lett.* **71**(25), 4127 (1993), Cat J.

[177] **kyhl:56**   R. L. Kyhl and H. F. Webster, Breakup of hollow cylindrical electron beams *IRE Trans. Electron Devices* **3**, 172 (1956).

[178] **lamb:83**   B. M. Lamb and G. J. Morales, Ponderomotive effects in non-neutral plasmas *Phys. Fluids* **26**, 3488 (1983), Cat A, B, C.

[179] **lars:86**   D. J. Larson, J. C. Bergquist, J. J. Bollinger, W. M. Itano, and D. J. Wineland, Sympathetic cooling of trapped ions: a laser-cooled two-species non-neutral ion plasma *Phys. Rev. Lett.* **57**, 70–73 (1986), Cat A, G, H, J.

[180] **lauk:86**   F. H. Laukien, The effects of residul spatial magnetic field gradients on fourier transform ion cyclotron resonance spectra *Int. J. Mass Spectrom. Ion Processes* **73**, 81 (1986), Cat E.

[181] **levy:65**   R. H. Levy, Diocotron instability in a cylindrical geometry *Phys. Fluids* **8**, 1288 (1965), Cat C, P.

[182] **levy:69**   R. H. Levy, J. D. Daugherty, and O. Buneman, Ion resonance instability in grossly non-neutral plasmas *Phys. Fluids* **12**, 2616 (1969), Cat B, C, P.

[183] **lund:93**   S.M. Lund, J.J. Ramos, and R.C. Davidson, Coherent structures in rotating nonneutral plasma *Phys. Fluids B* **B5**, 19 (1993), Cat A.

[184] **lund:93a**   S.M. Lund and R.C. Davidson, A class of coherent vortex structures in rotating nonneutral plasma *Phys. Fluids B* **B5**, 1421 (1993), Cat A.

[185] **malm:75**   J. H. Malmberg and J. S. deGrassie, Properties of a non-neutral plasma *Phys. Rev. Lett.* **35**, 577 (1975), Cat A B C O.

[186] **malm:77**   J. H. Malmberg and T. M. O'Neil, The pure electron plasma, liquid, and crystal *Phys. Rev. Lett.* **39**, 1333 (1977), Cat A G P.

[187] **malm:80**   J. H. Malmberg and C. F. Driscoll, Long-time containment of a pure electron plasma *Phys. Rev. Lett.* **44**, 654 (1980), Cat B O.

[188] **malm:82**   J. H. Malmberg, C. F. Driscoll, and W. D. White, Experiments with pure electron plasmas *Physica Scripta* **T2**, 288 (1982), Cat A, B, C, O.

[189] **malm:84**   J H. Malmberg, T. M. O'Neil, A. W. Hyatt, and C. F. Driscoll, The cryogenic pure electron plasma, in *Proc. of 1984 Sendai Symposium on Plasma Nonlinear Phenomena* p. 31, (1984), Cat A, G, O.

[190] **malm:88**   J. H. Malmberg, C. F. Driscoll, B. Beck, D. L. Eggleston, J. Fajans, K. Fine, X. P. Huang, and A. W. Hyatt, Experiments with pure electron plasmas, in C.W. Roberson and C.F. Driscoll, editors, *Non-Neutral Plasma Physics*, volume AIP 175 p. 28, New York (1988), American Institute of Physics, Cat A B C D O.

[191] **malm:92**   J. H. Malmberg, Some recent results with non-neutral plasmas, in K. Lackner and W. Lindinger, editors, *Plasma Physics 1992: Joint Conference of the 9th Kiev Intl. Conf. on Plasma Theory, 9th Intl. Congress on Waves and Instabilities in Plasmas, and 19th European Physical Society Conf. on Controlled Fusion and Plasma Physics*, volume 34 p. 1767, (1992), Cat A, B, C, D, O.

[192] **mich:89**   F. Curtis Michel, Nonneutral plasmas in the laboratory and astrophysics *Comments Astrophys.* **13**, 145 (1989), Cat F, H, K, N.

[193] **mitc:93**   T. B. Mitchell, *Experiments on electron vortices in a Malmberg-Penning trap*, PhD thesis, University of California, San Diego (1993), Cat A, C, O.

[194] **miti:93a**   T. B. Mitchell, C. F. Driscoll, and K. S. Fine, Experiments on stability of equilibria of two vortices in a cylindrical trap *Phys. Rev. Lett.* **71**, 1371 (1993), Cat A, D, O.

[195] **miti:93b**   T. B. Mitchell, *Experiments on Electron Vortices in Malmberg-Penning Trap*, PhD thesis, University of California, San Diego (1993), Cat A, C, O.

[196] **miti:94**   T. B. Mitchell and C. F. Driscoll, Symmetrization of 2d vortices by beat-wave damping *Phys. Rev. Lett.* **73**, 2196 (1994), Cat B, C, D, O.

[197] **mood:92**   J. D. Moody and J. H. Malmberg, Free expansion of a pure electron plasma column *Phys. Rev. Lett.* **69**, 3639 (1992), Cat B, C, O.

[198] **mora:88**   G. J. Morales, 2-d non-neutral plasmas on liquid helium, in *Non-Neutral Plasma Physics* p. 111, New York (1988), Cat A, K, C, H.

[199] **murp:90**   T. J. Murphy and C. M. Surko, Annihilation of positrons in xenon gas *Journal of Physics B* **23**, 727–32 (1990), Cat I, J, O.

[200] **murp:91**   T. J. Murphy and C. M. Surko, Annihilation of positrons on organic molecules *Physical Review Letters* **67**, 2954–7 (1991), Cat I, J, O.

[201] **murp:92**   T. J. Murphy and C. M. Surko, Positron trapping in an electrostatic well by inelastic collisions with nitrogen molecules *Physical Review A* **46**, 5696–705 (1992), Cat I, O.

[202] **niel:94**   J. S. Nielsen, J. S. Hangst, O. Poulsen, J. P. Schiffer, P. Shi, and D. Wannor, Laser cooling of 24mg+ in the astrid storage ring, in J. Bosser, editor, *Beam Cooling and Related Topics* p. 339, CERN 94-03, (1994), Cat G, H, I, K.

[203] **nott:92**   J. Notte, A. J. Peurrung, J. Fajans, R. Chu, and J.S. Wurtele, Asymmetric, stable equilibria of non-neutral plasmas *Phys. Rev. Lett.* **69**, 3056 (1992), Cat A, O, P.

[204] **nott:93a**   J. Notte, J. Fajans, R. Chu, and J.S. Wurtele, Experimental breaking of an adiabatic invarient *Phys. Rev. Lett.* **70**, 3900 (1993), Cat C, O, P.

[205] **nott:93b**   J. A. Notte, *The Effect of Asymmetries on Non-Neutral Plasmas*, PhD thesis, University of California, Berkeley (1993), Cat A, C, O.

[206] **nott:94**   J. Notte and J. Fajans, The effect of asymmetries on non-neutral plasma confinement time *Phys. Plasmas* **1**, 1123 (1994), Cat B, O.

[207] **onei:79**   T. M. O'Neil and C. F. Driscoll, Transport to thermal equilibrium of a pure electorn plasma *Phys. Fluids* **22**, 266 (1979), Cat A B P.

[208] **onei:80**   T. M. O'Neil, A confinement theorem for non-neutral plasmas *Phys. Fluids* **23**, 2216 (1980), Cat A P.

[209] **onei:80a**   T. M. O'Neil, Cooling of a pure electron plasma by cyclotron radiation *Phys. Fluids* **23**, 725 (1980), Cat B, C, P.

[210] **onei:80b**   T. M. O'Neil, Pure electron plasmas *Proc. Int. Conf. on Plasma Physics* **II**, 321 (1980), Cat A, G, P.

[211] **onei:80c**   T. M. O'Neil, Non-neutral plasmas have exceptional confinement properties *Comments Plasma Phys. Cont. Fusion* **5**, 231 (1980), Cat A, P.

[212] **onei:81**   T. M. O'Neil, Centrifugal separation of a multispecies pure ion plasma *Phys. Fluids* **24**, 1447 (1981), Cat A, P.

[213] **onei:83**   T. M. O'Neil, Collision operator for a strongly magnetized pure electron plasma *Phys. Fluids* **26**, 2128 (1983), Cat B P.

[214] **onei:85**   T. M. O'Neil and P. G. Hjorth, Collisional dynamics of a strongly magnetized pure electron plasma *Phys. Fluids* **28**, 3241 (1985), Cat B E P.

[215] **onei:87**   T. M. O'Neil, C. F. Driscoll, and D. H. E. Dubin, Like particle transport: A new theory and experiments with pure electron plasmas, in *Turbulence and Anomalous Transport in Magnetized Plasmas* pp. 293–308, Orsay (1987), Editions de Physique, Cat B C O P.

[216] **onei:88**   T. M. O'Neil, Plasmas with a single sign of charge, in C. W. Roberson and C. F. Driscoll, editors, *Non-neutral Plasma Physics* p. 1, New York (1988), American Institute of Physics, Cat A, B, G, P.

[217] **onei:90**   M. O'Neil T., P. G. Hjorth, B. Beck, J. Fajans, and J. H. Malmberg, Collisional relaxation of a strongly magnetized pure electron plasma: Theory and experiment, in S. Ichimaru, editor, *Strongly Coupled Plasma Physics* p. 313, Elsevier Science Pub. B.V./Yamada Science Foundation, (1990), Cat B, E, O, P.

[218] **onei:92**   T. M. O'Neil and R. A. Smith, Stability theorem for off-axis states of a non-neutral plasma column *Phys. Fluids B* **4**, 2720 (1992), Cat A C D P.

[219] **onei:94**   T. M. O'Neil, Plasmas with a single sign of charge *Physica Scripta* (1994), Cat A, B, C, D, G, O, P.

[220] **onei:94a**   M. O'Neil T. and R. A. Smith, Stability theorem for a single species plasma in a toroidal magnetic configuration *Phys. Plasmas* **1**, 2430 (1994), Cat A, K, P.

[221] **onel:85a**   T. M. O'Neil, A new theory of transport due to like particle collisions *Phys. Rev. Lett.* **55**, 943 (1985), Cat B, P.

[222] **pass:89**   A. Passner, C. M. Surko, M. Leventhal, and A. P. Mills, Jr., Ion production by positron-molecule resonances *Physical Review A* **39**, 3706–9 (1989), Cat I, J, O.

[223] **peti:87**   J.J. Petillo and R.C. Davidson, Kinetic equilibrium and stability properties of high-current betatrons *Phys. Fluids* **30**, 2477 (1987), Cat A, B, C, M.

[224] **peur:90**   A. J. Peurrung and J. Fajans, Non-neutral plasma shapes and edge profiles *Phys. Fluids B* **2**, 693 (1990), Cat A, B, P.

[225] **peur:92d**   A. J. Peurrung, *Imaging of Instabilities in a Pure Electron Plasma*, PhD thesis, University of California, Berkeley (1992), Cat A, B, C, D, O.

[226] **peur:93a**   A. J. Peurrung and J. Fajans, A pulsed, microchannel plate-based, non-neutral plasma imaging system *Rev. Sci. Instrum.* **64**, 52 (1993), Cat D, O.

[227] **peur:93b**   A. J. Peurrung and J. Fajans, Experimental dynamics of an annulus of vorticity in a pure electron plasma *Phys. Fluids A* **5**, 493 (1993), Cat C, D, O.

[228] **peur:93c**   A. J. Peurrung, J. Notte, and J. Fajans, Collapse and winding in an asymmetric annulus of vorticity *J. Fluid Mech.* **252**, 713 (1993), Cat C, D, O.

[229] **peur:93e**   A. J. Peurrung, J. Notte, and J. Fajans, Observation of the ion resonance instability *Phys. Rev. Lett.* **70**, 295 (1993), Cat B, C, O.

[230] **peur:93f**   A. J. Peurrung and J. Fajans, A limitation to the analogy between pure electron plasmas and 2-d inviscid fluids *Phys. Fluids B* **5** (1993), Cat D, O, P.

[231] **peur:94a**   A. J. Peurrung and R. T. Kouzes, Long-term coherence of the cyclotron mode in a trapped ion cloud *Phys. Rev. E* **49**, 4362 (1994), Cat E.

[232] **pill:94**   N. S. Pillai and R. W. Gould, Damping and trapping in 2d inviscid fluids *Phys. Rev. Lett.* **73**, 2849 (1994), Cat D, C, O.

[233] **pouk:81**   J. W. Poukey and J. R. Freeman, Diocotron instability in asymmetric beams *Phys. Fluids* **24**, 2376 (1981).

[234] **prad:93**   S.; K. Avinash Pradhan, Diocotron instability in curved magnetic field *Phys. Fluids B* **5**, 2334 (1993), Cat C, K.

[235] **pras:79**   S. A. Prasad and T. M. O'Neil, Finite length equilibria of a pure electron plasma column *Phys. Fluids* **22**, 278 (1979), Cat A, P.

[236] **pras:81**   S. A. Prasad, *Thermal Equilibria and Wave Properties of Finite Length Pure Electron Plasma Columns*, PhD thesis, University of California, San Diego (1981), Cat A, C, P.

[237] **pras:83**   S. A. Prasad and T. M. O'Neil, Waves in a cold pure electron plasma of finite length *Phys. Fluids* **26**, 665 (1983), Cat C, P.

[238] **pras:84**   S. A. Prasad and T. M. O'Neil, Vlasov theory of electrostatic modes in a finite length electron column *Phys. Fluids* **27**, 206 (1984), Cat C, P.

[239] **pras:85**   S. A. Prasad, G. J. Morales, and B. D. Fried, Cyclotron resonance in a non-neutral plasma *Phys. Rev. Lett.* **54**, 2336 (1985), Cat E.

[240] **pras:86**   S. A. Prasad and J. H. Malmberg, A nonlinear diocotron mode *Phys. Fluids* **29**, 2196 (1986), Cat A, C, D, P.

[241] **pras:87**   S. A. Prasad, G. J. Morales, and B. D. Fried, Cyclotron resonance phenomena in a non-neutral plasma *Phys. Fluids* **30**, 3093 (1987), Cat E.

[242] **pras:87a**   S. A. Prasad and G. J. Morales, Equilibrium and wave properties of two-dimensional ion plasmas *Phys. Fluids* **30**, 3475 (1987), Cat A, K, C, H.

[243] **pras:88**   S. A. Prasad and G. J. Morales, Nonlinear resonance of two- dimensional ion layers *Phys. Fluids* **31**, 562 (1988), Cat A, K, C, H.

[244] **pras:89**   S. A. Prasad and G. J. Morales, Magnetized equilibrium of a two-dimensional ion plasma *Phys. Fluids B* **1**, 1329 (1989), Cat A, K, C, H.

[245] **rafa:91**   Robert Rafac, John P. Schiffer, Jeffrey S. Hangst, Daniel H. E. Dubin, and David J. Wales, Stable configurations of confined cold ionic systems *Proceedings of the National Academy of Sciences* **88**, 483 (1991), Cat G, H, I, K.

[246] **rahm:86a**   A. Rahman and J. P. Schiffer, Structure of a one-component plasma in an external field: A molecular-dynamics study of particle arrangement in a heavy-ion storage ring *Phys. Rev. Lett.* **57**, 1133 (1986), Cat G, H, I, K, P.

[247] **rahm:88**   A. Rahman and J. P. Schiffer, A condensed state in a system of stored and cooled ions *Physica Scripta* **T22**, 133 (1988), Cat G, H, I, K.

[248] **raiz:92**   M. G. Raizen, J. C. Bergquist, J. M. Gilligan, W. M. Itano, and D. J. Wineland, Linear trap for high accuracy spectroscopy of stored ions *J. Mod. Opt.* **39**, 233–242 (1992), Cat G, H, J.

[249] **raiz:921**   M. G. Raizen, J. M. Gilligan, J. C. Bergquist, W. M. Itano, and D. J. Wineland, Ionic crystals in a linear Paul trap *Phys. Rev. A* **45**, 6493–6501 (1992), Cat G, H, J.

[250] **rama:93**   H. Ramachandran, G. J. Morales, and V. K. Decyk, Particle simulation of non-neutral plasma behavior *Phys. Fluids B* **5**, 2733 (1993), Cat B, C, G.

[251] **rasb:93**   S. Neil Rasband, Ross L. Spencer, and Richard R. Vanfleet, Exponential growth of an unstable l=1 diocotron mode for a hollow electron column in a warm fluid model *Phys. Fluids B* **5**, 669 (1993), Cat A, Q.

[252] **robe:88**   C.W. Roberson and C.F. Driscoll, *Non-Neutral Plasma Physics*, American Institute of Physics, New York, (1988), Cat A, B, C, D, E, F, G, H, I, J, K, L, M, N, O, P.

[253] **rose:87**   G. Rosenthal, G. Dimonte, and A. Y. Wong, Stabilization of diocotron instability in an annular plasma *Phys. Fluids* **30**, 3257 (1987), Cat C, D, H, O.

[254] **rose:90**   G. B. Rosenthal, *Experimental Studies of an Annular Non-Neutral Electron Plasma*, PhD thesis, University of California, Los Angeles (1990), Cat C, D, H, O.

[255] **schi:85**   J. P. Schiffer and P. Kienle, Could there be an ordered condensed state in beams of fully stripped heavy ions? *Z. Phys. A-Atoms and Nuclei* **321**, 181 (1985), Cat G, H, I, K.

[256] **schi:86**   J. P. Schiffer and O. Poulsen, Possibility of observing a condensed crystalline state in laser-cooled beams of atomic ions *Europhys. Lett.* **1**, 55 (1986), Cat G, H, I, K.

[257] **schi:88**   John P. Schiffer, Layered structure in condensed, cold, one-component plasmas confined in external fields *Phys. Rev. Lett.* **61**, 1843 (1988), Cat G, H, I, K.

[258] **schi:88a**   J. P. Schiffer and A. Rahman, Feasibility of a crystalline condensed state in cooled ion beams of a storage ring *Z. Phys.-Atomic Nuclei* **331**, 71 (1988), Cat G, H, I, K.

[259] **schi:89**   John P. Schiffer, Order in cold ionic systems: Dynamic effects, in *Proceedings of Workshop on Crystalline Ion Beams* p. 2, GSI-89-10 ISSN 0171-4546, (1989), Cat G, H, I, K.

[260] **schi:91**   J. P. Schiffer and J. S. Hangst, On the way towards crystallized beams: The transverse temperature of particle beams *Z. Phys. A - Hadrons and Nuclei* **341**, 107 (1991), Cat G, H, I, K.

[261] **schi:93**   J. P. Schiffer, Phase transitions in anisotropically confined ionic crystals *Phys. Rev. Lett.* **70**, 818 (1993), Cat G, H, I, K.

[262] **schi:93a**   J. P. Schiffer, Recoil-free absorption and scattering of light from confined crystalline ionic systems *Phys. Rev. A* **47**, 5193 (1993), Cat G, H, I, K.

[263] **schi:93b**   John P. Schiffer, Editorial *Nucl. Phys. News* **3**, 4 (1993), Cat G, H, I, K.

[264] **schi:94**   J. P. Schiffer and J. S. Hangst, Crystalline beams in alternating focusing fields, and for curved trajectories, in J. Bosser, editor, *Beam Cooling and Related Topics* p. 279, CERN 94-03, (1994), Cat G, H, I, K.

[265] **schi:94a**   J. P. Schiffer, Summary talk on beam crystallization, in J. Bosser, editor, *Proceedings of the Workshop on Beam Cooling and Related Topics, Montreux, Switzerland, October 4-8, 1993* p. 455, CERN 94-03, (1994), Cat G, H, I, K.

[266] **schu:88**   H. A. Schuessler, R. D. Knight, D. Dubin, W. D. Phillips, and G. Lafyatis, Summary of the physics in traps panel *Physica Scripta* **T22**, 228 (1988), Cat G, H, J, K, M, O, P.

[267] **smit:89**   R. A. Smith,   Phase-transition behavior in a negative-temperature guiding-center plasma *Phys. Rev. Lett.* **63**, 1479 (1989), Cat A, D, P.

[268] **smit:90**   R. A. Smith and M. N. Rosenbluth, Algebraic instability of hollow electron columns and cylindrical vortices *Phys. Rev. Lett.* **64**, 649 (1990), Cat A, C, D, P.

[269] **smit:90a**   R. A. Smith and T. M. O'Neil, Nonaxisymmetric thermal equilibria of a cylindrically bounded guiding center plasma or discrete vortex system *Phys. Fluids B* **2**, 2961 (1990), Cat A, D, P.

[270] **smit:91**   R. A. Smith, Maximization of vortex entropy as an organizing principle in intermittent, decaying two-dimensional turbulence *Phys. Rev. A* **43**, 1126 (1991), Cat A, D, P.

[271] **smit:92**   R. A. Smith, Effectively non-ergodic behavior of guiding-center and discrete-vortex systems, in I. Prigonine et al., editor, *Research Trends in Physics: Chaotic Dynamics and Transport in Fluids and Plasmas* p. 396, New York (1992), American Institute of Physics, Cat A, D, P.

[272] **smit:92a**   R. A. Smith, Effects of electrostatic confinement fields and finite gyro-radius on an instability of hollow electron columns *Phys. Fluids B* **4**, 287 (1992), Cat A, C, D, P.

[273] **smit:92b**   R. A. Smith, T. M. O'Neil, S. M. Lund, J. J. Ramos, and R.C. Davidson, Comment on the stability theorem of davidson and lund *Phys. Fluids B* **4**, 1373 (1992), Cat A, C, D, P.

[274] **spen:90**   Ross L. Spencer, The effect of externally-applied oscillating electric fields on the l=1 and l=2 diocotron modes in non-neutral plasmas *Phys. Fluids B* **2**, 2306–2314 (1990), Cat C, D, Q.

[275] **spen:92**   Ross L. Spencer and Grant W. Hart, Linear theory of non-neutral plasma equilibrium in a tilted magnetic field *Phys. Fluids B* **4**, 3507 (1992), Cat A, D, Q.

[276] **spen:93a**   Ross L. Spencer, S. Neil Rasband, and Richard R. Vanfleet, Axisymmetric non-neutral plasma equilibria *Phys. Fluids B* **5**, 4267 (1993), Cat A, L, Q.

[277] **spen:93b**   Ross L. Spencer and Grant W. Mason, Large amplitude l=1 coherent structures in non-neutral plasmas confined in a cylindrical trap *Phys. Fluids B* **5**, 1738–1745 (1993), Cat C, D, L, Q.

[278] **surk:86a**   C. M. Surko, M. Leventhal, W. S. Crane, and A. P. Mills, Jr., The positron trap—a new tool for plasma physics, in A. P. Mills, Jr., W. S. Crane, and K. F. Canter, editors, *Positron Studies of Solids, Surfaces, and Atoms. A Symposium to celebrate Stephan Berko's 60th Birthday* pp. 221–33, World Scientific, Singapore (1986), Cat I, O.

[279] **surk:86b**   C. M. Surko, M. Leventhal, W. S. Crane, A. Passner, and F. Wysocki, Use of positrons to study transport in tokamak plasmas *Rev. Sci. Instrum.* **57**, 1862–7 (1986), Cat I, O.

[280] **surk:86c**   C. M. Surko, M. Leventhal, A. Passner, and F. J. Wysocki, A positron plasma in the laboratory—how and why, in *Symposium on Non-Neutral Plasma Physics, AIP Conference Proceedings, No. 175* pp. 75–90, (1988), Cat I, O.

[281] **surk:88**   C. M. Surko, A. Passner, M. Leventhal, and F. J. Wysocki, Bound states of positrons and large molecules *Phys. Rev. Lett.* **61**, 1831–4 (1988), Cat I, J, O.

[282] **surk:89**   C. M. Surko, M. Leventhal, and A. Passner, Positron plasma in the laboratory *Phys. Rev. Lett.* **62**, 901–4 (1989), Cat I, O.

[283] **surk:90**   C. M. Surko and T. J. Murphy, Use of the positron as a plasma particle *Phys. Fluids B* **2**, 1373 (1990), Cat I, J, O.

[284] **surk:93**   C. M. Surko, R. G. Greaves, and M. Leventhal, Use of traps to study positron annihilation in astrophysically relevant media *Hyperfine Interactions* **81**, 239–52 (1993), Cat I, J, O.

[285] **tan:94**   J. N. Tan, J. J. Bollinger, and D. J. Wineland, Minimizing the time-dilation shift in Penning trap atomic clocks *IEEE Trans. Instrum. Meas. (to be published)* (1994), Cat A, H, J,

[286] **tang:92**   G. Tang, M. D. Tinkle, R. G. Greaves, and C. M. Surko, Annihilation gamma-ray spectra from positron-molecule interactions *Physical Review Letters* **68**, 3793–6 (1992), Cat I, J, O.

[287] **tink:94**   M. D. Tinkle, R. G. Greaves, C. M. Surko, R. L. Spencer, and G. W. Mason, Low-order modes as diagnostics of spheroidal non-neutral plasmas *Physical Review Letters* **72**, 352–5 (1994), Cat C, I, P.

[288] **tink:95**   D. Tinkle, R. G. Greaves, and C. M. Surko, Low-order longitudinal modes of single-component plasmas, submitted to Physics of Plasmas, (1995), Cat C, O.

[289] **tots:75**   H. Totsuji, Thermodynamic properties of surface layers of classical electrons *J. Phys. Soc. Japan* **39**(1), 253 (1975), Cat G, D, P.

[290] **tots:76**   H. Totsuji, Theory of two-dimensional classical electron plasma *J. Phys. Soc. Japan* **40**(3), 857 (1976), Cat G, D, P.

[291] **tots:78**   H. Totsuji, Numerical experiments on two-dimensional electron liquids. thermodynamic properties and onset of short-range order *Phys. Rev. A* **17**(1), 399 (1978), Cat G, D, P.

[292] **tots:79**   H. Totsuji, Cluster expansion for two-dimensional electron liquids *Phys. Rev. A* **19**(2), 889 (1979), Cat G, P.

[293] **tots:80**   H. Totsuji, Effect of electron correlation on high-frequency conductivity of electron liquids on a liquid helium surface *Phys. Rev. B* **22**(1), 187 (1980), Cat B, G, P.

[294] **tots:80a**   H. Totsuji and H. Kakeya, Dynamical properties of two-dimensional classical electron liquids *Phys. Rev. A* **22**(3), 1220 (1980), Cat G, C, P.

[295] **tots:81**   H. Totsuji, Distribution of charged particles near a charged hard wall in a uniform background *J. Chem. Phys.* **75**(2), 871 (1981), Cat G, P.

[296] **tots:81a**   H. Totsuji and T. Funahashi, Dynamical fluctuation spectra of two-dimensional classical electron liquids in magnetic fields *Phys. Lett. A* **84**(4), 185 (1981), Cat G, C, P.

[297] **tots:81b**   H. Totsuji, On the nature of transverse excitations in strongly coupled two-dimensional classical electron liquids *Phys. Lett. A* **85**(6/7), 349 (1981), Cat G, C, P.

[298] **tots:82**   H. Totsuji, Distribution of charged particles near a charged hard wall in a uniform background: Comparison with exact results *J. Chem. Phys.* **77**(7), 3772 (1982), Cat G, P.

[299] **tots:84**   H. Totsuji, Triplet correlation function in strongly-coupled three- and two-dimensional classical one-component plasmas *Phys. Rev. A* **29**(1), 314 (1984), Cat G, P.

[300] **tots:84a**   H. Totsuji and K. Tokami, Thermodynamic properties of classical plasmas in a polarizing background: Numerical experiments *Phys. Rev. A* **30**(6), 3175 (1984), Cat G, H, L, P.

[301] **tots:86**   H. Totsuji, Surface properties of classical one-component plasma *J. Phys. C: Solid State Phys.* **19**(26), 1573 (1986), Cat G, P.

[302] **tots:87**   H. Totsuji and H. Wakabayashi, Comparisons of solutions of integral equations for one-component plasma in supercooled liquid state *Phys. Rev. A* **36**(9), 4511 (1987), Cat G, P.

[303] **tots:88**   H. Totsuji and J.-L. Barrat, Structure of non-neutral classical plasma in a magnetic field *Phys. Rev. Lett.* **60**(24), 2484 (1988), Cat A, H, G, P.

[304] **tots:88a**   H. Totsuji, Madelung energy of a one-dimensional coulomb lattice *Phys. Rev. A* **38**(10), 5444 (1988), Cat A, H, G, P.

[305] **tots:92**   H. Totsuji, H. Shirokoshi, and S. Nara, Molecular dynamics of a coulomb system with deformable periodic boundary conditions *Phys. Lett. A* **162**(2), 174 (1992), Cat G, L, P.

[306] **tots:92a**   H. Totsuji, Spectrum of schottky noise in ion storage rings *Phys. Rev. A* **46**(4), 2106 (1992), Cat H, G, C, P.

[307] **tots:93**   H. Totsuji, Two-dimensional system of charges in cylindrical traps *Phys. Rev. E* **47**(5), 3784 (1993), Cat A, H, K, P.

[308] **turn:87**   Leaf Turner, Collective effects on equilibria of trapped charged plasmas *Phys. Fluids* **30**, 3196 (1987), Cat A, I, P.

[309] **turn:90**   Leaf Turner, Confinement of non-neutral plasma in unconventional geometries, in *Proceedings of the Fourth International Workshop on Slow-Positron Beam Technologies for Solids and Surfaces*, New York (1990), American Institute of Physics, Cat A, K, P.

[310] **turn:91**   Leaf Turner, Brillouin limit for non-neutral plasma in inhomogeneous magnetic fields *Phys. Fluids B* **3**, 1355 (1991), Cat A, K, P.

[311] **turn:92**   D. C. Barnes and Leaf Turner, Non-neutral plasma compression to ultrahigh density year=with d. c. barnes) *Phys. Fluids B* **4**, 3890 (1992), Cat A, K, M, P.

[312] **turn:93**   Leaf Turner and D. C. Barnes, The brillouin limit and beyond: A route to inertial-electrostatic confinement of a single species plasma *Phys. Rev. Lett.* **70**, 798 (1993), Cat A, K, P.

[313] **turn:93a**   D. C. Barnes, R. A. Nebel, and Leaf Turner, Production and application of dense penning trap plasmas *Phys. Fluids B* **5**, 3651 (1993), Cat A, B, K, M, P.

[314] **turn:93b**   D. C. Barnes, R. A. Nebel, Leaf Turner, and T. N. Tiouririne, Alternate fusion: Continuous inertial confinement *Plasma Phys. Control. Fusion* **35**, 929 (1993), Cat A, B, K, M, P.

[315] **turn:94**   T. N. Tiouririne, Leaf Turner, and A.W.C. Lau, Multipole traps for non-neutral plasmas *Phys. Rev. Lett.* **72**, 1204 (1994), Cat A, K, P.

[316] **turn:94a**   Leaf Turner andT. N. Tiouririne and A.W.C. Lau, Compressional oscillation frequency of an anharmonic oscillator: The spherical non-neutral plasma *J. Math. Phys.* **35**, 2349 (1994), Cat A, P.

[317] **turn:94b**   Leaf Turner and John M. Finn, Streaming instabilites of a non-neutral plasma with turning points, submitted to the Physics of Plasmas, October 1994, (1994), Cat A, C, K, M, P.

[318] **uech:92**   G. T. Uechi and R. C. Dunbar, Space charge effects on relative peak heights in fourier transform-ion cyclotron resonance spectra *J. Am. Soc. Mass Spectrom.* **3**, 734 (1992), Cat E.

[319] **uhm:78**   H.S. Uhm and R.C. Davidson, Influence of finite ion larmor radius and equilibrium self-electric fields on the ion resonance instability *Phys. Fluids* **20**, 579 (1978), Cat A, B, C.

[320] **uhm:80**   H.S. Uhm and R.C. Davidson, Stability properties of intense nonneutral ion beams for heavy ion fusion *J. Particle Accelerators* **11**, 65 (1980), Cat A, B, C, M.

[321] **uhm:80a**   H.S. Uhm and R.C. Davidson, Kinetic description of coupled transverse oscillations in an intense relativistic electron beam-plasma system *Phys. Fluids* **23**, 813 (1980), Cat A, B, C.

[322] **uhm:82**   H.S. Uhm and R.C. Davidson, Ion resonance instability in a modified betatron accelerator *Phys. Fluids* **25**, 2334 (1982), Cat A, B, C, M.

[323] **uhm:83**   H.S. Uhm and R.C. Davidson, Free electron laser instability for a relativistic solid electron beam in a helical wiggler field *Phys. Fluids* **26**, 288 (1983), Cat A, B, C, M.

[324] **wang:89**   M. Wang and A. G. Marshall, A screened electrostatic ion trap for enhanced mass resolution, mass accuracy, reproducibility, and upper mass limit in fourier transform ion cyclotron resonance mass spectrometry *Anal. Chem.* **61**, 1288 (1989), Cat K.

[325] **webs:55**   H. F. Webster, Breakup of hollow electron beams *J. Appl. Phys.* **26**, 1386 (1955).

[326] **weim:94**   C. S. Weimer, J. J. Bollinger, F. L. Moore, and D. J. Wineland, Electrostatic modes as a diagnostic in Penning trap experiments *Phys. Rev. A* **49**, 3842–3853 (1994), Cat C, J.

[327] **whit:82**   W. D. White, J. H. Malmberg, and C. F. Driscoll, Resistive wall destabilization of diocotron waves *Phys. Rev. Lett.* **49**, 1822 (1982), Cat C, O.

[328] **wine:85**   D. J. Wineland, Trapped ions, laser cooling, and better clocks *Science* **226**, 395–400 (1985), Cat J.

[329] **wine:85a**   D. J. Wineland, J. J. Bollinger, W. M. Itano, and J. D. Prestage, Angular momentum of trapped atomic particles *J. Opt. Soc. Am. B* **2**, 1721–1730 (1985), Cat A, H, J.

[330] **wine:87**   D. J. Wineland, J. C. Bergquist, W. M. Itano, J. J. Bollinger, and C. H. Manney, Atomic ion Coulomb clusters in an ion trap *Phys. Rev. Lett.* **59**, 2935–2938 (1987), Cat G, H, J.

[331] **wine:90**   D. J. Wineland, J. C. Bergquist, J. J. Bollinger, W. M. Itano, D. J. Heinzen, S. L. Gilbert, C. H. Manney, and M. G. Raizen, Progress at NIST towards absolute frequency standards using stored ions *IEEE Trans. Ultrasonics, Ferroelectrics, Frequency Control* **37**, 515–523 (1990), Cat H, J.

[332] **wine:93**   D. J. Wineland, C. S. Weimer, and J. J. Bollinger, Laser-cooled positron source *Hyp. Int.* **76**, 115–125 (1993), Cat I.

[333] **wyso:88**   F. J. Wysocki, M. Leventhal, A. Passner, and C. M. Surko, Accumulation and storage of low energy positrons *Hyperfine Interactions* **44**, 185–200 (1988), Cat I, O.

[334] **xian:93**   X. Xiang, P. B. Grosshans, and A. G. Marshall, Image charge-induced ion cyclotron orbital frequency shift for orthorhombic and cylindrical ft-icr ion traps *Int. J. Mass Spectrom. Ion Processes* **125**, 33 (1993), Cat E.

[335] **yin:92**   W. W. Yin, M. Wang, A. G. Marshall, and E. B. Ledford, Experimental evaluation of a hyperbolic ion trap for fourier transform ion cyclotron resonance mass spectrometry *J. Am. Soc. Mass Spectrom.* **3**, 188 (1992), Cat K.

[336] **zave:92**   P. Zaveri, P. I. John, K. Avinash, and P. K. Kaw, Low-aspect-ratio toroidal equilibria of electron clouds *Phys. Rev. Lett.* **68**, 3295 (1992), Cat A, K.

AUTHOR INDEX

A

Anderegg, F., 1, 139, 184

B

Barlow, S. E., 129
Barnes, D. C., 113
Barton, A. S., 215
Beck, B. R., 25
Bollinger, J. J., 215
Book, D. L., 7

C

Cluggish, B. P., 14
Corngold, N. R., 20
Cowan, T. E., 25
Crooks, S. M., 32

D

Davidson, R. C., 118, 149
Decyk, V. K., 155
Dholakia, K., 162
Driscoll, C. F., 1, 14, 38, 184
Dubin, D., 262

E

Eggleston, D. L., 54

F

Fajans, J., 25, 64, 87, 271
Fang, Z., 108
Fine, K. S., 38

G

Gopalan, R., 25
Greaves, R. G., 70, 229

H

Hansen, C., 87
Hartley, J. H., 25
Hernandez-Pozos, J. L., 162
Hoffnagle, J. A., 92
Holzscheiter, M. H., 113
Horvath, G. Zs. K., 162
Howell, R. H., 25
Huang, X.-P., 1, 38

I

Ichimaru, S., 252

K

Kaye, S. M., 118
Kervalishvili, N. A., 102
Kouzes, R. T., 129
Kwong, V. H. S., 108

M

McDonald, J. L., 25
Mitchell, T. B., 38, 113
Moore, D. A., 118
Morales, G. J., 124, 155

N

Neu, S. C., 124

293

AIP Conference Proceedings

		L.C. Number	ISBN
No. 170	Nuclear Spectroscopy of Astrophysical Sources (Washington, DC, 1987)	88-71625	0-88318-370-6
No. 171	Vacuum Design of Advanced and Compact Synchrotron Light Sources (Upton, NY, 1988)	88-71824	0-88318-371-4
No. 172	Advances in Laser Science—III: Proceedings of the International Laser Science Conference (Atlantic City, NJ, 1987)	88-71879	0-88318-372-2
No. 173	Cooperative Networks in Physics Education (Oaxtepec, Mexico, 1987)	88-72091	0-88318-373-0
No. 174	Radio Wave Scattering in the Interstellar Medium (San Diego, CA, 1988)	88-72092	0-88318-374-9
No. 175	Non-neutral Plasma Physics (Washington, DC, 1988)	88-72275	0-88318-375-7
No. 176	Intersections Between Particle and Nuclear Physics (Third International Conference) (Rockport, ME, 1988)	88-62535	0-88318-376-5
No. 177	Linear Accelerator and Beam Optics Codes (La Jolla, CA, 1988)	88-46074	0-88318-377-3
No. 178	Nuclear Arms Technologies in the 1990s (Washington, DC, 1988)	88-83262	0-88318-378-1
No. 179	The Michelson Era in American Science: 1870–1930 (Cleveland, OH, 1987)	88-83369	0-88318-379-X
No. 180	Frontiers in Science: International Symposium (Urbana, IL, 1987)	88-83526	0-88318-380-3
No. 181	Muon-Catalyzed Fusion (Sanibel Island, FL, 1988)	88-83636	0-88318-381-1
No. 182	High T_c Superconducting Thin Films, Devices, and Applications (Atlanta, GA, 1988)	88-03947	0-88318-382-X
No. 183	Cosmic Abundances of Matter (Minneapolis, MN, 1988)	89-80147	0-88318-383-8
No. 184	Physics of Particle Accelerators (Ithaca, NY, 1988)	89-83575	0-88318-384-6
No. 185	Glueballs, Hybrids, and Exotic Hadrons (Upton, NY, 1988)	89-83513	0-88318-385-4
No. 186	High-Energy Radiation Background in Space (Sanibel Island, FL, 1987)	89-83833	0-88318-386-2